Predictive Species and Habitat Modeling in Landscape Ecology

C. Ashton Drew • Yolanda F. Wiersma
Falk Huettmann
Editors

Predictive Species and Habitat Modeling in Landscape Ecology

Concepts and Applications

Editors
C. Ashton Drew
North Carolina State University
USA
cadrew@ncsu.edu

Yolanda F. Wiersma
Memorial University
St. John's, NL
Canada
ywiersma@mun.ca

Falk Huettmann
University of Alaska
Institute of Arctic Biology
Department of Biology & Wildlife
99775 Fairbanks Alaska
USA
fhuettmann@alaska.edu

ISBN 978-1-4899-8135-6   ISBN 978-1-4419-7390-0 (eBook)
DOI 10.1007/978-1-4419-7390-0
Springer New York Dordrecht Heidelberg London

© Springer Science+Business Media, LLC 2011
Softcover re-print of the Hardcover 1st edition 2011
All rights reserved. This work may not be translated or copied in whole or in part without the written permission of the publisher (Humana Press, c/o Springer Science+Business Media, LLC, 233 Spring Street, New York, NY 10013, USA), except for brief excerpts in connection with reviews or scholarly analysis. Use in connection with any form of information storage and retrieval, electronic adaptation, computer software, or by similar or dissimilar methodology now known or hereafter developed is forbidden.
The use in this publication of trade names, trademarks, service marks, and similar terms, even if they are not identified as such, is not to be taken as an expression of opinion as to whether or not they are subject to proprietary rights.

Printed on acid-free paper

Springer is part of Springer Science+Business Media (www.springer.com)

CAD: To my son, Matthew, for his love, patience, and laughter.

YFW: To my husband John, for support, and to my sons William and Xavier, for inspiration and with hopes for beautiful landscapes for them to explore their whole lives through.

FH: With gratitude to my long-term co-workers Sergey Resvy, Nikita Chernetsov, Yuri Gerasimov, Igor Dorogoy, Sasha Solovyow, Aleksey Antonov, Misha Markovets, Olga Valchuck, Katya and Sasha Matsyna, Dima Lisityn, Andrey Koroblyov, and Pavel Ktitorov. Without your dynamic support, skill, time, effort, discussions and fun, most of this research and the underlying thoughts would not have evolved.

Foreword

Chair of the NASA-MSU Program, Former President of US-IALE
Rachel Carson Chair in Sustainability, Center for Systems
Integration and Sustainability, Michigan State University
Jianguo (Jack) Liu

Landscape ecology is a relatively young but rapidly growing discipline. This phenomenal growth is partially due to the energetic "new blood" to the field – a young generation of dedicated scholars who are innovative thinkers, insightful theorists, skillful modelers and experimentalists, and enthusiastic promoters and practitioners of applying landscape ecology to natural resources management, planning, and policy-making processes.

The content of this book is an excellent example of scholarship under the leadership of some outstanding young landscape ecologists. Focusing on species distribution and habitat modeling, it contributes to landscape ecology theories, methods, and applications in many ways. The book integrates fundamental ecological theories into the model development processes, presents novel approaches to untangling the complexities of real-world systems, addresses variability and uncertainty in species–habitat associations across space and time, and discusses research frontiers. As species and habitats are important components of global biodiversity, models are particularly useful tools for predicting and explaining biodiversity distribution and dynamics. While the world is celebrating the International Year of Biodiversity (2010), and the United Nations is revising its 2010 biodiversity targets for the future, this timely book helps meet relevant global challenges, such as protecting biodiversity and providing sustainable ecosystem services for human well-being. Many valuable insights in the book have long-lasting impacts.

This book evolved from a symposium celebrating the tenth anniversary of the NASA–MSU Professional Enhancement Program. With support from the National Aeronautics and Space Administration (NASA) and Michigan State University (MSU), especially William Taylor, then the chair of the Department of Fisheries and Wildlife at MSU, I established the NASA–MSU Program to provide funds for some outstanding students to attend the 1998 meeting of US-IALE (US Regional Association of the International Association for Landscape Ecology). The awardees were selected based on their academic backgrounds, professional goals, abstracts for presentations, and financial needs. While at the meeting, they learned the latest developments in landscape ecology, contributed to the meeting by presenting and sharing their research, and established professional and personal networks for future collaborations. They were recognized at the conference banquet with award certificates. Also, a special dinner was organized for the awardees to interact with leaders in the field, such as Richard Forman, Frank Golley, Eugene Odum, Monica Turner, and John Wiens. The students benefitted enormously from interacting with those leaders and felt honored to meet scholars whose publications had influenced them. Meeting with leaders boosted students' confidence, further inspired their interest in landscape ecology, and helped their professional growth. After that meeting, John Wiens, then president of IALE, asked me to continue this program for the 1999 World Congress of Landscape Ecology that he was organizing. Since then, the NASA–MSU Program has become an integral part of the US-IALE annual meetings. The program has continued under three NASA managers (Anthony Janetos, Jon Ranson, and Garik Gutman); all of them have maintained a strong level of commitment to the program. There are now more than 270 NASA–MSU alumni around the world, and many of them have become leaders in landscape ecology and within US-IALE.

The symposium was organized by the three editors of this book – Ashton Drew, Yolanda F. Wiersma, and Falk Huettmann. They are NASA–MSU alumni and did a fantastic job in selecting an excellent group of speakers (most of whom were also NASA–MSU alumni). The room for the symposium was packed with a very engaged and enthusiastic audience. To benefit those who were not present, the symposium organizers decided to turn the symposium presentations into a book. Ten of the 13 chapters have a former NASA–MSU awardee as the lead author, and of all 28 chapter authors, 12 are NASA–MSU alumni. They are shining and rising stars of the field. With these and other promising scholars, landscape ecology continues to develop rapidly and play an increasing role in addressing important challenges to biodiversity and society around the world.

Acknowledgments

This book developed as a wide and international collaboration; consequently, it was supported by many "helpers" from all over the world. It presents the fruits from years of "thinking" and doing science. We are further grateful to all our chapter authors. Secondly, we would like to thank US-IALE for providing a fertile "breeding ground" for the development of the discipline and for the mentoring of young scientists and students. US-IALE kindly provided a professional platform and audience for our modeling symposium, and for this publication. Most of the book contributors are directly involved in IALE. We are grateful to Jack Liu for his work in creating and maintaining the NASA–MSU awards. This has created a nice cohort of Modeling Landscape Ecologists, many of whom came together at the Golley–Odum NASA–MSU symposium at the US-IALE meeting in 2007 (again, facilitated by Jack Liu). That symposium has formed the basis for the book, and many of the past NASA–MSU awardees who presented at the symposium are contributors here.

Production of this book would not be possible without the assistance of our many reviewers. All primary chapter authors provided a review of at least one, with some providing constructive feedback on as many as three chapters. In addition, many colleagues provided blind peer-review of individual chapters. Specifically, we would like to thank Thomas Albright, Ali Arab, Mike Austin, Janine Bollinger, Ray Dezzani, Alessandro Gimona, Thomas Gottschalk, Barry Grand, Joe Hightower, Jeff Hollister, Todd Jobe, Matthew Krachey, Cohn Loredo-Osti, Bruce Marcot, Jana McPherson, Rua Mordicai, Dirk Nemitz, Tom Nudds, Anatha Prasad, Sam Riffell, Kim Scribner, Mark Seamans, Darren Sleep, Erica Smithwick, John Wiens, Chris Wilke, and Miguel Zavala for providing peer-review of individual chapters. Thanks are also due to N. Dunn, who under supervision of YW and funding from a Memorial University Career Experience Program grant, assisted with copy editing and formatting of a number of chapters in this volume.

AD first recognizes the outstanding support and friendship of colleagues in the USFWS Eastern North Carolina and Southeastern Virginia Strategic Habitat Conservation Team, especially those that have served as experts and provided logistical and financial support for her postdoctoral research. She wishes to thank her supervisor J. Collazo at the NC Cooperative Fish and Wildlife Research Unit at NCSU for allowing her great flexibility to pursue her diverse academic and professional interests. Also within the Coop Unit, colleagues at the Biodiversity and

Spatial Information Center (BaSIC: A. McKerrow, S. Williams, C. Belyea, T. Earnhardt, M. Rubino, N. Tarr, A. Terando, J. White, M. Iglecia) have provided data and modeling assistance, invaluable advice, friendly reviews of very rough drafts, and generally an amazingly fun and productive work environment.

YW wishes to thank D. Urban and the group in the Landscape Ecology lab at Duke 2002–2003 for providing an excellent venue to be immersed in landscape ecology and habitat modeling. This "PhD sabbatical" was supported by the Fulbright Foundation. She is also grateful to her PhD supervisor and mentor T. Nudds for promoting thinking on philosophy of science and to colleagues M. Drever and D. Sleep for discussions on statistics. A final thanks to colleagues in the Biology Department at Memorial University, especially P. Marino for fantastic support of new faculty. Also thanks to J. Sandlos for unconditional love and support, and to my two sons for inspiration.

FH wishes to acknowledge the student power of his EWHALE lab at UAF, and his project collaborators in Alaska GAP (T. Gotthardt, S. Payare, A. Baltensberger), ArcOD (Arctic Ocean Diversity Project: B. Bluhm, C. Hopcroft, R. Hopcroft and R. Gradinger; funded by SLOAN Foundation), Global Mountain Biodiversity Project (GMBA, E. Spehn and C. Koerner; funded by DIVERSITAS), and the UAF Avian Influenza Center of Excellence (J. Rundstadler, G. Happ; funded by NIH). Further, FH likes to thank all GIS, data and modeling supporters and for brainstorming and feedback: T. Gottschalk, G. Humphries, M. Lindgren, S. Oppel, D. Nemitz, B. Dickore, E. Green, B. Best, GBIF, Eamonn O'Tuama, NCEAS, T. Peterson, B. Chen, M. Parson, T. Chapin, V. Wadley, J. Ausubel, T. Diamond, S. Cushman, S. Linke, A. Huettmann, I. Presse, Salford Systems Ltd team of D. Steinberg.

Finally, all three of the editors appreciate Springer NY Publishers, and especially J. Slobodien, for taking on this project and working with us toward completion. Final thanks to M. Higgs, our Production Manager at Springer, J. Quatela our Book Production Editor at Springer and to Marian John Paul of SPi Technology Services who oversaw production of the internal text.

Contents

1 **Introduction. Landscape Modeling of Species and Their Habitats: History, Uncertainty, and Complexity** 1
Yolanda F. Wiersma, Falk Huettmann, and C. Ashton Drew

Part I Current State of Knowledge

2 **Integrating Theory and Predictive Modeling for Conservation Research** 9
Jeremy T. Kerr, Manisha Kulkarni, and Adam Algar

3 **The State of Spatial and Spatio-Temporal Statistical Modeling** 29
Mevin B. Hooten

Part II Integration of Ecological Theory into Modeling Practice

4 **Proper Data Management as a Scientific Foundation for Reliable Species Distribution Modeling** 45
Benjamin Zuckerberg, Falk Huettmann, and Jacqueline Frair

5 **The Role of Assumptions in Predictions of Habitat Availability and Quality** 71
Edward J. Laurent, C. Ashton Drew, and Wayne E. Thogmartin

6 **Insights from Ecological Theory on Temporal Dynamics and Species Distribution Modeling** 91
Robert J. Fletcher Jr., Jock S. Young, Richard L. Hutto, Anna Noson, and Christopher T. Rota

Part III Simplicity, Complexity, and Uncertainty in Applied Models

7 **Focused Assessment of Scale-Dependent Vegetation Pattern** 111
Todd R. Lookingbill, Monique E. Rocca, and Dean L. Urban

8   **Modeling Species Distribution and Change Using Random Forest** 139
    Jeffrey S. Evans, Melanie A. Murphy, Zachary A. Holden, and Samuel A. Cushman

9   **Genetic Patterns as a Function of Landscape Process: Applications of Neutral Genetic Markers for Predictive Modeling in Landscape Ecology** 161
    Melanie A. Murphy and Jeffrey S. Evans

10 **Simplicity, Model Fit, Complexity and Uncertainty in Spatial Prediction Models Applied Over Time: We Are Quite Sure, Aren't We?** 189
    Falk Huettmann and Thomas Gottschalk

11 **Variation, Use, and Misuse of Statistical Models: A Review of the Effects on the Interpretation of Research Results** 209
    Yolanda F. Wiersma

12 **Expert Knowledge as a Basis for Landscape Ecological Predictive Models** 229
    C. Ashton Drew and Ajith H. Perera

Part IV   Designing Models for Increased Utility

13 **Choices and Strategies for Using a Resource Inventory Database to Support Local Wildlife Habitat Monitoring** 251
    L. Jay Roberts, Brian A. Maurer, and Michael Donovan

14 **Using Species Distribution Models for Conservation Planning and Ecological Forecasting** 271
    Josh J. Lawler, Yolanda F. Wiersma, and Falk Huettmann

15 **Conclusion: An Attempt to Describe the State of Habitat and Species Modeling Today** 291
    C. Ashton Drew, Yolanda F. Wiersma, and Falk Huettmann

Author Bios 299

Index 303

Abbreviations

AIC	Akaike's Information Criteria
AKN	Avian Knowledge Network
ANN	Artificial neural networks
AUC	Area Under Curve
BAS	British Antarctica Service
BBS	Breeding bird survey
BIC	Baysian Information Criterion
BRT	Boosted regression trees
CART	Classification and Regression Tree
DiGIR	Distributed Generic Information Retrieval
DIS	Digital Information Science
DOP	Dilution of precision
EIR	Entomological inoculation rate
EML	Ecological metadata language
FGDC	Federal Geographic Data Committee
GAM	Generalized additive model
GAP	Gap Analysis Program
GARP	Genetic Algorithms for Rule-set Production
GBIF	Global Biodiversity Information Facility
GCM	General Circulation Model
GDM	Generalized dissimilarity models
GEE	Generalized estimating equations
GIGO	"Garbage in, garbage out"
GIS	Geographic Information Systems
GLM	Generalized Linear Model
GPDD	Global Populations Dynamics Database
GPS	Global Positioning System
HSI	Habitat suitability index
IDE	Integro-Difference Equations
IPY	International Polar Year
ITIS	Integrated Taxonomic Information System
KML	Keyhole Markup Language
LTER	Long-term ecological research

MARS	Multivariate Adaptive Regression Spline
MAUP	Modifiable aerial unit
MDM	Mechanistic distribution models
MIGAP	Michigan Gap Analysis Program
NBII	National Biodiversity Information Infrastructure
NCEAS	National Centre for Ecological Analysis and Synthesis
NGO	Nongovernment organization
NIH	National Institutes of Health
NRLMP	Northern Region Landbird Monitoring Program
NSF	National Science Foundation
OBIS	Oceanic Biogeographic Information System
OOB	Out-of-bag
OPS	Overall prediction success
PCA	Principal component analysis
PCC	Percent correctly classified
PDA	Personal Digital Assistant
PDE	Partial Differential Equations
PDOP	Positional dilution of precision
PRISM	Parameter Elevation Regressions on Independent Slopes Model
RDBMS	Relational databases management system
ROC	Receiver Operator Characteristic
RPART	Recursive Partitioning
SML	Sensor Metadata Language
USFS	US Forest Service
USGS	US Geological Survey
XML	Extensible Markup Language

Chapter 1
Introduction. Landscape Modeling of Species and Their Habitats: History, Uncertainty, and Complexity

Yolanda F. Wiersma, Falk Huettmann, and C. Ashton Drew

1.1 Where Do We Come from?

From the start, the discipline of MODERN Landscape Ecology has focused on the interaction between spatial pattern and ecological processes. One area of focus has been to better understand how the patterns of environmental features, habitats, and resources (e.g., gradients, patches) influence patterns of species distribution. Some of the earliest predictive modeling papers in the journal *Landscape Ecology* dealt with predicting vegetation patterns based on topography (e.g., Bolstad et al. 1998; Ostendorm and Reynolds 1998). One of the early tools developed for wildlife management was "Habitat Evaluation Procedures" (HEP) (Schamberger 1982; Urich and Graham 1983; Mladenoff et al. 1995).

A suite of qualitative and quantitative models to predict habitat and species distributions followed. Landscape ecologists were also aware of the power of computer-based modeling to predict other ecological processes. The early 1990s saw the development of models to predict ecological processes in space and time. These included predicting forest stand dynamics (Keane et al. 1990) and the HARVEST model for predicting forest pattern as a result of disturbance (Gustafson and Crow 1996). There has been an increase in studies carried out to model species distribution and other ecological processes since the mid-1990s, largely the result of increased access to large data sets, advances in Geographic Information Systems (GIS) and remote sensing technology, new statistical models, and, most importantly, dramatic increases in computing power (see Fig. 11.1; Chap. 11). Models exist for two main reasons: to infer and explain ecological processes, and to predict future conditions or distributions and patterns in locations that have not yet been sampled. Some of the literature to date has focused on the latter function of models for prediction in space and/or time.

Landscape ecologists must make assumptions about how organisms experience and make use of the landscape. This is true whether models are being used to code

---

Y.F. Wiersma (✉)
Department of Biology, Memorial University, St. John's NL, A1B 3X9, Canada
e-mail: ywiersma@mun.ca

behavioral rules for dispersal of simulated organisms through simulated landscapes, or for designing the sampling extent of field surveys and experiments in real landscapes. These convenient working postulates allow modelers to project the model through time and space, yet rarely are these predictions explicitly considered. Out of necessity, the early years of landscape ecology focused on the evolution of effective data sources, metrics, and statistical approaches that could truly capture the spatial and temporal patterns and processes of interest. Many texts already exist which deal with predictive modeling for both species distribution (Manly et al. 2002; Scott et al. 2002; Cushman and Huettmann 2010) and ecological processes, (Dale 2002) or in relation to resource management and conservation planning (Bissonette and Storch 2003; Millspaugh and Thompson 2009). The available literature offers detailed "how to" sections, with an emphasis on the statistical tools used to develop models (Franklin 2009). However, little of the available literature includes detailed information on the explicit link between underlying ecological theories and the structure of the models; or on the actual role of data, scale, and autocorrelation. Now that tools and techniques are well established, we are prompted to reflect on the ecological theories that underpin the assumptions commonly made during species distribution modeling and mapping.

This book will (1) highlight how fundamental ecological theories are being explicitly integrated into the model building processes for a robust inference approach, (2) offer practical examples of how modelers are addressing the conflict between the complexity of ecological systems and the relative simplicity of their modeled systems, and (3) present novel prediction methods and underlying philosophies to identify and quantify sources of uncertainty and variability in species–habitat associations in time and space, and to contribute progressive management applications.

1.2 Where Are We Going?

As recently completed PhDs who all had a modeling focus in their dissertations, it seemed to us that ecological theory was being largely ignored. We feel that modelers must take a moment to consider existing "first principles" of the ecology of an organism or process that they are trying to model. That motivated us to organize a special symposium at the 2007 meeting of the United States chapter of the International Society of Landscape Ecology (US-IALE) in Arizona. This book is a result of that international symposium, and many of the chapter authors were presenters at that meeting.

We set out to make the symposium and the resultant book more than a collection of individual presentations and case studies. To provide additional progress on the field of predictive modeling in space and time, we asked authors to reflect on the role of ecological theory in modeling, where they felt the field of modeling was going, and to identify some of the gaps that needed to be addressed to move the field forward. We do not intend this book to be a "how to model" guide (there are many excellent books and articles that already do this), but rather, we wanted participants in the symposium to offer their own thoughts and insights on potential pitfalls, highlight areas of confusion, and provide commentary on how they felt the field of modeling could be improved. Authors were encouraged to share potential solutions to ongoing modeling problems, and to identify those areas requiring further research. This volume is an attempt to promote thinking about the current challenges in predictive species/habitat modeling, including issues about dynamic ecological systems, stochasticity, complexity, data management, public use of models and data, and the need to include explicit ecological mechanisms.

This book is written primarily for graduate students and professionals and/or practitioners, especially those considering the use and/or application of models for the first time – although we feel that our review of ideas and concepts is of value to veteran modelers as well. For the new modeler, the field may seem daunting with its technical language, seemingly arcane debates over statistical methods, and overwhelming data sets. However, we remind readers that models are all around us. Nearly everyone interacts with models daily when they consult the local weather forecast. Weather forecasting is based on a complex set of models and carries with it uncertainties and variation in predictive accuracy. Yet, most of us feel comfortable with the uncertainty inherent in a 5-day weather projection and are content to be largely oblivious to the myriad meteorological models underlying the report we consult. On a larger scale, the International Panel on Climate Change (IPCC) is the international body that reports on climate models and makes projections regarding the impact that varying climate change scenarios may have on the earth's ecosystems and societies. These models are much less trusted by the lay public. The recent "climategate" scandal, wherein the emails of some IPCC scientists at East Anglia University were leaked to the media, highlighted the mistrust that many still have in models. While fellow scientists recognized and were largely comfortable with the uncertainty in the models expressed in the leaked emails, some elements of the media promoted the idea to the lay public that the scientific uncertainty was a sign that climate models were completely unreliable. When models are – or are perceived to be – unreliable, it can be difficult to use them for decision making, management, and policy formation.

Several key themes emerge through this volume. In some cases there is congruence between chapters, in others there is conflict. We hope the conflicts and apparent contradictions do not discourage prospective landscape modelers. Rather, we feel the disagreements in perspective between the authors of some chapters are part of the scientific discourse and reflect a healthy, dynamic, and still-growing field. These areas of disagreement or uncertainty represent diverse areas that are ripe for more research and offer up new hypotheses for testing.

1.3 Key Themes

One of the key themes, and the one which prompted us to organize the symposium that was the foundation for this book, is the issue of mechanistic versus predictive models and the tradeoffs inherent in choosing one over other. Mechanistic models have traditionally been perceived as useful for testing ecological hypotheses, but whether models need to be mechanistic to make accurate predictions is an open question.

The tradeoff between simplicity (parsimony) and complexity for increased predictive performance has emerged as a key debate in the modeling literature, and also in science as a whole. This book is focused on *predictive* habitat–species distribution models. Prediction in general is a key tenet of the scientific method; prediction allows one to generalize, and forms the very basis of statistical hypothesis testing. Thus, according to scientific standards, models (i.e., hypotheses) should yield inferences and results that can be generalized to other points in space and time; that is, they should predict well, or to the best possible degree. Principles of parsimony suggest that simplicity in models will yield better generalization (Burnham and Anderson 2002).

Thus, a key question that remains is whether the intent of models is to be explanatory, predictive, or both. While the focus of this volume has been on predictive models, and notwithstanding the importance of prediction as part of the scientific process, it is entirely possible to have accurate predictive models with no underlying biological or mechanistic explanation.

The available data also influence the models and methods of accuracy assessment – another key theme in this volume. To adequately assess predictive power, accuracy assessment, and the use of alternative data is essential. Management decisions based on models should probably not be done without quantifying model uncertainty. The available data are the essential foundation for choosing the method of accuracy assessment to use (e.g., reserving validation data vs. leave-one-out methods such as bootstrapping). When using models in management and decision-making, it is also important to consider situations where models are created based on and assessed with existing models. Errors add up, and can behave in a chaotic fashion. Such concepts are known to landscape ecologists from other applications (e.g., percolation theory, landscape metrics) but have not been widely discussed in the context of predictive habitat modeling.

Several other themes throughout this volume hinge on process and technical issues. These are areas which are still quite open to debate, but we feel that highlighting them opens up avenues for further research, provides guidelines for starting a modeling exercise, and highlights some of the important considerations managers and end-users must take into account when implementing the results of modeling exercises. These issues include statistical and model selection decisions (see Chap. 2 for an overview of available tools), and considerations around open access data sharing and metadata (see Chap. 3). Debates are not always resolved. Whereas chapter authors do not always agree on statistical tools, the authors have clarified their thought process around particular statistical choices made in their case examples so that readers may apply these considerations in their own decision-making. Most authors agree that the statistical choices should reflect the question being asked, and that suitability is somewhat predicated on predictive accuracy. That is, if you can build a model that will predict well using a standard statistical model, there is no need to apply a more recent or apparently more sophisticated model. On the other hand, there is valid argument for work to explore better models and to use data, software, and computer power to build on and further improve existing models.

A related consideration is the issue of sampling design. Sampling considerations are important for data management and accuracy assessment. Many statistical textbooks that deal with sampling considerations still assume that samples are used for frequentist-based testing. We are not aware of any texts that address sampling design for predictive modeling which employ alternative statistical methods (e.g., neural networks or random forests).

As illustrated by some of the controversy around climate modeling, model literacy is important. Increased literacy around the field of predictive species/habitat modeling will also increase the ability of the scientific community and the public to offer constructive critiques of models as they appear. Here we have identified some key elements necessary to improve the use of models (i.e., comprehensive data management, appropriate sampling and statistical methods, clear and transparent processes, robust accuracy assessment, online delivery, and effective communication of model results including model uncertainty). Models are likely to be at the forefront of legislative and judicial processes in the future; thus, transparent and robust models that are based on the best available science will be the most effective. It is our hope that this volume will contribute to more of such models.

1.4 Organization of the Book

This book is organized into four sections. The first section outlines the current state of knowledge in the field of modeling. Chapter 2 by Kerr et al. provides a perspective on the ecological issues germane to these types of models. In Chap. 3, Hooten outlines the statistical landscape, and Zuckerberg et al. offer some thoughts on the challenges of acquiring and managing large data sets in Chap. 4.

The second section, "*Integration of Ecological Theory into Modeling Practice*," outlines how we can ensure that ecological first principles remain at the forefront of modeling. Laurent et al. provide a thoughtful overview of some of the common ecological assumptions inherently made in modeling in Chap. 5. In Chap. 6, Fletcher et al. look at how predictive models can also be tests of ecological theory dealing with temporal processes.

The third section of the book, "*Simplicity, Complexity, and Uncertainty in Applied Models*," highlights some key issues that the new modeler may want to consider. This is the most "how to" section of the book, and probably of most interest to those attempting to develop models for the first time. Although the chapters are not structured as instructive "recipes," they highlight specific issues, offer suggestions and references for further reading, and through case studies, illustrate how some of these issues might be addressed. In Chap. 7, Lookingbill et al. address pattern-based issues through a discussion highlighting how focused assessment techniques can help identify scale-dependent vegetation pattern. Evans et al. deal with process-based issues and provide detail on machine-learning approaches through a treatise on "Random Forests" in Chap. 8. In Chap. 9, Murphy and Evans follow with a second chapter addressing process-based issues, this time looking at the process of gene flow across landscapes. Huettmann and Gottschalk look at temporal and accuracy issues and engage in a debate on simplicity versus complexity in Chap. 10. Wiersma provides some perspective on various statistical issues in Chap. 11, and the implications of choosing one statistical model over another through a meta-analysis of statistical models applied in the literature over a 10-year period. Finally, in Chap. 12, Drew and Perera round out this section with a discussion on expert opinion and how it can be integrated with modeling to address knowledge and data issues.

The final section of the book, "*Designing Models for Increased Utility*," provides two examples. Roberts et al. (Chap. 13) show how models have been applied to the Gap Analysis Program (GAP) in Michigan to support management decision making, and Lawler et al. look at how models have been used in strategic conservation planning in Chap. 14. The book concludes with a chapter by the editors synthesizing key findings, and pointing to areas for future research. Although the fields of landscape ecology and predictive species/habitat modeling are rapidly evolving, it is our hope that many of the chapters in this volume will remain timeless, and help to set new standards. It is our wish that both the new and veteran modeler and landscape ecologist will find valuable insights in the reflections on the state of the practice of predictive modeling that we and our collaborators and colleagues offer in the pages of this book for advancing the science of landscape ecology world-wide.

References

Bissonette JA, Storch I (eds) (2003) Landscape ecology and resource management: linking theory with practice. Island Press, Washington DC.
Bolstad PV, Swank W, Vose J (1998) Predicting Southern Appalachian overstory vegetation with digital terrain data. Landsc Ecol 13:271–283.
Burnham KP, Anderson DR (2002) Model selection and multimodel inference: a practical information-theoretic approach, 2nd edition. Springer, New York.
Cushman S, Huettmann F (eds) (2010) Spatial complexity, informatics, and wildlife conservation. Springer, Tokyo.
Dale VH (2002) Ecological modeling for resource management. Springer, New York.
Franklin J (2009) Mapping species distributions: spatial inference and prediction. Cambridge University Press, New York.
Gustafson EJ, Crow TR (1996) Simulating the effects of alternative forest management strategies on landscape structure. J Environ Manage 46:77–94.
Keane RE, Arno SF, Brown JK, Tomback DF (1990) Modelling stand dynamics in whitebark pine (*Pinus albicaulis*) forests. Ecol Modell 51:73–95.
Manly BFJ, McDonald LL, Thomas DL, McDonald TL, Erickson WP (2002) Resource selection by animals: statistical design and analysis for field studies, 2nd edition. Kluwer Academic Publishers, Dordrecht.
Millspaugh JJ, Thompson FR Jr (2009) Models for planning wildlife conservation in large landscapes. Academic Press, Burlington MA.
Mladenoff DJ, Sickley TA, Haight RG, Wydeven AP (1995) A regional landscape analysis and prediction of favorable gray wolf habitat in the Northern Great Lakes. Conserv Biol 9:279–294.
Ostendorf B, Reynolds JF (1998) A model of arctic tundra vegeation derived from topographic gradients. Landsc Ecol 13:187–201.
Schamberger M (1982) Habitat: a rational approach to assessing impacts of land use changes on fish and wildlife. Env Pro 4:251–259.
Scott JM, Heglund PJ, Morrison MM, Haufler JB, Raphael MG, Wall WA, Samson FB (eds) (2002) Predicting species occurrences: issues of accuracy and scale. Island Press, Washington.
Urich DL, Graham JP (1983) Applying habitat evaluation procedures (HEP) to wildlife area planning in Missouri. Wildl Soc Bull 11:215–222.

Part I
Current State of Knowledge

Chapter 2
Integrating Theory and Predictive Modeling for Conservation Research

Jeremy T. Kerr, Manisha Kulkarni, and Adam Algar

2.1 Introduction

The need for effective techniques to predict how global changes will alter biological diversity has never been greater and continues to increase (Buckley and Roughgarden 2004; Thomas et al. 2004). Although accelerating climate and land use changes loom especially large, extinction rates have risen as a result of other types of threats as well – such as overkill and pollution. Individually, each of these perils is serious, but it is through their additive and sometimes synergistic interactions that the world is now in the midst of a sixth mass extinction (Wake and Vredenburg 2008).

Predicting responses of individual species or biological communities to the shifting environmental conditions with which they are increasingly confronted is the business of models (see Scott et al. 2002). Those predictions may focus on many different biological responses, including population fluctuations, shifting distributions, or collective changes in the species richness of communities (Woodward 1987; Root 1988; Parmesan 2005; White and Kerr 2006). Models may adopt spatial or temporal perspectives, or combine both. Regardless of the purpose, application, or spatio-temporal perspective intrinsic to models, the criteria for determining their utility can be narrowed down to two, alternative questions: did the model provide successful predictions and, if not, did it fail in interesting ways? In the former case, models are more likely to be helpful for policy-makers but much can be learned in the latter.

Although many modeling systems exist to build these predictions, few that incorporate extensive mechanistic justification are in wide use. There are at least two good reasons for this state of affairs. First, predictive models that lack strong mechanistic backup have proven surprisingly capable (Araújo and New 2007). Second, complex models with detailed biological underpinnings may include terms that are impractical to measure. So, higher model sophistication comes with a cost

---

J.T. Kerr (✉)
Canadian Facility for Ecoinformatics Research, Department of Biology,
University of Ottawa, 30 Marie Curie, Ottawa ON Canada, K1N 6N5
e-mail: jkerr@uottawa.ca

in terms of ease of application but also can bring benefits, particularly in the form of greater reliability as environmental conditions change (Buckley 2008). Models built on such detailed theoretical underpinnings are not necessarily more effective, nor are models without such foundations more likely to be ineffective. For any model, from the theoretically derived to the purely phenomenological, utility is defined by the ability to test and the resultant tested predictions that may be used to inform policy (e.g., Dillon and Rigler 1974).

Here, we consider theoretical perspectives on modeling the distributions of species across landscapes and regions. Predicting spatial distributions of species and how those distributions change over time has been possible without using detailed mechanistic models; however, theoretical contributions to these models may improve their reliability as environmental conditions change. Different degrees of theoretical support for predictive models suggest an array of strategies to maximize model reliability for policy application (Kerr et al. 2007). We bring these together in a single framework that addresses the role of theory in model construction and the reliability of applications that result. Many examples demonstrate the importance of theoretical contributions to predictive modeling but that those models can often be successful even in the absence of such contributions.

2.2 Ecological Theory and a Framework for Predictive Modeling

2.2.1 *Integrating Data, Testing, and Theory for Predictive Modeling*

As the theoretical complexity of models increases, it becomes increasingly difficult to test their reliability across a range of environmental conditions. Collecting new data to test elaborate hypotheses about biological processes can quickly become impractical, whereas an able theoretician faces fewer constraints in pursuing increasingly sophisticated models able to detail mechanisms (McLean and May 2007). This does not suggest that very detailed models fail to illuminate potentially important biological processes, but if those models cannot be tested, their application to real world situations will be hampered. The point at which increasing model complexity, via the introduction of untestable mechanisms that may not actually be operating in the system in question, actually begins to *decrease* the reliability of real-world predictions could, perhaps, be viewed as the point beyond which modeling efforts serve a reduced practical purpose. Conservatively speaking, that point could be reached when data can no longer be collected to test additional model details.

The interaction between theoretical and empirical approaches to modeling is frequent and mutually indispensable (Fig. 2.1). In fact, most researchers work somewhere in the space between these extremes and use theory (or experimental observations) to inform data collection, and field observations to refine existing theory. The objective of this interplay is to produce reliably predictive models. It is possible to make progress on purely empirical grounds, or similarly on purely theoretical grounds, but rapid advances made based on such grounds may decrease the chances that scientific progress can reliably be translated into policy action.

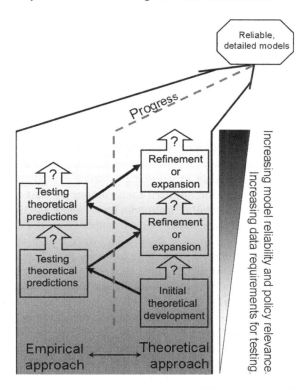

Fig. 2.1 One view of the interplay between theoretical and data-driven (empirical) approaches to predictive modeling (Box 1979; Austin 2002). In this framework, an initial theoretical model yields predictions that can be tested empirically. Results from those tests can then be used to refine (possibly refute) or expand models in terms of theoretical content, which may mean specification of additional biological processes. Scientific "progress" is expanded capacity to predict species responses to environmental change, across some spatial gradient (e.g., of habitat fragmentation) or through time. Discoveries do not require this interplay: progress can be driven entirely by empirical or theoretical perspectives, but such approaches may be less certain. With increased model detail and reliability, it may be possible to contribute to the development and implementation of policy, which are most likely to be implemented according to adaptive management principles, and not as a one-way process. Data requirements are likely to increase with increasing model detail, which can inhibit hypothesis testing. Hypothesis (model) testing can take many forms, including experimentation and simulations, but real-world use of models should include real-world tests wherever possible

2.2.2 Case Study: Butterfly Models Using Mechanistic Knowledge

In recent years, much work has focused on the impacts of observed climate changes on species distributions (Thuiller 2003; Peterson et al. 2004; Luoto et al. 2005; Guisan et al. 2008). Butterflies have proven to be a useful focal taxon in this kind of global change research (Dennis 1993; Wilson et al. 2005; Menendez et al. 2006). Because most butterflies, along with most other animals, are non-migratory, they must be able to tolerate a full year's range of environmental conditions or face local extinction. In northern environments, where climates are changing rapidly, butterflies cannot use behavior to avoid extreme winter conditions. Instead, they must be freeze tolerant. Keeping in mind that models predicting how changing climates will alter species distributions would be valuable, it would be helpful to know the mechanism limiting the northern distributions of these species. Possible mechanisms, on the ecology side, include limits imposed by host plant distributions or temperature-dependent foraging, whereas lab-based physiological research has demonstrated the molecular basis for freeze tolerance. If temperatures drop too far below the supercooling point of an individual, its antifreezes are no longer effective and its cells lyse as they freeze solid. This mechanism suggests an empirical step toward predicting climate change effects on butterfly distributions. Kukal et al. (1991) took a step toward quantifying this empirical step, and provided field experimental evidence – supported by molecular observation – that suggested the northern limits among at least two swallowtail butterflies (the Tiger Swallowtail, *Papilio machaon*, and Canadian Swallowtail, *P. canadensis*) were limited by minimum winter temperatures. If those temperatures drop sufficiently, mortality rates increase sharply. This work, in turn, supports the theoretical mechanism linking temperature to range limits for these butterflies, leading to further prediction on the empirical side of the equation: warming temperatures cause range expansion among butterflies. Species distribution models (Anderson et al. 2003; Phillips et al. 2006) can then be constructed. Consequently, range expansion among butterflies in Canada has been shown to track observed climate changes in the twentieth century (White and Kerr 2006; Kharouba et al. 2009). Species distribution models were validated with data that were spatially and temporally independent from data used to construct models, increasing their reliability for policy application.

From such examples, it is possible to generalize one approach to the modeling process beginning with a theoretically derived hypothesis at the outset, leading to predictions that are tested empirically, and an expansion or modification of the starting hypothesis which leads to further empirical evaluation, and so on (Fig. 2.1). The goal in this process is to construct reliable, predictive models for a biological phenomenon of interest, such as how climate changes will alter species' distributions. At each step in this process, the requirements for data increase, but so does the likelihood that modeling results up to that point could reasonably be used for policy decisions.

2.3 Models Missing Mechanisms

Predictive models of species distributions across landscapes and regions commonly exclude detailed mechanistic information, but their predictions have often proven quite accurate. Some of the most influential observations that species are actually shifting their ranges and phenological timing in the direction expected given climate change include relatively little environmental measurement (Parmesan 1996; Parmesan et al. 1999; Root et al. 2003). Although these efforts did not use predictive modeling, they provided critical evidence that biological responses to shifting environmental conditions were likely. Recent efforts link observed changes to environmental drivers (Rosenzweig et al. 2008).

2.3.1 Case Study: Madagascan Chameleons

Predictive models have had some surprising successe even when mechanisms are not directly considered. Raxworthy et al. (2003) used a common niche modeling technique (Genetic Algorithms for Rule-set Production; GARP) to estimate possible geographic ranges for chameleon species in Madagascar (Fig. 2.2). In this study, point observations for chameleon species were available in much of Madagascar, but GARP niche models consistently predicted some chameleon species could be present in a disjunct area from which no data had been collected. When that area was surveyed, several new chameleon species were discovered that were sister species to those that had been modeled. Missing from these GARP models were any considerations of the mechanisms that might be necessary for speciation – including isolation, the presence of dispersal barriers, mutation rates, or selection pressures. Predictive models can clearly succeed under some circumstances even without consideration of mechanism, although these models are strongly driven by expert knowledge of the niche requirements of chameleons.

Fig. 2.2 The modeled distribution (in *gray*) of *Furcifer pardalis*, a chameleon in Madagascar, included a large area in the eastern half of the country. This area (shown in *white*) was predicted to be suitable for the species and contained all historical observations. The model, developed using GARP, also predicted that this species should be present in a previously unexplored area (indicated by a *question mark* within a *black-shaded ellipse*) to the west of its known range (Raxworthy et al. 2003). When that unexplored area was surveyed, new chameleon species were discovered, including a sister species to *F. pardalis*. Although the GARP models linking environmental characteristics to chameleons' distributions in Madagascar did not specify biological processes or mechanisms that would generate the species' distribution, they provide a strong indication that such models can still generate surprising, useful, and correct predictions

It is even possible to predict the spatial distributions of species without any biological information at all. The simple reason this is possible is that environmental factors are spatially autocorrelated. This fact has long been recognized by ecologists, who have responded to the problems (and opportunities) that spatial autocorrelation presents with a series of increasingly sophisticated statistical tools, including adjustments to make probability tests more conservative, and the control of biases in coefficient estimation in regression models (see Legendre and Legendre 1998; Koenig 1999; Diniz-Filho and Hawkins, 2003). Commonly, spatial autocorrelation of environmental factors is considered after first investigating the potential, main effects of environmental variables thought to affect a biological response of interest, such as a species' distribution. Bahn and McGill (2007) take the reverse approach, simulating a series of species' ranges and modeling purely spatial predictors of those ranges, then testing whether adding environmental variables, after applying purely spatial variables, improved predictions of species ranges. Their result was striking: spatial structure did a better job of predicting simulated species ranges than mechanistic approaches. In other words, it is possible that spatial models attempting to predict the distribution of species may sometimes do so purely because of the underlying spatial structure of the species' range, not because the model successfully captures any biological mechanisms.

The view that prediction of a species' range requires no biological information is incorrect. Bahn and McGill (2007) argue that these models have little capacity to predict distributional shifts in changing environments. In fact, this is not strictly true: spatial models that do not depend on environmental conditions predict that distributions will remain relatively constant when those conditions change. Clearly, this proposition is refuted by many observations. Environmentally based models have made accurate predictions of distribution expansions and contractions, at least for some taxa (e.g., Kharouba et al. 2009). However, Bahn and McGill's warning should be heeded by those who predict distributional shifts. Observed correlations between environmental variables and species distributions may arise in whole or in part from spatial autocorrelation rather than mechanistic links. For example, Algar et al. (2009) have shown that accounting for potential confounding effects of spatial autocorrelation while training environment-diversity models results in more accurate predictions of diversity shifts in response to twentieth century climate change than similar models that ignore spatial autocorrelation.

2.4 Testing Spatial Models through Time

Of course, the search for mechanisms goes on, not least because of two problems created by ignoring them. First, if predictions succeed, the reason of "why" is a mystery in the absence of some mechanistic understanding. Second, it is impossible to distinguish correlation and causation. These problems reduce expectations that a model developed in one place may be applied in another geographic location or that model predictions will be continue to be effective as environmental conditions change. It is probably self-evident that strong theoretical thinking should support predictive modeling efforts, especially when models include a temporal component. Unfortunately, predictions of change through time often rely on models observing change across space.

2.4.1 The "Space-for-Time" Assumption

Ecologists recognize the perils of the "space-for-time" assumption; that is, using purely spatial models to predict changes through time. The reason many authors have been forced to make this risky assumption is the dearth of long-term data that can be used to build predictive models that include temporal components. In the case of species distribution modeling, the absence of historical species observations and/or environmental data often limits model construction to the purely spatial. Spatial environmental gradients within the species' range are then assumed to indicate how that species responds to environmental differences. Expectations of environmental change into the future can then be translated into predictions of biological impacts. Modeling techniques have proven highly capable for fitting spatial relationships between environmental data and observational data of species on the ground, and statistical assessments make them seem very accurate. When run through time, however, model behavior can be grossly unreliable. Using extensive datasets on the environmental factors that South African Protea are known to respond to, Pearson et al. (2006) developed highly accurate spatial models predicting each species' range using nine different modeling methods. When anticipated environmental conditions for 2030 were substituted into the different models, they returned wildly divergent predictions of biological responses for the same species, ranging from 92% loss of current modeled range to more than 300% expansions. Species-by-species modeling approaches might be more prone to return idiosyncratic results, but variability for individual species might diminish in relative importance when large species assemblages are considered. However, one study focusing on an assemblage of about 150 of the most thoroughly studied butterfly species in Canada found that the relationship between human population density and butterfly species richness was positive spatially, but negative when tested temporally (White and Kerr 2006). The message for models with only partial mechanistic support appears to be that temporal tests may be essential.

2.4.2 Global Change as a Pseudo-experiment

Fortunately, climate and land use changes during the twentieth century provide many pseudo-experimental opportunities to test otherwise spatial links between environmental factors and the distribution of species. Butterflies in Canada again provide very effective examples for a few reasons. First, long-term data are available for observations of many of these species, as are historical environmental data for factors that physiological ecology literature (e.g., Kukal et al. 1991) suggests should influence species ranges. Importantly, climate and land uses have changed in much of Canada during the twentieth century. In short, it is possible to test whether models

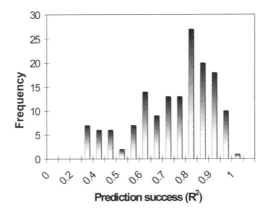

Fig. 2.3 Spatial models predicting species distributions may not work when used to predict distribution shifts through time (i.e., the "space-for-time" assumption may not hold), but a study of 139 butterfly species in Canada demonstrates that such predictions can be reliable (Kharouba et al. 2009). This histogram shows the distribution of conditional autoregressive R^2 values comparing predicted range shift to observed range shift during the twentieth century. For most species, models were sufficiently accurate to advise policy, but some predictions were erratic. Using large species assemblages reduces the impact of noisy predictions and creates a strong signal on which to base decisions. Observed, spatial correlations between environment and species distributions appear causative in this case

remain accurate through time using known climate and land-use changes as a kind of pseudo-experiment. That is, historical distributions can be estimated using models (in this case, Maximum Entropy). As environmental conditions change, the effects on species distributions can then be forecast using the same models, and actually tested using independent data collected from the present day. Models projected through time using this framework are generally accurate (Fig. 2.3), as would be expected if their spatial analogs specify environmental determinants of range correctly. Models developed for some species fail, on the other hand, suggesting that these techniques should only be applied to large species assemblages. Inferences around mechanisms are possible from purely spatial models, but adding a temporal component to them demonstrates whether projections of future conditions have any likelihood of being accurate. Macroecological research has few enough experimental opportunities: when they arise, it is essential to use them.

2.5 Theoretical Perspectives on Predictive Modeling

2.5.1 Niches, Neutrality and Predictive Models

Theoretical contributions to predictive modeling can include nearly any biological ingredient that affects the likelihood that a species will be present in a particular locality. Specific mechanisms may include physiological tolerances to the range of local environmental conditions (e.g., temperature, moisture, wind) and the capacity to survive negative biotic interactions (e.g., competition, parasitism, predation). Dispersal capabilities determine whether species can colonize suitable habitats (Thomas 2000) or rescue failing populations that are already present there (Brown and Kodric-Brown 1977). Species differ in all of these respects, but these differences may not always be essential to the development of models that successfully predict species distributions and responses to environmental change (Bell 2001). A conceptually simple modeling approach that could be applied to assemblages of species in a region that rely on similar trophic strategies is to ignore admitted differences among species entirely. Despite such a massive simplifying assumption, neutral theory is able to generate predictions for biological parameters that are essential for landscape ecologists – including extinction rates, abundance patterns, and species richness. Where neutral theory fails, additional complexity around biological mechanisms or species-specific data may be necessary, which is an area for hybrid theories that incorporate both neutral and niche-based perspectives. More sophisticated still are detailed mechanistic models that rely explicitly on taxon-specific physiology, dispersal, and trophic strategies to generate detailed predictions of the dynamics of species across landscapes. Other models for single species, which draw from metapopulation dynamics and habitat selection, are detailed elsewhere in this volume (*see* Chapter 6).

The niche concept, which lies at the core of ecological research, argues that each species has a unique, n-dimensional array of environmental tolerances and resource needs (Hutchinson 1957). Consequently, predictive models are strongly and justifiably influenced by the mechanisms that species' niches imply, such as differences in dispersal rates, physiological tolerances, and details of metabolism. Models often begin with substantial sophistication because such mechanisms certainly exist and are thought to determine species abundances and distributions. The distributions of individuals and species among habitat patches in fragmented landscapes, for instance, result from such niche-based predictions.

Neutral theory ignores differences in dispersal, birth, and death rates among all individuals within a community, a perspective imported from population genetics (Bell 2001). Beginning with the observation that biotic communities are saturated with individuals, neutral theory shows that differences in the properties of those communities through time can arise through stochastic processes of ecological drift rather than niche-based determinism (*see* Hubbell 2001 for a detailed discussion). Then, with information on the number of species and individuals within the regional meta-community, the spatiotemporal dynamics of that community can be predicted with remarkable accuracy. Neutral theory provides a theoretical framework predicting rank abundance distributions among biotic communities that range from log-series (highly inequitable, with nearly all individuals in the community drawn from a single species) to broken stick (relatively equitable distribution of individuals among species). It also provides a dispersal-based explanation for differences in the slope of species-area relationships observed in mainland areas relative to islands (i.e., species richness increases with area more slowly on mainlands because dispersal rates are higher in such areas, leading to lower beta diversity) and for the positive correlation between range-size and abundance. Long-term modeling can incorporate speciation, which led Hubbell (2001) to observe a fundamental biodiversity number that emerges repeatedly from his extensive mathematical proofs and that provides a kind of index to emergent properties of meta-communities:

$$\theta = 2J\upsilon,$$

where θ (theta) is the fundamental biodiversity number, J is the number of individuals in the meta-community, and υ (nu) is the speciation rate. By predicting a number of widely observed patterns, such as differences in species-area or range-abundance relationships, neutral theory startled ecologists into a revolutionary reconsideration of many widely supported hypotheses.

2.5.2 A Test of Neutral Theory in Fragmented Landscapes

Experimental evidence from fragmented tropical forests suggests neutral theory is able to predict some, but not all, aspects of community dynamics. The Biological Dynamics of Forest Fragments project in central Amazonia has collected long-term data for an array of changes in forest communities following experimental fragmentation in 1979 (Lovejoy et al. 1983; Laurance et al. 2002). Neutral theory predicts that forest fragmentation should alter species diversity, composition, and extinction rates among tree species in these forest communities. In particular, in an isolated forest patch, random birth–death processes (ecological drift) should favor those tree species that are initially most abundant, while simultaneously making least common species relatively extinction prone. Using observed mortality rates for tree species in experimentally fragmented patches, neutral theory predicts extinction rates among rare species (i.e., that were represented by one or two individuals after experimental fragmentation) and, less accurately, overall extinction rates (Gilbert et al. 2006). Neutral theory underestimated species turnover in fragments, likely as a result of differences among species in their tolerances to rapidly changing environmental conditions within fragments – such as increased light availability and desiccation. In other words, neutral theory fails because niche differences can influence the trajectory of biological communities over relatively short periods of time. Although it is remarkable that predictive models derived from neutral theory could predict some aspects of community dynamics, rapid environmental changes may push communities beyond thresholds within which neutral assumptions remain valid. The result is that ignoring differences among species or individuals is risky, especially in the context of rapid, human-induced environmental change.

2.5.3 Adding Niche into Neutral Models: Stochastic Niche Theory

Including differences among species for resource competition and/or individual growth rates can improve predictions of species and community-level responses to environmental change (Tilman 2004; Zhou and Zhang 2008). Stochastic niche theory (Tilman 2004) seeks to predict how biotic communities are assembled and change over time, focusing on plant communities (although this reasoning could be adapted for animal communities). New invaders within a system, such as a plant arriving in a habitat fragment, can establish if and only if they survive to maturity, which in turn requires their ability to tolerate existing resource concentrations. Thus, the probability that a propagule grows to maturity depends on its chances of mortality through time, m, and the total time taken to reach maturity, y:

$$p = (1-m)^y, \text{where } y = \ln(B_a / B_s)/U$$

U is the organism's growth rate, which depends on resource availability, and B_a/B_s is the ratio of adult body size to seed size, which is an index of how much growth (and therefore resource consumption) must take place before the organism reaches maturity. In this approach, p depends strongly on Tilman's R^* concept, the concentration to which a species reduces a potentially limiting resource. Stochastic niche theory makes survival to adulthood for invaders an improbable chance event and it is able to explain additional biological observations that are not consistent with neutral theory, including correlations among species traits, their abundances, and environmental conditions, all of which are relevant to predictive modeling efforts. Similarly, heterogeneous environments having higher diversity are more likely to contain combinations of resources that cannot be fully exploited by species already present in the regional species pool, making such regions more prone to invasion. Thus, stochastic niche theory offers an explanation for the paradox of invasion, that high diversity locales are less prone to invasion, whereas high diversity regions are more prone to invasion.

Neutral theory or its hybrid niche-neutral cousins may unify much of biogeography and community ecology on the basis of mechanistic, but not necessarily deterministic, processes. This field is evolving rapidly.

2.5.4 *Mechanistic Distribution Models – Early Developments*

Considerably greater mechanistic detail can be resolved for very well-studied species, or assemblages. These mechanistic models take a different approach to modeling species' distributions than the approaches already discussed. Rather than beginning with a set of environmental variables and species occurrence records, mechanistic distribution models (MDMs) begin with a set of processes or mechanistic links between an organism's physiology, ecology, and environment. Such models may include biophysical and ecophysiological relations between environment, behavior, body temperatures, and water requirements (Kearney and Porter 2004); effects of temperature on fecundity or survivorship (Kearney and Porter 2004; Crozier and Dwyer 2006); or interactions between temperature, prey abundance, and metabolic costs of foraging (Buckley 2008). The goal of these studies is to derive an estimate of a species' fundamental niche from first principles as well as lab and field experiments. This can then be used to predict its distribution, or as a basis for more complex models of realized niches that incorporate biotic interactions or dispersal limitation (Buckley 2008).

The most complete example of an MDM is Buckley's (2008) model for the eastern fence lizard (*Sceloporus undulatus*), a territorial sit-and-wait predator. Buckley's model merges an individual-based energetics model with the familiar logistic model of population dynamics to predict the abundance of *S. undulatus* across the continental United States. The model includes two components. The first predicts the energetic yield ($E(d)$) of an individual lizard foraging within a radius (d):

$$E(d) = \left(e_i - e_w t_w(d) - e_p t_p(d)\right) / \left(t_w(d) + t_p(d)\right)$$

where e_i is the energy per prey individual, and e_w and e_p are the per unit time energetic costs of waiting and pursuing, respectively. Lastly, t_w and t_p are time spent waiting and pursuing and are functions of distance (d), prey abundance and lizard velocity. Buckley links this energetic model to population dynamics by the difference equation:

$$\Delta N = [bE(d) - \lambda]N$$

where b is the per capita reproductive rate (number of offspring) per unit of energetic yield and λ encompasses both mortality and the cost of metabolism while not foraging. Since foraging radius for a given transect length is a declining function of N, $E(d)$ decreases with increasing N, resulting in density-dependent population growth.

Buckley parameterized her model using data from experimental and field studies and pre-established relationships between parameters usually based on published regression equations. For example, the cost of metabolism was a function of operating temperature and mass; operating temperature was determined from a biophysical model, whereas mass was calculated from snout–vent length. Energy per prey item was calculated from estimates of average prey mass (converted from average prey length), estimates of joules per unit mass and the percentage of this energy available to lizards, and lizard digestive efficiency which was temperature dependent. Estimates of prey abundance were based on field observations and snout–vent lengths were measured from several distinct populations. The only free parameter in Buckley's model was the proportion of insects consumed by lizards within the foraging radius.

After parameterization, Buckley's MDM predicted the range of *S. undulatus* quite well, especially when population-specific body sizes and life-history parameters were included. In fact, Buckley's model made slightly better overall predictions than a purely empirical distribution model (based on temperature and generated using DesktopGarp). The MDM correctly predicted presences (i.e., model sensitivity) slightly less well than the GARP model, however it was substantially superior at predicting absences (specificity). These results are quite striking and suggest that Buckley's model, despite the necessary level of abstraction and parameter estimation, accurately captures temperature-mediated mechanisms that are highly relevant in determining the distribution of *S. undulatus*.

MDMs have several advantages and disadvantages relative to correlative distribution models, which are exemplified by Buckley's model. The most obvious is that they can directly predict abundances and not just occurrence. Other advantages include the capacity to separate fundamental and realized niches and to include geographic variability in species' traits (Buckley 2008). The primary disadvantage is the extensive level of species-specific information on behavior, diet, prey availability, and thermoregulation that is necessary to construct and parameterize the model. The required level of detail to apply even a relatively simple MDM is high, and obtaining sufficient information – especially for large assemblages of species – may range from extremely challenging to impractical. Furthermore, MDMs increase quickly in complexity as additional niche axes are considered. For example, consider that Buckley's MDM did not include water requirements. The model could be expanded to include these requirements by determining species-specific rates of evaporative water loss as determined by operating temperature, relative humidity, metabolic rate, body shape, activity hours, and water gained from prey following Kearney and Porter (2004). Metabolic costs of having to obtain supplemental water through drinking could then be incorporated via estimates of water availability and foraging time lost. As in any modeling endeavor, the increased realism achieved by including additional processes must be weighed against a loss of generality or practicality and the potential for unnecessary complexity. Detailed knowledge of the focal species' ecology, physiology and behavior will be invaluable in guiding these efforts, but the ultimate judge of a model's success should be its ability to predict changes in a species' abundance and distribution as environments change.

A recognized (Helmuth et al. 2005; Buckley 2008), but unexplored and potentially exciting application of MDMs is their potential ability to explore evolutionary implications of changing environments. Models lacking mechanistic detail are unlikely to address such questions effectively. However, models that incorporate population dynamics could be expanded to include population-based variance in traits, such as mass, foraging ability, or digestive efficiency that improve energetic yield. Such measures, combined with estimates of selection pressure and heritability could allow MDMs to predict when species, or populations, will adapt or decline to extinction in response to changing environmental conditions.

A model's failures can be as informative as its successes. This is especially true of MDMs. MDMs can separate fundamental and realized niche determinants (Buckley 2008). Under-prediction of a species' observed range indicates the model's failure to capture the limits of the fundamental niche along the modeled axis. Alternately, over-prediction may indicate the necessity of including additional fundamental niche axes, or could indicate constraints on the realized niche – such as dispersal limitation, competition, or predation pressures. For example, Buckley (2008) found that her MDM greatly over-predicted the range of *S. graciosus*, a species similar to *S. undulatus* but with a predominantly disjunct range. This result suggests that, whereas the distribution of *S. undulatus* is determined by fundamental niche limits, *S. graciosus* inhabits a significantly reduced portion of its fundamental niche space, possibly as a result of biotic interactions of dispersal limitation. Each of these processes can be included in subsequent models and the improvement in prediction their inclusion conveys can be quantified.

One of the attractions of MDMs for predicting species distributions arises from the perspective that only a mechanistic explanation of species ranges can produce trustworthy predictions (Kearney and Porter 2004; Helmuth et al. 2005). The basis of this perspective is that correlations between environment variables and occurrences may not be causal and thus are likely to fail, especially when predictions are made under novel conditions and in the absence of historical calibration (Willis et al. 2007; Kharouba et al. 2009); alternately, because MDMs use mechanistic interactions to explain species distributions, their predictions should be valid regardless of the novelty of the projected environmental conditions (Kearney and Porter 2004). However, if the goal of distribution modeling is to make predictions under changing environments, then predictive ability, rather than explanation, determines model success. Mechanistic models, even if they capture the critical processes governing current distributions, make assumptions and exclude numerous processes. If changing environments result in the violation of the model's assumptions, or result in a change in the processes limiting species distributions, then the predictions of MDMs will fail when projected through time. Therefore, just as with empirical models, direct tests of a MDMs ability to predict temporal changes in species distributions are essential. Such tests are currently lacking. If such tests find that MDMs make better predictions than empirical models, especially for those species for which empirical models are known to fail, then ecologists will have a powerful tool that not only predicts, but helps satisfy the broadly held scientific desire for explanation and understanding.

2.5.5 Integrating Detailed Mechanisms and Predictions: A Case Study for Malaria

The need for mechanistic models is considerable but there are probably no examples from conservation in which mechanistic details are so well understood as for malaria transmission. There are good reasons for the intensive work on this disease: malaria, which is transmitted by mosquitoes, affects between 350 and 500 million people worldwide each year, with the greatest burden of morbidity in sub-Saharan Africa (WHO 2008), particularly among children.

Epidemiological models may follow the same underlying principles and processes as those used in conservation biology or biogeography. In particular, the ecological techniques applied to species distribution modeling lend themselves to modeling vector-borne diseases, because the insect vectors that transmit diseases such as malaria are strongly influenced by environmental conditions. It is generally considered that models that account for complex biological mechanisms in disease transmission are more robust than those based on one or two *a priori* assumptions, although the latter approach may sometimes suffice to explain large-scale patterns in disease transmission. Climate-based models of malaria risk for the African continent have been developed that apply generalized biological limits for vector and parasite survival to coarse resolution temperature and precipitation data (Craig et al. 1999; Snow et al. 1999). Whereas these simplified models serve as a basis for understanding large-scale temporal and spatial patterns in malaria transmission, they mask small-scale variations which are essential for disease control. Malaria transmission can vary widely over short distances (Hay et al. 2000), and the capacity of health systems is often limited in resource-poor, malaria-endemic countries (WHO 2008); therefore, it is important to target control strategies to increase effectiveness and maximize the use of resources.

Spatial and temporal prediction of vector abundance is a key goal of more complex malaria models because vector abundance is the main factor contributing to the entomological inoculation rate (Bodker et al. 2003). The entomological inoculation rate (EIR) is a measure of malaria transmission intensity defined as the number of infectious bites received by a person in one year. The annual EIR is the product of the sporozoite rate (the proportion of vectors carrying the infectious stage of malaria parasite) and the human-biting rate (the number of bites received per person per year). EIR is the best estimate of the risk of mosquito-borne malaria infection (Smith et al. 2004). Ecological models are limited in their ability to directly predict vector abundance (Rogers et al. 2002); however, the proximity to suitable vector habitat provides an index of this critical parameter (Bogh et al. 2007). In contrast to purely climate-based models, high-resolution models that incorporate satellite-derived measures of land cover and/or topography as indices of vector habitat are better able to identify local variations in malaria risk (e.g., Balls et al. 2004; Bogh et al. 2007). These may serve as practical tools in the development of targeted malaria control strategies; however, it may be difficult to extrapolate these models to other geographic areas unless the underlying ecological and biological mechanisms can be taken into account. Niche modeling methods that have emerged in the field of conservation biology, in particular those based on maximum entropy algorithms (Phillips et al. 2006), may provide a robust means to extrapolate detailed, mechanistic models across broader areas.

The niche modeling approach may prove particularly useful in the context of malaria transmission, where the mechanisms have been studied in detail (see Killeen et al. 2000; Depinay et al. 2004). For example, temperature-dependent effects on the rate of mosquito larval development and the duration of the gonotrophic cycle largely determine the seasonal dynamics of vector populations and the frequency of blood-feeding (Teklehaimanot et al. 2004). Vector breeding is also dependent on precipitation, because mosquito larval stages are aquatic, although paradoxical associations between precipitation and malaria transmission may arise due to washing out of breeding sites by excessive rainfall (Lindsay et al. 2000). Among these are adult mosquito dispersal capabilities: anophelene mosquitoes that transmit the particularly dangerous form of malaria caused by *Plasmodium falciparum* are able to disperse several kilometers, or possibly farther, from breeding habitat. Factors affecting dispersal patterns in mosquitoes will affect disease transmission rates even in areas that are unsuitable mosquito habitat but that are within dispersal distance of suitable breeding sites. High adult dispersal rates among these disease vectors also

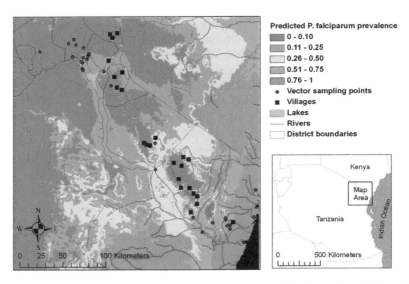

Fig. 2.4 Map of north eastern Tanzania (East Africa) showing the predicted prevalence of Plasmodium falciparum malaria in children. The malaria risk model algorithm is based on field-derived parasitological data, and incorporates the predicted presence of suitable malaria vector habitat, derived from maximum entropy niche models of the three major malaria vector species in Tanzania, and altitude. High resolution risk maps can assist malaria control efforts by identifying potential hot spots of malaria transmission. This is particularly important in the epidemic-prone highland areas of East Africa which have experienced a recent resurgence in malaria. Predicted malaria prevalence in these areas matches observed prevalence unexpectedly closely ($R^2 = 0.73$).

increases the likelihood that vectors will successfully shift in response to regional warming in malaria-affected regions, including up mountain slopes into areas that have traditionally experienced low malaria transmission rates. Thus, critical survival thresholds for mosquitoes are useful to define coarse ecological limits for vector species, but finer gradients in habitat suitability can be predicted on the basis of more complex ecological and biological mechanisms.

To demonstrate a novel ecological approach to malaria modeling, a maximum-entropy method was used to model malaria vector distributions of species in a highland area of northeastern Tanzania, and the resulting models were used to generate a high-resolution map of malaria risk. Niche models incorporated climatic indices of temperature and precipitation – including means, critical levels, and measures of seasonality, human population density, elevation, and land cover classified from Landsat TM imagery – to predict areas suitable for the three major vector species: *Anopheles gambiae*, *An. arabiensis,* and *An. funestus*. Species were modeled individually to account for differences in larval and adult ecology. The amount of suitable vector habitat within a 1.5 km radius buffer area, corresponding to a maximum potential dispersal distance (Takken et al. 1998), was defined for each 30 m pixel. By combining this measure of vector habitat, derived from vector species' niche models, with altitude-dependent effects on the malaria parasite (see Drakeley et al. 2005), the model successfully predicted 73% of the local variation in observed malaria prevalence in children (Fig. 2.4). Models such as this may be used to identify hot spots of local transmission, which is of particular importance in highland areas that are prone to seasonal malaria epidemics.

The capacity to predict actual malaria prevalence among children in areas of sub-Saharan Africa depends not on the sophistication of the modeling technique, but on the knowledge of the mechanisms causing disease transmission. In the specific case of malaria models, it is the inclusion of the mechanistic or theoretical detail that enables successful prediction. But, the best proof that these, or any, models have "an objective grip on reality" (Gottfried and Wilson 1997) is that they are able to predict successfully, a lesson that applies equally to modeling activities in global change biology, conservation biology, or epidemiology.

Acknowledgments All authors would like to acknowledge research support from the Natural Sciences and Engineering Research Council, as well as infrastructure and research support from the Canadian Foundation for Innovation and the Ontario Ministry of Research and Innovation. We are grateful to three anonymous reviewers for their assistance in improving this work.

References

Algar AC, Kharouba HM, Young EM, Kerr JT (2009) Predicting the fu
ture of biodiversity: direct tests of alternate forecasting methods. Ecography 32:22–33.
Anderson RP, Lew D, Peterson AT (2003) Evaluating predictive models of species' distributions: criteria for selecting optimal models. Ecol Modell 162:211–232.
Araújo MB, New M (2007) Ensemble forecasting of species distributions. Trends Ecol Evol 22:42–47.
Austin MP (2002) Spatial prediction of species distribution: an interface between ecological theory and statistical modelling. Ecol Model 157:101–118.
Bahn V, McGill BJ (2007) Can niche-based distribution models outperform spatial interpolation? Glob Ecol Biogeogr 16:733–742.
Balls MJ, Bodker R, Thomas CJ, Kisinza W, Msangeni HA, Lindsay SW (2004) Effect of topography on the risk of malaria infection in the Usambara Mountains, Tanzania. Trans R Soc Trop Med Hyg 98:400–408.
Bell G (2001) Neutral macroecology. Science 293:2413–2418.
Bodker R, Akida J, Shayo D, Kisinza W, Msangeni HA, Pedersen EM, Lindsay SW (2003) Relationship between altitude and intensity of malaria transmission in the Usambara Mountains. Tanzania J Med Entomol 40:706–717.
Bogh C, Lindsay SW, Clarke SE, Dean A, Jawara M, Pinder M, Thomas CJ (2007) High spatial resolution mapping of malaria transmission risk in The Gambia, West Africa, using Landsat TM satellite imagery. Am J Trop Med Hyg 76(5):875–881.
Box GEP (1979) Some problems of statistics and everyday life. J AM Stat Assoc 74:1–4.
Brown JH, Kodric-Brown A (1977) Turnover rates in insular biogeography: effect of immigration on extinction. Ecology 58:445–449.
Buckley LB, Roughgarden J (2004) Biodiversity conservation: effects of changes in climate and land use. Nature 430:2.
Buckley LB (2008) Linking traits to energetics and population dynamics to predict lizard ranges in changing environments. Am Nat 171:E1–E19.

Craig MH, Snow RW, le Sueur D (1999) A climate-based distribution model of malaria transmission in sub-Saharan Africa. Parasitol Today 15:105–111.

Crozier L, Dwyer G (2006) Combining population dynamic and ecophysiological models to predict climate-induced insect range shifts. Am Nat 167:853–866.

Dennis RLH (1993) Butterflies and climate change. Manchester University Press, Manchester.

Depinay J-M, Mbogo CM, Killeen G, Knolls B, Beier J, Carlson J, Dushoff J, Billingsley P, Mwambi H, Githure J, Toure AM, McKenzie FE (2004). A simulation model of the African *Anopheles* ecology and population dynamics for the analysis of malaria transmission. Malar J 3: 29.

Dillon PJ, Rigler FH (1974) The phosphorus-chlorophyll relationship in lakes. Limnol Oceanogr 19:767–773.

Diniz-Filho JAF, Bini LM, Hawkin BA (2003) Spatial autocorrelation and red herrings in geographical ecology. Glob Ecol Biogeogr 12:53–64.

Drakeley C, Carneiro I, Reyburn H, Malima R, Lusingu JP, Cox J, Theander TG, Nkya WM, Lemnge M, Riley EM (2005) Altitude-dependent and -independent variations in *Plasmodium falciparum* prevalence in northeastern Tanzania. J Infect Dis 191:1589–1598.

Gilbert B, Laurance W, Leigh Jr. E, Nascimento H (2006) Can neutral theory predict the responses of Amazonian tree communities to forest fragmentation? Am Nat 168:304–317.

Gottfried K, Wilson KG (1997) Science as a cultural construct. Nature 386:545–547.

Guisan A, Zimmermann NE, Elith J, Graham CH, Phillips S, Peterson AT (2008) What matters for predicting the occurrences of trees: techniques, data, or species' characteristics? Ecol Monogr 77:615–630.

Hay SI, Rogers DJ, Toomer JF, Snow RW (2000) Annual *Plasmodium falciparum* entomological inoculation rates (EIR) across Africa: literature survey, internet access and review. Trans R Soc Trop Med Hyg 94:113–126.

Helmuth B, Kingsolver JG, Carrington E (2005) Biophysics, physiological ecology and climate change: does mechanism matter? Ann Rev Physiol 67:177–201.

Hubbell SP (2001) The unified neutral theory of biodiversity and biogeography. Princeton University Press, Princeton.

Hutchinson GE (1957) Concluding remarks. Cold Spring Harb Symp Quant Biol 22:415–427.

Kearney M, Porter WP (2004) Mapping the fundamental niche: physiology, climate, and the distribution of a nocturnal lizard. Ecology 85:3119–3131.

Kerr JT, Kharouba HM, Currie DJ (2007) The macroecological contribution to global change solutions. Science 316:1581–1584.

Kharouba HM, Algar AC, Kerr JT (2009) Historically calibrated predictions of butterfly species' range shift using global change as a pseudo-experiment. Ecology 90:2213–2222.

Killeen GF, McKenzie FE, Foy BD, Schieffelin C, Billingsley PF, Beier JC (2000) A simplified model for predicting malaria entomological inoculation rates based on entomologic and parasitologic parameters relevant to control. Am J Trop Med Hyg 62(5):535–544.

Koenig WD (1999) Spatial autocorrelation of ecological phenomena. Trends Ecol Evol 14:22–26.

Kukal O, Ayres MP, Scriber JM (1991) Cold tolerance of the pupae in relation to the distribution of swallowtail butterflies. Can J Zool 69:3028–3037.

Laurance WF, Lovejoy TE, Vasconcelos HL, Bruna HM, Didham RK, Stouffer PC, Gascon C, Bierregaard RO, Laurance SG, Sampiao E (2002) Ecosystem decay of Amazonian forest fragments: a 22-year investigation. Conserv Biol 16:605–618.

Legendre P, Legendre L (1998) Numerical ecology, 2nd edn. Elsevier Science, Amsterdam.

Lindsay SW, Bodker R, Malima R, Msangeni HA, Kisinza W (2000) Effect of 1997–98 El Nino on highland malaria in Tanzania. Lancet 355:989–990.

Lovejoy TE, Bierregaard Jr. RO, Rankin JM, and Schubart HOR (1983) Ecological dynamics of forest fragments. In: Sutton SL, Whitmore TC, and Chadwick AC, (eds) Tropical rain forest: ecology and management. Blackwell Scientific, Oxford, United Kingdom 377–384.

Luoto M, Pöyry J, Heikkinen RK, Saarinen K (2005) Uncertainty of bioclimate envelopemodels based on geographical distribution of species. Glob Ecol Biogeogr 14:575–584.

McLean A, May RM (2007) Introduction. In: May RM, McLean A (eds) Theoretical ecology: principles and applications. Oxford University Press, Oxford.

Menendez R, Gonzalez Megias A, Hill JK, Braschler B, Willis SG, Collingham Y, Fox R, Roy DB, Thomas CD (2006) Species richness changes lag behind climate change. Proc R Soc B 273:1465–1470.
Parmesan C. 1996. Climate and species' range. Nature 382:765–766.
Parmesan C (2005) Biotic response: range and abundance changes. In: Lovejoy TE, Hannah L (eds) Climate change and biodiversity. Yale University Press, New Haven.
Parmesan C, Ryrholm N, Stefanescus C, Hill JK, Thomas CD, Descimon H, Huntley B, Kaila L, Kullberg J, Tammaru T, Tennent WJ, Thomas JA, Warren M (1999) Poleward shifts in geographical ranges of butterfly species associated with climate change. Nature 399:579–583.
Pearson RG, Thuiller W, Araújo MB, Martinez-Meyer E, Brotons L, McClean C, Miles L, Segurado P, Dawson TP, Lees DC (2006) Model-based uncertainty in species range prediction. J Biogeogr 33:1704–1711.
Peterson AT, Martinez-Meyer E, Gonzalez-Salazar C, Hall PW (2004) Modeled climate change effects on distributions of Canadian butterfly species. Can J Zool 82:851–858.
Phillips SJ, Anderson RP, Schapire RE (2006) Maximum entropy modeling of species geographic distributions. Ecol Modell 190:231–259.
Raxworthy CJ, Martinez-Meyer E, Horning N, Nussbaum RA, Schneider GE, Ortega-Huerta A, Peterson AT (2003) Predicting distributions of known and unknown reptile species in Madagascar. Nature 426:837–841.
Rogers DJ, Randolph SE, Snow RW, Hay SI (2002) Satellite imagery in the study and forecast of malaria. Nature 415:710–715.
Root TL (1988) Environmental factors associated with avian distributional boundaries. J Biogeogr 15:489–505.
Root TL, Price JT, Hall KR, Schneider SH, Rosenzweig C, Pounds JA (2003) Fingerprints of global warming on wild animals and plants. Nature 421:57–60.
Rosenzweig C, Karoly D, Vicarelli M, Neofotis P, Wu Q, Casassa G, Menzel A, Root TL, Estrella N, Seguin B, Tryjanowski P, Liu C, Rawlins S, Imeson A (2008) Attributing physical and biological impacts to anthropogenic climate change. Nature 453:353–357.
Scott JM, Heglund PJ, Morrison ML, Haufler JB, Raphael MG, Wall WA, Samson FB (2002) Predicting species occurrences: issues of accuracy and scale. Island Press, Washington, DC.
Smith DL, Dushoff J, McKenzie FE (2004) The risk of a mosquito-borne infection in a heterogeneous environment. PLoS Biol 2:e368.
Snow RW, Craig MH, Deichmann U, le Sueur D (1999) A preliminary continental risk map for malaria mortality among African children. Parasitol Today 15:99–104.
Takken W, Charlwood JD, Billingsley PF, Gort G (1998) Dispersal and survival of *Anopheles funestus* and *A. gambiae*s.L. (Diptera: Culicidae) during the rainy season in southeast Tanzania. Bull Entomological Res 88:561–566.
Teklehaimanot H, Lipsitch M, Teklehaimanot A, Schwartz J (2004) Weather-based prediction of *Plasmodium falciparum* malaria in epidemic-prone regions of Ethiopia I. Patterns of lagged weather effects reflect biological mechanisms. Malar J 3:41.
Thomas CD (2000) Dispersal and extinction in fragmented landscapes. Proc R Soc B 267: 139–145.
Thomas CD, Cameron A, Green RE, Bakkenes M, Beaumont LJ, Collingham LC, Erasmus BFN, Ferreira de Siqueira M, Grainger A, Hannah L, Hughes L, Huntley B, van Jaarsveld AS, Midgley GF, Miles L, Ortega-Huerta MA, Peterson AT, Phillips OL, Williams SE (2004) Extinction risk from climate change. Nature 427:145–148.
Thuiller W (2003) BIOMOD – optimizing predictions of species distributions and projecting potential future shifts under global change. Glob Change Biol 9:1353–1362.
Tilman D (2004) Niche tradeoffs, neutrality, and community structure: a stochastic theory of resource competition, invasion, and community assembly. Proc Natl Acad Sci USA 101:10854–10861.
Wake DB, Vredenburg VT (2008) Are we in the midst of a sixth mass extinction? A view from the world of amphibians. Proc Natl Acad Sci USA 105:11466–11473.
White PJ, Kerr JT (2006) Contrasting spatial and temporal global change impacts on butterfly species richness during the 20th century. Ecography 29:908–918.

Willis KJ, Araujo MB, Bennett KD, Figueroa-Rangel B, Froyd CA, Myers N (2007) How can a knowledge of the past help to conserve the future? Biodiversity conservation and the relevance of long-term ecological studies. Philos Trans R Soc B 362:175–186.

WHO (2008) World malaria report 2008. World Health Organization/UNICEF, Geneva/New York.

Wilson RJ, Gutierrez D, Gutierrez J, Martinez D, Agudo R, Monserrat VJ (2005) Changes to the elevational limits and extent of species ranges associated with climate change. Ecol Lett 8:1138–1146.

Woodward FI (1987) Climate and plant distribution. Cambridge University Press, Cambridge.

Zhou SR, Zhang DY (2008) A nearly neutral model of biodiversity. Ecology 89:248–258.

Chapter 3
The State of Spatial and Spatio-Temporal Statistical Modeling

Mevin B. Hooten

3.1 Introduction

The purpose of this chapter is to provide an overview of how statistical analyses have been used for studying ecological processes on landscapes and where the field of statistics is headed in general. Various approaches to the statistical analysis of spatial and spatio-temporal problems are presented and discussed; also, references for several suggested readings, containing further information and examples, are provided at the end of each section.

3.1.1 Why Statistics?

Scientific endeavor owes a great debt of gratitude to pioneers in the field of statistics, a relatively young area of study that has undergone significant change (including paradigm shifts, as well as both splits and merges in philosophical underpinnings) since its inception (Stigler 1990; Brown 2000). Historical flux in the field of statistics aside, one thing can be said with certainty: much of the scientific progress made in the past century would not have been possible without it (Salsburg 2001). Contemporary statistical analyses now encompass an incredibly wide range of methods, some of which can be used to study very complicated natural systems while still following the original basic tenet of statistics; that is, formally addressing and characterizing uncertainty when using data to learn about natural phenomena in an inverse fashion. Inverse modeling is the act of using data explicitly to learn about the underlying causal process; this is in contrast to

M.B. Hooten (✉)
USGS Colorado Cooperative Fish and Wildlife Research Unit, Department of Fish, Wildlife, and Conservation Biology, Colorado State University, 201 JVK Wagar Bldg, 1484 Campus Delivery, Fort Collins, CO 80523-1484
e-mail: mevin.hooten@colostate.edu

forward modeling, where models are constructed to simulate possible future observations of the process under study. In light of this distinction, the act of fitting statistical models to data is inherently stochastic inverse modeling. So, one may ask, "where do statistical prediction and forecasting fit in?" These are often thought of in a forward modeling context, but chronologically, predictions and forecasts depend on the model fit.

Landscape ecology is a field concerned with the study of natural processes over large spatial extents. Many introductory statistical methods require strong assumptions that can be difficult for scientists and managers to justify in practice. One of the most commonly required assumptions is that of independence among the observations, however, most spatial, temporal, and spatio-temporal data are explicitly dependent because of latent spatial and temporal autocorrelation. The term "autocorrelation" refers to data or residuals (depending on the context) that are correlated with themselves rather than independent. This can present significant challenges in the development and implementation of appropriate statistical analyses. On the other hand, these forms of dependence are invaluable for making scientifically meaningful predictions and forecasts. Rigorous statistical approaches (as opposed to *ad hoc* approaches) then, allow one to formally quantify the inherent uncertainty in predictions and forecasts. The difference between statistical and *ad hoc* approaches is often described as "optimality;" that is, rather than incorporate stochasticity in a haphazard manner, statistical methods provide the best estimates and predictions, using the available data. This is accomplished by ensuring that the estimates have good properties (e.g., unbiasedness, minimum variance among all other estimates). Therefore, statistical estimates and predictions with high quality properties lead to the best possible scientific inference given the available data.

3.1.2 Main Types of Data

As data will be discussed in detail in the following chapter, only a general overview is presented here. Statistical analyses of all kinds require quantitative measurements of the process under study. Qualitative observations certainly have their place in the scientific process (usually in the development of hypotheses and interpretation of inference), but statistical methods are not currently equipped to use them directly. Therefore, those measurements useful for statistics come in two broad varieties: discrete, most commonly in the form of counts or quantified categories, and continuous, often measurements of mass in some form. Adequately accommodating various types of data is one of the chief concerns in statistics. To guarantee important properties of statistical quantities one must pay careful attention to the many characteristics of data (e.g., orientation, scale, measurement error, dependence) as well as modeling assumptions (Hilborn and Mangel 1997).

3.2 Statistical Models

3.2.1 Parameters: Fixed or Random?

In general, data are considered as observed random variables; the probability distribution they arise from is generally of interest for making inference. When the form of the distribution is specified as part of the statistical model, the parameters in the distribution often become the subject of interest and thus the approach is labeled "parametric" statistical modeling. Conversely, non-parametric statistical modeling seeks to loosen the distributional assumptions made *a priori*. Non-parametric models are discussed in more detail in Sect. 3.2.5.

In parametric modeling, parameters are predominantly treated as fixed but unknown population quantities to be estimated using data. It should be noted, however, that some of the earliest statistical models [e.g., Laplace's model for astronomical quantities and Bayes' model for billiard balls (Stigler 1990)] considered parameters to be random variables, the probability distribution of which was to be estimated using the data (Carlin and Louis 2000; Salsburg 2001). The difference between the two views leads to a subtle but fundamentally different implementation and interpretation of the results (Clark 2007; Cressie et al. 2009). That is, both forms of modeling are still considered statistical because they serve to formally help learn about natural phenomena in an inverse fashion while accommodating uncertainty (that is, they work backwards from the data toward the parameters, rather than forward from the parameters to the data) (Clark 2007); the primary difference then, is in the resulting inference. If one believes that the true parameters governing the process are indeed fixed quantities then a carefully designed experiment and accompanying frequentist statistical analysis is in order. In this case, inference will be made in terms of long-run frequencies, and thus statements such as, "if the experiment were conducted a large number of times, we would expect to make the same decision approximately 95% of the time," are used to convey the results of the analysis.

On the other hand, the treatment of model parameters as random variables can be useful if data are observational or if measurements were obtained in such a manner that does not guarantee the assumption of fixed parameters holds; in this case, either a frequentist or Bayesian approach (i.e., methods with specification and inference based on conditional probability) may be taken. For example, if a set of measurements is collected over a period of time (which is nearly always the case in ecological studies) then one has to ask themselves if the unknown "population parameter" varied during that time. The same analogy could be applied to measurements collected over space, and thus it is often most appropriate to treat such parameters as random and employ methods that characterize the manner in which they are random (e.g., estimate the inherent stochasticity via a probability distribution). Another situation where random terms can be useful in a statistical model commonly arises in analysis of variance (Neter et al. 1996). To illustrate this, consider the situation where, out of 100 study sites, a random sample of 10 is selected. At each of the selected study sites, a sample of stationed field technicians collects data at their site. In this situation, if the researcher wishes to make general inference about the whole set of one hundred study sites, rather than each of the ten selected sites individually, they could let the ten sites constitute ten levels of a random factor in their analysis.

Frequentist and Bayesian statistical methods can both be employed to help learn about random parameters. Common frequentist approaches to dealing with random parameters are often in conjunction with fixed parameters and take the form of mixed models (Neter et al. 1996, pgs. 978–981) and state-space models (Chap. 6 of Shumway and Stoffer 2006). Many times Bayesian methods are preferred for modeling random parameters (Cressie et al. 2009) because of their flexibility, ease in specification and implementation for complex models, and the ability to directly incorporate prior scientific information (e.g., conclusions resulting from different data or historically documented quantities in the literature). The frequentist approach to parameter estimation is still preferred when frequency-based inference (e.g., confidence intervals) or objectivity in the parameters is desired (Lele et al. 2007), though numerous objective Bayesian methods exist for fitting various statistical models to data (Gelman et al. 2004). Given that statistical analysis can proceed in either fashion, an important question in the model construction phase is whether model parameters should be treated as fixed, random, or some combination of both (i.e., a mixed model).

3.2.2 Naïve Models

Conventionally, the dominant type of statistical model used to study natural processes is specified in such a manner that its form facilitates implementation and capitalizes on the rigor of study design in controlled experiments. For example, the linear regression model is often a model of choice for linearly linking a response variable to a set of covariates (Neter et al. 1996), such as in linking coyote abundance to a set of environmental variables like canopy openness, distance to nearest house, or distance to paved road, as discussed by Kays et al. (2008). If specified with independent additive Gaussian error, numerous beneficial properties of the estimated regression coefficients and predictions can be exploited for inference (Christensen 2002). In fact, rather than specify a more scientifically meaningful model (e.g., a nonlinear model with multiplicative error for example), it is a common practice to perform various transformations of the response and/or covariate data to justify necessary assumptions and the use of a linear model. In other words: when you've got a hammer, every problem starts to look like a nail.

It should be noted, however, that a simplified analysis does not imply a useless analysis. That is, naïve models such as linear models with continuous response variables and additive error can be fit easily, and in situations where statistical assumptions hold, they can be readily used to make valid inference about underlying natural processes of interest. Such models can be thought of as "structurally parsimonious," in the sense that they are sparse on model structure, but still useful under certain circumstances. Also, linear models are not always trivial. Consider the common situation in landscape ecological studies where a continuous response variable is measured over a landscape (e.g., soil moisture) and the researcher wishes to investigate its relationship to other important environmental features (i.e., covariates) of the landscape (e.g., slope, aspect, elevation, percent vegetation cover) and also possibly utilize that information to make predictions. In order to appropriately employ multiple linear regression analysis for making inference about the natural process, several model assumptions must be justified. A critical assumption that is often overlooked in such analyses is that the additive model errors are independent. In fact, assessing and accommodating possible spatial dependence in the errors is the premise of geostatistics (Cressie 1993; Diggle and Ribeiro 2007). Erroneous inference is one consequence of failing to account for dependence in the error when it is present (Chap. 9 in Waller and Gotway 2004). This fact is easily shown, and often used as an early exercise in a course on spatial statistics. In short, if residual spatial dependence exists, parameter estimates can be both biased and have incorrect precision (Chap. 6 in Schabenberger and Gotway 2005).

Generalizing the linear model specification used in regression analysis to accommodate residual spatial dependence is relatively simple yet adds significant complexity to the fitting procedure. That is, rather than assume observations can be modeled by a large-scale trend (involving spatial covariates and an associated set of regression coefficients) plus some independent measurement error, we wish to allow for possible dependence in these additive errors. In this way, any potential residual autocorrelation beyond what can be explained by the large-scale trend may be accounted for. In most cases, such autocorrelation in the errors can be characterized via variogram estimation and modeling (Chap. 2 in Cressie 1993), and then incorporated into the linear model for parameter estimation and prediction. On the surface, the regression model still looks the same (i.e., response=covariate effects+error), though the incorporation of correlated error necessitates a slightly more complicated estimation procedure for the regression coefficients (i.e., generalized least squares rather than ordinary least squares).

Once the residual spatial autocorrelation is taken into account, the prediction of continuous spatial processes is referred to as Kriging (Chap. 3 in Cressie 1993) and can be employed, under certain distributional assumptions, with relative ease; in fact, this can often be accomplished at the click of a button in many geographic information systems (GIS) and statistical software. Many types of Kriging have been developed for spatial prediction in various circumstances. For example, Ver Hoef et al. (2006) provide a method for extending Kriging from the standard Euclidean setting to stream networks.

When the response variable of interest has discrete support (e.g., presence/absence or counts of an organism at various locations across a landscape) similar naïve models can be useful for linking the observed natural process to a set of covariates. These models are referred to as generalized linear models and many specifications require similar assumptions about independence of errors, but can also be modified to accommodate correlated errors if necessary (Chap. 6 in Schabenberger and Gotway 2005). Some of the most common generalized linear models are for binary data (i.e., logistic and probit regression) and can be used for studying presence/absence or occupancy (Royle and Dorazio 2008). For example, Hooten et al. (2003) and Gelfand et al. (2006) present similar approaches for modeling vegetation abundance on a landscape using generalized linear models and binary data. In the former, Hooten et al. (2003) use presence/absence data on forest understory legumes (i.e., *Desmodium glutinossum* and *D. nudiflorum*) collected over a large number of plots spread across a Southern Missouri watershed as part of the Missouri Ozark Forest Ecosystem Project. In this study, large-scale spatial predictions of these plant distributions were desired. Thus, a generalized linear mixed model was specified to explicitly accommodate the binary data while characterizing the underlying probability of presence in terms of a set of spatial covariates (i.e., aspect, elevation, land type, and soil depth) and latent spatial autocorrelation. This model allowed for the prediction of probability of presence across the entire study area as well as provided maps of prediction standard deviation as a measure of uncertainty in the predictions.

In situations where boundless counts of organisms are the response variable of interest, a Poisson regression approach can be taken (see Royle and Dorazio 2006 for an example of avian abundance modeling). Another study, by Royle and Wikle (2005), discusses a generalized linear model for predicting avian abundance across the Eastern United States using North American Breeding Bird Survey (BBS) data. In their study, they assumed that BBS route counts of species followed a Poisson distribution. They incorporated covariate effects and spatial autocorrelation in the much same manner as Hooten et al. (2003), but in this case, rather than probability of presence; they linked these spatial effects to the log of the Poisson intensity parameter. This allowed Royle and Wikle to make large-scale predictive maps for bird abundance (specifically for Carolina Wren in this study) as well as maps of prediction uncertainty.

Numerous significant scientific findings have benefited from the use of a naïve model structure such as the linear model, but new tools have come to light with advancements in statistical theory and the advent of high performance personal computers.

3.2.3 Scientific Models

In this section, "scientific" is used to describe those statistical models that explicitly incorporate mathematical and/or physical processes. Such specifications are often most useful for studying time-evolving natural processes because they can incorporate explicit dynamic behavior (Hilborn and Mangel 1997). Because landscape ecology involves the study of spatial systems, relevant statistical models with a temporal component are termed spatio-temporal.

The study of dynamical systems has a long history in both pure and applied mathematics but only recently has it become prominent in statistics. As with static systems such as the spatial-only examples of the last section, naïve statistical models can be employed for studying temporal systems. In these cases, the "dynamics" (i.e., the components of the model controlling the change in the system being studied over time) are expressed in a general form that may be flexible but lacks a direct scientific interpretation. The temporal autoregressive specification is an example of a naïve time-series model where the form contains a distinct dynamic component but in most cases is over-simplified (Chap. 9 in Clark 2007). Employed in an ecological setting, such models may capture dynamic behavior and can often be useful for making inference but are not built on formal principles of ecological theory. Hooten and Wikle (2007) provide an example of a naïve spatio-temporal model that was used for studying the changes in dynamics of forest growth. In this study, a vector autoregressive model is used to analyze the differences in a reduced dimensional dynamical system (representing the spatio-temporal growth in shortleaf pine forests) before and after an anthropogenically created change-point and in response to climatic fluctuation. Their findings included a notable acceleration in the temporal evolution of shortleaf pine growth after a massive clearing of forest at the turn of the twentieth century and also in response to periods of drought. This model can be considered naïve because, although it is dynamic, the dynamics are represented by a simple autoregressive evolution equation where the estimated parameters have no inherent scientific meaning or interpretation.

In contrast to naïve models, scientific models for studying ecological systems on a landscape over time explicitly incorporate meaningful physical processes. For example, the diffusion (i.e., dispersal in ecological terms) of a natural phenomenon through a medium can be expressed using a number of different mathematical models such as:

- *Integro-Difference Equations (IDE)*: Wikle (2001) models a dynamic atmospheric process by integrating the product of two functions; one describing the increase in cloud intensity over time and the other represented by a spatial redistribution kernel that describes cloud spread over time. The distinguishing characteristic in IDE models is that the dynamical component operates via integration.
- *Partial Differential Equations (PDE)*: The mathematical opposite of integration is differentiation and this can also be a reasonable way to describe some natural dynamic processes. For example, Wikle (2003) places a spatio-temporal PDE model into a statistical framework for describing the spread of an invasive species over the North American continent using BBS count data. These models are distinguished from IDE models by the fact that the change in the underlying process of interest (e.g., bird abundance) is expressed in terms of spatial and temporal derivatives.
- *Markov Matrix Equations*: Another approach to modeling both spreading and growing phenomena is through the use of matrix models (Caswell 2001). Though typically employed to study changes in population demographics, matrix models can also be placed in a statistical spatio-temporal context. For example, Hooten et al. (2007) use a spatio-temporal matrix model to characterize and forecast the invasion of the Eurasian Collared-Dove in North America. While PDE and IDE models are inherently continuous in time and space, matrix models are derived explicitly in a discrete spatio-temporal setting.

- *Agent-Based Models*: Though the class of agent-based models is quite large, many consider them to include individual-based models. In general, agent-based models can be thought of as bottom-up models (as opposed to top-down), and are constructed by specifying how a small scale process behaves and then scaling them up to examine their larger-scale properties (Grimm and Railsback 2005). Given that this is how many believe all natural systems work, agent-based modeling has great potential. Hooten and Wikle (2010) construct a bottom-up statistical model to describe a spreading epidemiological process. Specifically, they specify a spatio-temporal cellular automata model that is capable of characterizing the complex dynamical behavior of the rabies epidemic as it spreads through raccoon populations in Connecticut.

3.2.4 Hierarchical Models

Many of the naïve models discussed earlier can be specified hierarchically (and are in many of the references provided), though complicated scientific models (including spatial and spatio-temporal dynamic models) can often be formulated with ease using a hierarchical framework. It is important to note here that the term "hierarchical," though always similar in spirit, is used differently across disciplines. In statistics, a hierarchical model is one that specifies a complicated joint probability distribution in terms of a set of simpler conditional distributions using well-known results from probability (see Cressie et al. 2009 for an excellent overview). In essence, this allows the modeler to break up a large intractable problem into a set of simpler problems that can be readily solved. Though the details are technical in nature, the basic premise is intuitive. Bayesian methods are particularly useful for specifying and fitting hierarchical models and thus have become very popular recently in complicated statistical analyses. When specifying a Bayesian hierarchical model, one can generally consider three main components (Berliner 1996): The data model (i.e., likelihood), the process model, and the parameter model (i.e., prior distribution). The product of these three appropriately scaled models (i.e., probability distributions) yields the "posterior distribution," a joint distribution of the model parameters and process given the data. This distribution is generally unknown and analytically intractable, hence the need to specify it in terms of a set of simpler conditional models. The first and second components (i.e., the likelihood and process model) by themselves have been considered from a traditional perspective in statistics, where it is the third component (i.e., the prior) that is both necessary and useful in a Bayesian implementation. In principle, the prior distribution contains all of the information about the model parameters that is available before the current data were collected. In practice, it is sometimes the case that little or no prior information exists, or if so, cannot be specified in terms of a probability distribution; thus, in such cases, a non-informative or vague distribution is then used for the parameter model so that any statistical learning is forced to come from the data rather than from an exogenous source. The posterior distribution is then used to make inference and can be thought of as the distribution of the process and parameters that has been updated (from the prior) using the data.

For further detail and examples of statistical hierarchical modeling in ecology see Banerjee et al. (2004), Zhu et al. (2005), Clark and Gelfand (2006), Arab et al. (2007), and Royle and Kery (2007). Note that although the methodological details may be technical and custom software is often necessary, many statistical packages, tutorials, and open-source code for fitting such models are readily available. One caveat is that, in high-dimensional settings, the algorithms used to implement hierarchical Bayesian models can be computationally cumbersome. Ecological science is currently transitioning from being data-poor to data-rich. With GIS layers, automated monitoring devices, and remotely sensed data becoming more prevalent, the field of statistics is rapidly adapting to meet such challenges. New methods for using statistical models with massive datasets are being regularly developed for the purposes of obtaining predictions in high-dimensional settings. Furrer et al. (2006) and Shi and Cressie (2007) have developed methods for modeling high-dimensional spatial datasets and illustrate their utility using examples pertaining to satellite data. Specifically, Shi and Cressie (2007) employed fixed rank kriging (Cressie and Johannesson 2006) to obtain global predictions of atmospheric aerosols (an important forcing component for climate models) using massive remotely sensed datasets resulting from the MISR sensor on NASAs Terra satellite. These rigorous statistical methods are shown by Shi and Cressie (2007) to be superior over the previously used *ad hoc* approaches.

3.2.5 Semi- and Non-parametric Models

In the current scientific era, statistics is transforming from a field that sought to get as much information as possible out of a small amount of data into a field that needs to reduce the dimensionality of the data before gleaning any useful information. Where scientific modeling seeks to explicitly introduce information about the underlying physical process into the model (which could be thought of as supervised model building), the statistical sub-discipline of machine learning (i.e., data mining) seeks to uncover naturally occurring relationships in data rather than build in predefined ones (Hastie et al. 2001).

Non-parametric statistical methods generally take a distribution-free approach to modeling. That is, they seek to make as few a priori assumptions about the data as possible. This can be incredibly valuable when data do not conform to conventional modeling assumptions and/or when no scientific modeling approach is obvious or available. Many non-parametric methods are able to fit data and make predictions extremely well, often better than any other technique; the only downside is that they sacrifice scientific interpretability. That is, because they do not explicitly build scientific information into the model itself, they must rely on *post hoc* scientific interpretations and inference. In fact, even though methods for machine learning involve elegant mathematical theory, they are still often treated as mere "black box" models by their users; however, the same could be said about many conventional statistical methods. See Cutler et al. (2007) and Chap. 8 in this volume for an example of a promising new non-parametric method, called "random forests," applied to invasive plant classification, rare lichen presence, and nesting site preference by cavity nesting birds.

Semi-parametric statistical models include both parametric and non-parametric components (Chap. 9 in Hastie et al. 2001). Various smoothing splines or wavelets are often used to implement such models, and in some specific cases, spatial predictions using these methods end up being equivalent to those via Kriging (Nychka 2000). Additive semi-parametric models are generally specified in a regression-style framework and can be useful for accommodating more complicated nonlinear relationships between response variables and covariates (Efron and Tibshirani 1991). For example, Holan et al. (2008) present an approach to modeling site-specific crop response to varying treatments (e.g., irrigation) in a spatial setting using semi-parametric relationships between the response and predictors.

3.3 Optimal Design

The notion of optimal sampling is not new; however, extensions of this concept to the spatial and spatio-temporal setting are being proposed with more regularity (Olea 1984; Cressie 1993). In this setting, the term "optimal" implies that the selected sampling design performs the best with regard to some design objective (Le and Zidek 2006). The choice of the design objective can vary depending on the goals of the study. For example, if the study involves the collection of spatial data at a finite set of locations across a landscape with the hopes of learning about a natural process at unobserved locations within the study area, then a design criterion based on spatial prediction error may be reasonable. In that case, one would examine all possible sampling designs to find the one design that minimizes prediction error variance (Stevens and Olsen 2004). In other situations with more vague study goals (e.g., data are being collected for many purposes and/or future modeling efforts), one may wish to reduce uncertainty in general; entropy-based design methods have been proposed and successfully used in these cases (Le and Zidek 2006). Regardless of the design criterion chosen, optimality allows one to get the most "bang for their buck" out of the data collected. That is, by exploiting properties pertaining to the manner in which the natural process is observed, it is possible to reduce uncertainty in the information gleaned from the data in a more efficient manner rather than just collecting more data. A simple example of how an optimal design could be useful is when more efficient estimation of the spatial structure in the process is desired. In such a case, both a regularly spaced sampling scheme and a fully random sampling scheme are inefficient (i.e., sub-optimal); however, a combination of the two can perform substantially better by providing better coverage of the study area (and thus being more representative) while still capturing small scale spatial structure in the process (Stevens and Olsen 2004; Zhu and Stein 2006; Zimmerman 2006).

Optimal design methods have also been proposed for spatio-temporal settings (Wikle and Royle 1999). That is, if one seeks to characterize a dynamically evolving system, specific forms of uncertainty can be reduced if care is taken in the construction of the sampling or monitoring design over time. In the current age of data collection, with remote wireless sensors constantly measuring features of the environment and mobile roving sensors capable of adaptively sampling a spatial domain, dynamic optimal monitoring designs will become more prevalent and useful. As an example, in a dynamic setting where a process is being monitored repeatedly over time and the data collected are to be used for various purposes including estimation of the dynamics of the system and statistical forecasting, the design objective may be to reduce uncertainty in the forecast (possibly via prediction error variance). In this case, several monitoring approaches could be taken to observe the system. If the system is evolving dynamically (i.e., changing over time) it is sensible to allow the design to change over time as well. This way, the monitoring scheme can adapt to capture important aspects of the behavior in the system. Allowing for roving monitors over time can help to avoid re-sampling redundant behavior and instead move to where the action is occurring. Ultimately this can maximize the power of the collected data given the available resources. An example of this kind of optimal design implemented to study plant community dynamics is presented by Hooten et al. (2009).

3.4 Conclusion

Statistics is an ever-changing field, constantly adapting to the new developments and needs of the scientific disciplines that depend on it. In this new and exciting computer era, we are witnessing a blurring of the lines between previously distinct areas of study. Methods for statistically analyzing spatial processes are built into GIS software, where spatial data manipulation has been occurring for decades; GIS tools are also being built into statistical software to aid in exploratory analyses and visualization of modeling results on spatial and spatio-temporal domains. In the face of mountains of data, machine learning and data mining methods are becoming more prominent, as is the need for formal data management skills. We are also seeing the direct integration of information formerly restricted to applied mathematical and physical studies into rigorous statistical analyses. Likewise, it can be said that new statistical methods (e.g., hierarchical Bayesian models) are being readily used in the applied mathematics literature. Statistics is growing and changing and we are rapidly approaching a time where every scientific problem, no matter how complex, can be considered naturally in a statistical framework. Landscape ecology, as a field concerned with multidimensional systems and an abundance of data, stands to benefit greatly from new statistical methods for analyzing spatial and spatio-temporal data.

Acknowledgments The author would like to thank A. Arab, C. Wikle, and two anonymous reviewers of this chapter for their helpful comments and suggestions.

References

Arab A, Hooten MB, Wikle CK (2007) Hierarchical spatial models. In: Encyclopedia of geographical information science. Springer, New York.
Banerjee S, Carlin BP, Gelfand AE (2004) Hierarchical modeling and analysis for spatial data. Chapman & Hall/CRC, Boca Raton, FL.
Berliner LM (1996) Hierarchical Bayesian time series models. In: Hanson K, Silver R (eds), Maximum entropy and Bayesian methods, pp. 15–22. Kluwer Academic Publishers, New York.
Brown LD (2000) An essay on statistical decision theory. J Am Stat Assoc. 95:1277–1281.
Carlin BP, Louis TA (2000) Bayes and empirical Bayes methods for data analysis, Second Edition. Chapman & Hall/CRC, Boca Raton, FL.
Caswell H (2001) Matrix population models: construction, analysis, and interpretation. Sinauer Associates, Inc., Sunderland, MA.
Christensen R (2002) Plane answers to complex questions. Springer-Verlag, New York.
Clark, JS (2007) Models for ecological data, an introduction. Princeton University Press, Princeton, NJ.
Clark JS, Gelfand AE (2006) Hierarchical modelling for the environmental sciences. Oxford University Press, New York.
Cressie NAC (1993) Statistics for spatial data, Revised Edition. John Wiley & Sons, New York.
Cressie NC, Calder C, Clark JS, Ver Hoef JM, Wikle C (2009). Accounting for uncertainty in ecological analysis: the strengths and limitations of hierarchical statistical modeling. Ecol Appl 19:553–570.
Cressie N, Johannesson G (2006) Fixed rank kriging for large spatial datasets. Technical Report No. 780, Department of Statistics, The Ohio State University, Columbus, OH.
Cutler DR, Edwards TC, Beard KH, Cutler A, Hess KT, Gibson J, Lawler JJ (2007) Random forests for classification in ecology. Ecology 88:2783–2792.
Diggle PJ, Ribeiro PJ, Jr (2007) Model-based geostatistics. Springer, New York.
Efron B, Tibshirani R (1991) Statistical analysis in the computer age. Science 253:390–395.
Furrer R, Genton MG, Nychka D (2006) Covariance tapering for interpolation of large spatial datasets. J Comput Graph Stat 15:502–523.
Gelfand AE, Silander JA, Wu S, Latimer A, Lewis PO, Rebelo AG, Holder M (2006) Explaining species distribution patterns through hierarchical modeling. Bayesian Anal 1:41–92.
Gelman A, Carlin JB, Stern HS, Rubin DB (2004) Bayesian data analysis: second edition, Chapman and Hall/CRC, Boca Raton, FL.
Grimm V, Railsback SF (2005) Individual-based modeling and ecology. Princeton University Press, Princeton, NJ.
Hastie T, Tibshirani R, Friedman J (2001) The elements of statistical learning. Springer, New York.
Hilborn R, Mangel M (1997) The ecological detective, confronting models with data. Princeton University Press, Princeton, NJ.
Holan S, Wang S, Arab A, Sadler J, Stone K (2008) Semiparametric geographically weighted response curves with application to site-specific agriculture. J Agric Biol Environ Stat 13:424–439.
Hooten MB, Wikle CK (2007) Shifts in the spatio-temporal growth dynamics of shortleaf pine. Environ Ecol Stat 14:207–227.
Hooten MB, Wikle CK (2010) Statistical agent-based models for discrete spatio-temporal systems. J Am Stat Assoc 105:236–248.
Hooten, MB, Larsen DR, Wikle CK (2003) Predicting the spatial distribution of ground flora on large domains using a hierarchical Bayesian model. Landsc Ecol 18:487–502.
Hooten MB, Wikle CK, Dorazio RM, Royle JA (2007) Hierarchical spatiotemporal matrix models for characterizing invasions. Biometrics 63:558–567.
Hooten MB, Wikle CK, Sheriff S, Rushin J (2009) Optimal spatio-temporal hybrid sampling designs for monitoring ecological structure. J Veg Sci 20:639–649.
Kays RW, Gompper ME, Ray JC (2008) Landscape ecology of eastern coyotes based on large-scale estimates of abundance. Ecol Appl 18:1014–1027.

Le ND, Zidek JV (2006) Statistical analysis of environmental space-time processes. Springer, New York.
Lele SR, Dennis B, Lutscher F (2007) Data cloning: easy maximum likelihood estimation for complex ecological models using Bayesian Markov chain Monte Carlo methods. Ecol Lett 10:551–563.
Neter J, Kutner MH, Nachtsheim CJ, Wasserman W (1996) Applied linear statistical models. WCB/McGraw-Hill, Boston.
Nychka D (2000) Spatial-process estimates as smoothers. In: Schimek, MG (ed) Smoothing and regression: approaches, computation, and application. John Wiley & Sons, New York.
Olea RA (1984) Sampling design optimization for spatial functions. Math Geol 16:369–392.
Royle JA, Dorazio RM (2006) Hierarchical models of animal abundance and occurrence. J Agric Biol Environ Stat 11:249–263.
Royle JA, Dorazio RM (2008) Hierarchical modeling and inference in ecology: the analysis of data from populations, metapopulations, and communities. Academic Press, London.
Royle JA, Kery M (2007) A Bayesian state-space formulation of dynamic occupancy models. Ecology 88:1813–1823.
Royle JA, Wikle CK (2005) Efficient statistical mapping of avian count data. Environ Ecol Stat 12:225–243.
Salsburg D (2001) The lady tasting tea: how statistics revolutionized science in the twentieth century. Henry Holt and Company, New York.
Schabenberger O, Gotway CA (2005) Statistical methods for spatial data analysis. Chapman & Hall/CRC, Boca Raton, FL.
Shi T, Cressie NAC (2007) Global statistical analysis of MISR aerosol data: a massive data product from NASA's Terra satellite. Environmetrics 18:665–680.
Shumway R, Stoffer DS (2006) Time series analysis and its applications. Springer, New York.
Stevens DL, Olsen AR (2004) Spatially balanced sampling of natural resources. J Am Stat Assoc 99:262–278.
Stigler SM (1990) The history of statistics: the measurement of uncertainty before 1900. Harvard University Press, Cambridge.
Ver Hoef JM, Peterson E, Theobald D (2006) Spatial statistical models that use flow and stream distance. Environ Ecol Stat 13:449–464.
Waller LA, Gotway CA (2004) Applied spatial statistics for public health data. John Wiley & Sons, Inc. Hoboken, NJ.
Wikle CK, Royle JA (1999) Space-time models and dynamic design of environmental monitoring networks. J Agric Biol Environ Stat 4:489–507.
Wikle CK (2001) A kernel-based spectral approach for spatio-temporal dynamic models. Proceedings of the 1st Spanish workshop on spatio-temporal modelling of environmental processes (METMA), Benicassim, Castellon (Spain), 28–31 October 2001, pp. 167–180.
Wikle CK (2003) Hierarchical Bayesian models for predicting the spread of ecological processes. Ecology 84:1382–1394.
Zhu J, Huang H-C, Wu C-T (2005) Modeling spatial-temporal binary data using Markov random fields. J Agric Biol Environ Stat 10:212–225.
Zhu Z, Stein M (2006) Spatial sampling design for prediction with estimated parameters. J Agric Biol Environ Stat 11:24–44.
Zimmerman DL (2006) Optimal network design for spatial prediction, covariance estimation, and empirical prediction. Environmetrics 17:635–652.

Part II
Integration of Ecological Theory into Modeling Practice

Chapter 4
Proper Data Management as a Scientific Foundation for Reliable Species Distribution Modeling

Benjamin Zuckerberg, Falk Huettmann, and Jacqueline Frair

4.1 Introduction

Data management, storage, curation, and dissemination are mainstays of computer modeling. Indeed, a traditional view of computer modeling has perpetuated the notion of *"garbage in, garbage out"* (GIGO), which serves as a constant reminder that, no matter how sophisticated the analysis, computers will "unquestioningly process" whatever type of data are provided regardless of its quality or suitability (Pearson 2007). In ecology, the datasets used in computer modeling are inherently complex and often characterized by missing values, dynamic environmental variables, and other factors leading to numerous data anomalies (Michener et al. 1997; Michener and Brunt 2000). Ecologists have long recognized, however, that although data quality is undoubtedly important, using different types of data, even messy ones, can still prove informative, and facilitates new questions, methods, and synergies in science and society. For instance, historical data often have greater error and biases due to shifting practices in data collection and cataloging, but research continues to explore the effects of these biases and ultimately issues identified from past sampling approaches are remedied in contemporary data collection programs (Graham et al. 2004). Without these past efforts we could not have achieved the high standards currently practiced in remote sensing, modeling and related disciplines (Manly et al. 2002; Scott et al. 2002; Kadmon et al. 2004; Jochum 2008). Even so, predictive species distribution modeling relies on quantifying complex relationships between a large amount of biological data and environmental predictors, and certainty in our inferences may depend as much upon data management, as with the analytical approaches.

There have been numerous publications reviewing the various statistical approaches used in species distribution modeling (Guisan and Zimmermann 2000; Segurado and Araújo 2004; Austin 2006), but relatively few have addressed data per se. Data

---

B. Zuckerberg (✉)
Cornell Lab of Ornithology, 159 Sapsucker Woods Road, Ithaca, NY 14850-1551, USA
e-mail: bz73@cornell.edu

management and data curation in ecology have received increasing attention (Michener et al. 1997; Michener and Brunt 2000; Huettmann 2005, 2007; Jan 2006; Karasti and Baker 2008), even resulting in online resources for managing ecological data (www.ecoinformatics.org), but there is little guidance on the role of data management in species distribution modeling. As an example, species distribution modeling usually benefits greatly from collecting, storing and incorporating geographic information on where a species or groups of species occur across a landscape. This spatial information is an inherent part of ecology and predictive modeling, and leaving it out can produce misleading results and inferences (Fortin and Dale 2005). The practicalities of conducting scientific studies, however, require that geographic information can be re-used and updated, making the maintenance, documentation, and dissemination of these data a key component of predictive and spatial modeling.

Predictive species distribution modeling is a large and diverse component of landscape ecology and the data issues involved in predictive modeling might be best considered in light of a large, focused area of research. Therefore, we use species distribution modeling to review the inherent advantages, disadvantages and management of commonly used data in landscape ecology applications. Data management touches on many components including choice of biological data, sampling design, spatial extent and resolution, sources of error and bias, relational databases, maintenance, metadata, storage, industrial technology, funding schemas and online data delivery to a global audience. All of these components play important roles when developing sound results and conclusions through predictive modeling.

4.2 Management Challenge and Theoretical Framework

The objectives of almost any predictive model are influenced by the nature and type of data upon which it is built. In a review of ecological theory and statistical modeling, Austin (2002) defined a "data model" conceptually as any decision-making system concerning how data are collected, measured, estimated and managed (Fig. 4.1). The success of any modeling effort therefore relies on the interaction of the data model with other models of ecological theory and statistical approaches (Austin 2002). The data model consists of several related components ranging from research design to data storage and dissemination. With an increasing reliance of data dissemination through the internet, a current challenge of predictive modeling is the need to understand the strengths and weaknesses of an ever-expanding diversity and availability of biological and environmental data.

4.2.1 The Practical Aspects of Research and Sampling Design

Sampling design has important implications for predictive modeling and data management. A comprehensive review of biological sampling for research projects

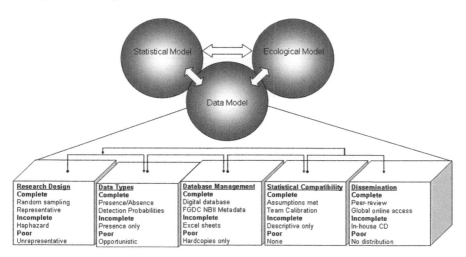

Fig. 4.1 In predictive modeling, the data model interacts with models of statistical and ecological theory. The data model, however, is unique and consists of a decision-making system focused on research design, data types, database management, statistical compatibility, and then topics such as efficiency, convenience, speed, costs, and result dissemination. We believe that the data model presents a foundation in species distribution modeling and care needs to be taken in its design. Here we offer examples of each of these components for complete, incomplete, and poor modeling initiatives

is beyond the scope of this chapter (for details see Thompson et al. 1998; Sutherland 2000, 2006; Elzinga 2001; Manly et al. 2002; Thompson 2004; Vesley et al. 2006). However, a critical consideration for predictive modeling is ensuring that the sampling strategy covers the full range of gradients thought to influence the species' distribution (Hirzel and Guisan 2002; Edwards et al. 2006). A sampling design that does not address these gradients accurately may cause the distribution of species responses to be biased or truncated and not cover the full range of the environmental limits of that species. Hirzel and Guisan (2002) performed a test, so far rare in the discipline, on the effects of different sampling strategies for modeling species distributions. They found that although sample size was the most important component in any sampling design, equal-stratified and uniform designs performed strongest for predictive modeling. With multi-agency, multi-year, or broad-scale data collection efforts, however, the development of an effective sampling design or the most appropriate database structure are inherently more complex. Many digital databases provide data collected using a wide variety of designs ranging from controlled systematic surveys to opportunistic sampling, and commonly datasets contain information from multiple sampling strategies. As an example of a spatially controlled survey, the North American Breeding Bird Survey (BBS) (Table 4.1), a digital database of roughly 3,700 roadside surveys conducted throughout the United States and Canada, incorporates several features in its design to achieve representative sampling, including the random location of survey routes in habitats that are representative of the entire region, consistent methodology and observer expertise, and large sample sizes to average local variations and reduce the effects

Table 4.1 Examples of publicly accessible online databases containing information on biological data for species' distribution modeling

Database type	Database description	Online source
Large databases and webportals	Global Biodiversity Information Facility (species occurrence data warehouse)	www.gbif.org
	Avian Knowledge Network (observational database on birds)	www.avianknowledge.net/content
	International Biogeography Society Database (updated list of biogeographical resources)	www.biogeography.org/databases.htm
	Global Population Dynamics Database (collection of animal and plant population data)	www3.imperial.ac.uk/cpb/research/patternsandprocesses/gpdd
	Biological Inventories of the World's Protected Areas (species inventories of plants and animals reported from the world's protected areas)	www.ice.ucdavis.edu/bioinventory/bioinventory.html
	Geogratis Canada (geospatial data for Canada)	www.geogratis.cgdi.gc.ca
Species occurrences	North American Breeding Bird Survey (population data and analyses on >400 birds)	www.pwrc.usgs.gov/bbs
	Christmas Bird Count (circular count areas on wintering birds)	www.infohost.nmt.edu/~shipman/z/cbc/homepage.html
	Modern and Fossil Pollen Database (pollen counts and related information)	www.ncdc.noaa.gov/paleo/pollen.html
	Lower Colorado Basin GIS Fish Data (fish records based on presence/absence within the Lower Colorado Basin)	www.peter.unmack.net/gis/fish/colorado/
	Ocean Biogeography Information System (marine biogeographic information system)	www.iobis.org
Natural history collections	Faunmap (database for the late Quaternary distribution of mammal species in the United States)	www.museum.state.il.us/research/faunmap/
	Paleobiology Database (data about fossil plants and animals)	www.Paleodb.org/
	World Data Center for Paleoclimatology (past climate and environment derived from proxies such as tree rings and ice cores)	www.ncdc.noaa.gov/paleo/
Range maps	NatureServe (distributional data on >70,000 plants, animals, and ecosystems of the US and Canada)	www.NatureServe.org
	Tree Species Distributions (range maps of tree species in N. America)	www.esp.cr.usgs.gov/data/atlas/little/
	GAP Analysis Program (distribution maps for native species in US, mostly based on expert knowledge)	www.gapanalysis.nbii.gov
Atlases	United Kingdom	www.bto.org/birdatlas
	North American Breeding Bird Atlas Viewer (atlas viewer and links to bird atlases in N. America)	www.mbirdims.fws.gov/nbii_bba
	Australia Natural Resource Atlas	www.anra.gov.au/index.html
	Birds of Australia Atlas	www.birdata.com.au/atlasstats.do

of sampling error (Sauer et al. 2007). In contrast, the Oceanic Biogeographic Information System (OBIS) (Table 4.1) collects occurrence data for marine species world-wide from a wide variety of sources and sampling designs, but provides less guidance on the actual research design and its description. Although OBIS imposes a number of data quality filters (e.g., authoritative sources, spatial error checking), this data collection includes a myriad of sampling strategies for data collected since 1850. Both of these databases are extremely valuable for purposes of predictive modeling, and may be the best available data in certain geographic areas; however, data that are collected in a more-standardized fashion and following a consistent sampling scheme and description are likely to produce predictive models that are more accurate, more precise, and allow for stronger conclusions.

When considering the use of a digital database, two particularly important features influencing predictive modeling are the spatial extent and resolution of the data (Fig. 4.2; Huettmann and Diamond 2006; Guisan et al. 2007; Meyer 2007). Spatial extent refers to the area over which data are collected (e.g., research station, state, region) and resolution is the size of a sampling unit (e.g., survey plot, atlas block) (Fig. 4.2). Austin (2006) suggested that if the purposes of the study are to model the environmental realized niche of a species, the spatial extent of the sampling design should range beyond the observed environmental limits of the species. As an example,

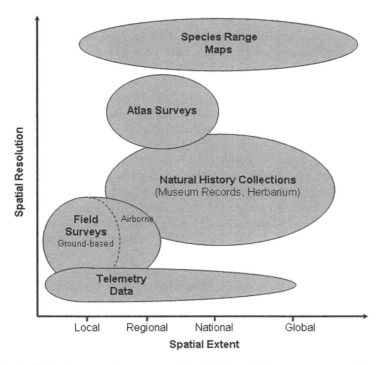

Fig. 4.2 Available data types for purposes of predictive modeling encompass a wide range of spatial resolutions and extents. There is a generally positive correlation between spatial resolution and extent where datasets covering a large extent tend to have a coarser resolution

the Lower Colorado Basin GIS Fish Databases (Table 4.1) contain many datasets that facilitate the analysis of native and exotic fish records based on presence–absence records for specified watersheds within the Lower Colorado Basin. Although these data offer a number of avenues for species distribution modeling, care must be taken in the selection of study species due the spatial limitations imposed by the extent of the study area. That is, any species–environmental associations used for purposes of predictive modeling would obviously be constrained to the artificial extent and environmental variation contained within the sampling area, thus limiting the inference of the model. Similar to extent, resolution can affect what species–environmental associations can be assessed. For example, target species occurrences derived from fine-grained survey plots can be combined with occurrence data for other species to model patterns of interspecific interactions (Ferrier and Guisan 2006). These types of interactions are difficult to model using biological surveys of a coarse resolution (e.g., atlases) (see Araújo and Luoto 2007) because coarse-grained surveys may not capture locally relevant variation affecting the occurrence of multiple species. Indeed, coarse-grained surveys, which can have resolutions of 25–2,500 km^2, are most often used for modeling relationships between species distributions and climate change, precipitation gradients, and habitat resource availability (Venier et al. 1999, 2004; Brotons et al. 2004).

Improperly designed sampling strategies can contain specific biases (e.g., locational error, differences in effort), some of which may be overcome statistically, and are often related to spatial extent and resolution (Kadmon et al. 2004; Barry and Elith 2006). Datasets that are merged, compiled, or collected in an ad hoc fashion may be biased by geographical constraints, differences in sampling intensity, and site accessibility. One of the largest public data sources on biodiversity in the world, the Global Biodiversity Information Facility (GBIF; Table 4.1), suffers from many of these problems. For predicting species distribution, sampling biases may be most egregious when sampling fails to adequately cover strong environmental gradients in the multivariate sense (Elith et al. 2006). The most glaring implication for predictive species distribution modeling is that modeled relationships will be dominated by patterns associated with the sampling effort, as opposed to true species–environmental associations. If sampling is neither random nor representative, or when pseudoreplication occurs, model predictions can be biased toward the best sampled rather than most biologically important conditions. These sampling biases are often associated with coarse-scaled sampling efforts with relatively large extents and geographic variability in observer effort (McGowan and Zuckerberg 2008). In addition, species differ in their detectability due to various factors such as elusive behavior and their habitat associations. Methods for dealing with sampling biases include estimation of detection probabilities from repeated surveys (MacKenzie 2005a, b; Mackenzie and Royle 2005) and distance sampling (Buckland 2001), rarifying to standardize sampling intensity or undertaking additional targeted sampling to fill data gaps (Araújo and Guisan 2006), standardizing for differences in effort (McGowan and Zuckerberg 2008), and using other species as surrogate variables for sampling biases (Barry and Elith 2006). The effects of sampling bias will likely be specific to the underlying database being used, and is often magnified when dealing with multiple surveyors, large geographic regions, varying sample sizes and historical datasets.

The spatial extent and resolution of a study will also affect the number of records included in a database. Small sample sizes are often associated with research having both a small grain size and small spatial extent (Manel et al. 2001; Stockwell and Peterson 2002; Barry and Elith 2006; Hernandez et al. 2006). When assessing digital databases for distribution modeling, it is thus typical to find areas that are entirely undersampled (usually a prime reason for carrying out predictive modeling in the first place). For example, the Information Center for the Environment's Biological Inventory databases contains taxonomically standardized inventories of plants and animals reported from the world's protected areas (Table 4.1). As part of the Biosphere Reserve Integrated Monitoring program, this database houses information on reserve names, characteristics, contact personnel, research datasets, and a PC microcomputer application (BioMon) used to create and query standardized databases of species occurrence information. Although this database contains biological information for many reserves, the small sample that result from the extent of the surveys or the biology of the target species is an important consideration for modeling applications. Importantly, sample size does not necessarily reflect the total number of survey sites and a large sample size does not necessarily equal a representative sample (Manel et al. 2001; Barry and Elith 2006; Pearson 2007). For occurrence data, sample size typically relates to the number of presence records available for the species of interest (i.e., prevalence) (Manel et al. 2001). Consequently, a large sampling effort may not necessarily result in a large number of presence records for rare species. Although low prevalence of recorded presences may be inadequate for a given modeling exercise, statistical advances (e.g., data mining) have led to successful modeling of species occurrence using less than 25 occurrence events (Hernandez et al. 2006; Pearson 2007; Pearson et al. 2007; Craig and Huettmann 2009; Jochum 2008).

How many species records are needed for predictive modeling, and how big does the database need to be? Although this question is typically asked when choosing a compatible statistical approach and setting up a field sampling regime, the expected minimum number of records can have important implications in deciding what database design to use. There is currently no rule of thumb of how sample size exactly increases the predictive performance of models. Elith et al. (2006) found that sample size was not consistently related to modeling success, but this could also be a result of the large number of localities used in this particular study. Using logistic regression, Coudun and Gégout (2006) suggest a general minimum value of 50 occurrences for species to derive acceptable ecological response curves. However, if the biological signal is very clear (i.e., large effects sizes and low variability), then for endemic species, even tiny sample sizes could suffice. Other important factors include the complexity of the pattern being modeled, the number of environmental predictors, possible spatial autocorrelation, and the desired precision and power of predictions. Burnham and Anderson (2002) offer another rule of thumb that provides a statistical model limit for sample size: $n \geq k \times 10$, where n is sample size and k is the number of parameters in the model. Issues associated with small sample sizes can be partially mitigated through approaches such as community modeling, habitat suitability modeling (HSI), and statistical methods (Araújo and Guisan 2006; Hernandez et al. 2006; Pearson 2007).

Importantly, species distribution modeling is not necessarily a static exercise because species–environment associations may vary over time. Consequently, issues of scale are not limited to space. If the objective of a study is to model the relationships between species' distributions and a changing environment, then the ability to collect data on species occurrences and environmental conditions over time plays a critical role that must be incorporated into database design. For example, collecting daily data, such as that available from automated sensors, will quickly lead to an exponential increase in database size. Thus, more hard drive space and data handling time will be required.

4.2.2 Data Types

4.2.2.1 Occurrence Data

The most common form of data used for predicting species' distributions is occurrence data (i.e., presence/absence data) that are collected from plots of a given size for a known geographic location (Austin 2006; Guisan et al. 2006; Pearson 2007). These occurrences represent the detection (confirmed) or non-detection (hopefully confirmed absence) of a species throughout a study region. They are often integrated into a Geographic Information System (GIS) as points, or polygons (many point locations get buffered to overcome spatial inaccuracy as we discuss later) allowing for association with environmental data (e.g., soil type, elevation, land cover, temperature gradients) (Table 4.2). Depending on the GIS, either flat files (i.e., non-relational) can be used following x, y (coordinate pair), and z (species and other attribute information) format. Because the biological input data are usually binary (e.g., detection=1 and non-detection=0 or carrying text labels) they are amendable to any categorical analysis techniques that can accommodate a binomial error distribution (e.g., GLMs or Classification and Regression Trees). The relationships analyzed in these models are usually correlative (as opposed to mechanistic) because they relate known species locations to a suite of environmental variables to infer environmental conditions suitable for species occurrence (Manly et al. 2002). In theory and in practice, the suite of environmental variables should be chosen carefully and should represent those factors most likely affecting the species' physiology, distribution, and probability of persistence. Often multiple environmental attributes may be required to explain complex patterns of species presence and absence. Given the growing array of available environmental and species occurrence datasets (Table 4.2), care is needed in order to avoid identifying spurious species–environmental associations from "data mining" (Anderson et al. 2001; Anderson 2008). Data mining may be useful for exploring patterns and potential avenues for future analysis (Craig and Huettmann 2009), and maximizing the predictive power of a model (Breiman 2001a, b), although potentially at the expense of reliable ecological insight. Anderson (2008) promotes that data analysts with clear hypotheses are better served by estimating a small set of candidate species occurrence models based on assumed habitat requirements, and using model selection approaches

Table 4.2 Examples of accessible online databases containing information on environmental data for species' distribution modeling

Data type	Database description	Online source
Human impact	Atlas of the Biosphere (GIS data on humans, land use, ecosystems, and water resources)	www.sage.wisc.edu/atlas
	United Nations' Environment Program Geo Data Portal (large database with numerous themes)	www.geodata.grid.unep.ch
	Global Resource Information Database	www.grid.unep.ch
	Center for International Earth Science Information Network (CIESIN) (global human interaction datasets and viewer)	www.ciesin.org
Climate	National Climate Data Center (climatological datasets)	www.ncdc.noaa.gov/oa/ncdc.html
	WORLDCLIM (global climate layers; 1 km^2 resolution)	www.worldclim.org
	IPCC Data Distribution Center (climate data and scenario models)	www.ipcc-data.org
	Parameter-elevation Regressions on Independent Slopes Model (PRISM) (climatic data, temperature and precipitation)	www.prism.oregonstate.edu
	Climate Research Unit (CRU)	www.cru.uea.ac.uk/cru/data/
Land cover	National Land Cover Data 1992	www.landcover.usgs.gov/natllandcover.php
	National Land Cover Data 2001	www.mrlc.gov/nlcd_multizone_map.php
	National Land Cover Database 1992/2001 Retrofit Land Cover Change	www.mrlc.gov/multizone.php
	Global Land Cover Facility (satellite imagery)	www.glcf.umiacs.umd.edu/data
	Landsat	www.landsat.gsfc.nasa.gov
	NASA (MODIS Data)	www.modis.gsfc.nasa.gov
Elevation	Earth Resources Observation and Science	www.edc.usgs.gov
	ETOPO2	www.ngdc.noaa.gov/mgg/fliers/01mgg04.html
	USGS National Elevation Dataset	www.ned.usgs.gov
	USGS Shuttle Radar Topography	www.srtm.usgs.gov

to infer the weight of evidence behind a given model. These approaches, however, can suffer from lack of wider generalization (Elith et al. 2006). Assuming no prior knowledge of a system, data mining may be useful for exploring patterns in the data and generating hypotheses that can be tested subsequently (Breiman 2001a; Hochachka et al. 2007), but data mining comes with two critical caveats in database management: First, cross-validation (i.e., the dividing of a dataset into parts, and use of one part to build a model and the other to assess the accuracy of this model) needs to be used in order to guard against the risk that models are identifying spurious patterns by over-fitting a specific sample of data (Hastie et al. 2001). Second, when exploratory analyses are used to identify hypotheses, testing such hypotheses with subsequent statistical analyses requires a meaningful independent set of data, which should be part of the research design. Data mining approaches are powerful with large and inherently complex databases, such as those created by sensor networks and coarse-scale monitoring projects, making the process of data quality checking inherently more critical (Huettmann 2007). The reality of wildlife management offers us many such datasets that qualify in this regard. Regardless of the analytical approach used, the central premise is that observed patterns in a species occurrences provide critical information on the environmental requirements of those species and it is incumbent upon the researcher to ensure an unbiased estimate of those requirements.

Occurrence data are available in multiple spatial resolutions and extents ranging from local field projects to regional and global initiatives (Fig. 4.2). Data collected from field surveys of species occurrences have a number of advantages that make them amendable for predictive modeling at landscape scales. The advantages of most field surveys is their ability to maintain consistency in survey methods and an a priori identification of important environmental gradients across which sampling can be planned (Fig. 4.2). In some cases, a more intensive sampling of species occurrences across a landscape can provide information on the co-occurrence with other species, useful for modeling biotic interactions (e.g., inter-specific competition, predation) or treating more abundant species as surrogates for species that are rare or difficult to detect (Ferrier and Guisan 2006; Araújo and Luoto 2007; Heikkinen et al. 2007; Royle and Dorazio 2008). In addition, the inclusion of reliable absence records generally improves model performance and flexibility (Brotons et al. 2004). In any occurrence dataset, however, false absences may be recorded when the species was not successfully detected, although it was present (MacKenzie et al. 2006). In some instances, confirmed absence is a function of survey effort. When possible, these details should be clearly specified in the database or accompanying publication so they may be dealt with properly in later analyses. Otherwise, the predictive model will interpret the absence as denoting unsuitable environmental conditions, even though this may not be the case (Pearson 2007). False absences bias estimates of prevalence, and may lead to identification of erroneous environmental associations if detection probability varies along environmental gradients. Thus, every effort needs to be made to correct for imperfect detection of organisms, ideally through use of repeated surveys and similar features of the field-based surveys to estimate detection probabilities (Mackenzie and Royle 2005). The database design needs to reflect these criteria both in sampling design (e.g., repeated visits) and database structure.

Although targeted field surveys allow for a more rigorous study design and control, this does not come without a price. In landscape-scale studies, there is typically a tradeoff between intensity of local sampling effort and the ability to conduct representative sampling across a wider area (Brennan et al. 2002; Travaini et al. 2007). That is, a more spatially intensive sampling effort generally corresponds with fewer presence–absence records or a limited spatial extent and it may not capture the full range of environmental conditions occupied by the species (Austin 2006;

Pearson 2007). In addition, occurrence data collected over relatively short time periods should not be confused with long-term persistence, especially for species with longer generation times. This is because individuals or populations may be occupying "sink" habitats and are dependent on immigrants from viable populations occupying "source" habitats for continued persistence (Pulliam 2000); such sinks can be identified through examination of long-term persistence at individual sites (Hames et al. 2001) or through collection of pertinent demographic data.

Grid-based sampling schemes, such as atlases, provide data on a species occurrence that can cover entire regions of interest, are often repeated, and are increasingly available in digital formats (e.g., online maps, relational databases) (Donald and Fuller 1998; Gibbons et al. 2007). Atlases provide a large spatial extent and an opportunity to explore species–environmental associations over a wider range of environmental variation. These data are increasingly used and re-used for purposes of conservation planning and policy (Bishop and Myers 2005), assessing species–habitat associations (Trzcinski et al. 1999), selecting areas for preserves (Araújo et al. 2002), and predicting a species distribution (Venier et al. 1999, 2004; Brotons et al. 2004). Breeding bird atlases (BBA) are the most common grid-based schemes, and there are approximately 411 ornithological atlases from nearly 50 countries with an increasing number being produced as searchable databases on the internet [*see* BBA Explore (www.pwrc.usgs.gov/bba) for avian atlases in the United States; Magness et al. (2008) and Nemitz (2009) for GRID data]. Online BBA Manager software (www.pwrc.usgs.gov/bba) is increasingly popular for online data management and dissemination, and is used by the atlasing projects of many states. Using these online systems, atlas participants can log into a BBA Manager site and directly enter their field data, which can be subsequently reviewed by coordinators. After the review process, the interim data become available for the public and can be mapped using online interfaces such as BBA Explorer or BBA Viewer (Table 4.1). Similar concepts are currently applied by Birds Australia in their Atlas (Table 4.1). The spatial extents of these atlases ranges from 12 to nearly 10,000,000 km^2 with resolutions of 0.06–14,400 km^2 (Gibbons et al. 2007). Although many of the advantages and disadvantages common in field surveys are also common in atlas surveys, atlases are more likely to suffer from variations in effort and biases associated with accessibility (i.e., areas farther away from roads) (Donald and Fuller 1998; Bibby et al. 2000; McGowan and Zuckerberg 2008).

Biotic interactions, with competitors or predators, may explain why species are not observed in otherwise suitable habitats. Additional data might be available to explicitly model biotic interactions (Ferrier and Guisan 2006; Araújo and Luoto 2007); however, such information is difficult to acquire and to make accessible in standard databases. Nevertheless, recent approaches linking species occurrence and demographic models to predict the fitness consequences of species use of a given area (see Nielsen et al. 2006; Aldridge and Boyce 2007; Frair et al. 2007), have led to greater ecological insight into processes governing species distributions. Moreover, community-level modeling has emerged as a technique for combining data from multiple species and producing information on the spatial distribution of biodiversity at a collective community level instead of the level of individual species (Ferrier and Guisan 2006; Royle and Dorazio 2008). Suitable datasets for community-level analysis typically consist of a large number of rare and common species and their geographic locations. In general, community-level modeling is less sensitive to small numbers of detections per species because it relies on using detections for a collection of species. This is especially useful for studies with small sample sizes for rare species because the data from more common species help guide the modeling of rare species. This approach may also be useful for incorporating information on competition and predation in species distribution modeling. Yet the more information available at a given point in space or collected in time, the more complicated the task of data management.

4.2.2.2 Occurrence Database Management

We cannot overemphasize the importance of proper data management and documentation, following international standards, such as those of the International Standardization Organization (ISO; www.iso.org). Data documentation often comes in the form of "metadata" which represent a set of documentation that describe the content, context, quality, structure, and accessibility of the data (Michener et al. 1997; Michener and Brunt 2000; Huettmann 2005). For a database to be useful for modeling species' distributions, the database records with accompanying metadata should describe: *what* (the identity of the organism), *how many* (were the units of observation individual organisms or colonies, was presence or actual counts recorded, and how many units were recorded), *where* (the location at which the organism was recorded and what coordinate system was referenced), *when* (the date and time of the recording event), *how* (what sort of record is represented and other details of data collection protocols, e.g., a sight record from 5-min point count, collected specimen in variable effort netting, what is known of errors associated with the data), and *who* (the person responsible for managing the data). Each of these components represents an important aspect of data collection and future use. For example, information on *how* a recording event was made allows someone disjunct from the data collection decisions to properly account for variation in effort and detection probability, deal with data from multiple protocols, and determine whether data are from multiple species or single-taxon records.

Given these necessary database characteristics, occurrence databases should include some basic and standardized inputs and descriptions (Huettmann 2005; Jan 2006). The Darwin Core is a simple data standard that is commonly used for taxon occurrence data (e.g., specimens, observations of living organisms) (www.wiki. tdwg.org/DarwinCore) (Table 4.1). The Darwin Core standard specifies several database components including record-level elements (e.g., record identifier), taxonomic elements (e.g., scientific name), locality elements (e.g., place name), and biological elements (e.g., life stage). In addition to describing the Darwin Core standard, Jan (2006) provides an excellent example of the primary structure of an observational database. Using the terminology of Jan (2006), sampling information relates to field site visits, and each of these visits is considered a *Gathering*.

Each *Gathering* event should record the occurrence of a species and additional site information including site name, the period of time, the name of the collector, the method of collection, and geography. The geography field should use country codes using ISO standards, accepted names, and it should have an attribute telling if this information is currently valid because political boundaries and names change over time (e.g., new countries form, their names can change) (Jan 2006). Geospatial data can be stored under the heading of *GatheringSite* and includes coordinate data (e.g., latitude and longitude, altitude), gazetteer data (e.g., political or administrative units), geo-ecological classifications (e.g., geomorphological types), and ecological sites. It is important that this field allows for high-resolution geo-referencing for subsequent integration with a GIS (e.g., using at least five significant digits for latitude and longitude coordinates). The *Unit* field includes organisms observed in the field, herbarium specimens, field data, taxonomic identifications, or descriptive data. An *Identifications* field supports the application of a name to a *Unit* (specimen, observation, etc.). Names should include species common name, species scientific name, and a species code (using the Integrated Taxonomic Information System, [it is]; www.itis.gov). *Identifications* can then be connected to a taxon database using a *TaxIdRef* field. The organization of any species occurrence database should have these necessary information fields (although field names may vary) and will likely require the use of a digital database for storage and manipulation.

Digital databases and their metadata are now considered an invaluable and commonly used tool for documenting and locating species occurrences. Even in remote field sites, researchers are now using mobile Global Positioning System (GPS) and Personal Digital Assistant (PDA) units to record georeferenced census tracks and species observations (Travaini et al. 2007). Using any laptop computer, these data can then be quickly integrated into database management software such as CyberTracker (www.cybertracker.org), Microsoft Excel (www.office.microsoft.com/en-us/excel), or Microsoft Access (www.office.microsoft.com/en-us/access). A key component to any digital database is the ability to georeference census points for later integration into a GIS, such as ArcGIS (www.esri.com/software/arcgis/). Travaini et al. (2007) provide an excellent review and application of a field-based database framework for mapping animal distributions in remote regions. A key advantage to recording data into a digital database during the collection event itself is the ability to develop and maintain multiple databases and a smoother integration for online data management and access.

Database managers often use online and electronic databases because they can be readily linked to other databases for greater functionality. Connecting multiple databases is best done in a relational databases management system (RDBMS), which allows for inter-database queries, and as such, the structure and frameworks of many large databases are increasingly more sophisticated. Thus, linking species occurrence records to data on demographic rates, predators, and other geographic information might be readily accommodated. The Global Populations Dynamics Database (GPDD) (Table 4.1) contains time-series data on animal and plant population data throughout the world, and is an excellent example of a relational database framework. The GPDD consists of a standard relational database of six sub-tables of data that are connected through a common record ID number. The *Main* table contains information on the sampling units, the units used for data presentation, the years of the study, and notes on sampling design. Other tables include the *Taxon* table (information about the organism sampled in each dataset; see ITIS www.itis.org for globally accepted species names), the *Biotope* table (habitat of the organism), the *Location* table (geographical details of the monitoring site), the *Data source* table (reference to the original source of the data), and the actual *Data* table (original population data). In this case, a relational database and a common record identifier enable the user to perform multiple queries based on species, taxonomic group, habitats, areas, latitudes or countries. In practice, developing, maintaining and retrieving data from a RDBMS often requires knowledge of Structured Query Language (SQL), a widely used database filter language that is specifically designed for management, query, and use of RDBMS. SQL is a standardized language with a huge user community that is recognized both by the American National Standards Institute (ANSI) and ISO (www.iso.org/iso/home.htm). It is implemented in many database software products. Popular and robust relational database management systems include Informix (www.ibm.com/software/data/informix), Oracle (www.oracle.com/index.html), SQL Server (www.microsoft.com/sql/default.mspx), MySQL (www.mysql.com), Microsoft Access (www.microsoft.com), and PostgretSQL (www.postgresql.org).

Databases on species occurrences should be designed with a clear focus on long-term use, public online dissemination, and compatibility with other databases in your own location and elsewhere, now and in the future. The ability to coordinate digital databases across the world provides unique opportunities for collaborative research and predictive modeling. As an example, the Avian Knowledge Network (AKN) is an online partnership and repository of observational data on bird populations throughout the world. Currently, the network consists of 43 partners contributing over 45,000,000 observations from more than 305,000 locations and 4,500 taxa. This includes data from bird-monitoring, bird-banding, and broad-scale citizen-based bird-surveillance programs. Of note is the unique data structure and data dissemination capability of the AKN. The functionality of the AKN relies on a technical infrastructure that allows access to the data through a federated data grid environment. Data providers can map their existing data bases to the Bird Monitoring Data Exchange (a metadata extension of Darwin Core; Table 4.1), and then use DiGIR (Distributed Generic Information Retrieval) client/portal software (www.specifysoftware.org/Informatics/informaticsdigir/) to merge distributed datasets into a primary data warehouse located within the AKN. In addition, a separate warehouse of over 200 environmental covariates is available including landcover, US Census data, and weather data. An application interface layer allows analysis or data visualization tools and provides access to a variety of third party websites. The AKN is directly linked and uploaded weekly to the GBIF and OBIS (Table 4.1). The metadata for AKN partners is also shared with the National Biodiversity Information Infrastructure (NBII) where it can be viewed in the NBII Metadata Clearinghouse (www.nbii.gov). Obviously, the AKN and other online databases are a clear example of how globally standardized, powerful digital databases are allowing for the dissemination of data for an unprecedented number of species and geographic regions.

4.2.2.3 Presence-only Data

Recording true absence events increases the depth and breadth of insights gained from predictive modeling efforts (Brotons et al. 2004; Elith et al. 2006). Yet in some cases, absences are either unreliable, lack a sound research design, or are altogether unavailable such as when using museum specimens or natural history collections to document species occurrence. Such collections are unique in providing "massive information resources," with voucher specimens accompanying date and location information, and data on the historical distribution of species spanning hundreds of years (Graham et al. 2004). Analyses of these "presence-only" data are informative for predicting species occurrences, even if less powerful than analyses using presence–absence data (Soberón et al. 2000; Graham et al. 2004; Elith et al. 2006). Despite their sampling biases, these data have greatly contributed to the "explosion" of species distribution models, online data delivery and progress in spatial modeling, science, management and society. Translating analog records to electronic catalog systems has been slow, with perhaps as little as 5–10% of specimens worldwide being digitally available (Graham et al. 2004). Nevertheless, digital collections provide a wealth of readily accessible data and are increasingly available for use over the internet (Table 4.1).

Fundamentally, records from museum and natural history collections include three primary elements: (1) taxonomic species identity, (2) geographic description where the specimen was collected, and (3) date of collection. Of these, species identity and spatial records are usually most prone to error, and the geographic description in particular may be too general to be useful (e.g., altitude information is missing or location accurate to a county or township level only). Thus, substantial effort to clean up specimen records by identifying, and where possible, fixing errors and provide reliable information is required (see Graham et al. 2004 for details). Other unique issues with specimen collections include rare species tending to be more abundant than common species (the exact opposite of what one tends to find in standardized field sampling programs), and certain locations being sampled more frequently such as roadsides or in close proximity to human centers (Lutolf et al. 2006). As a result, museum and natural history collections are often incomplete and biased in relation to the true spatial or environmental distributions of a species, which may, in turn, produce spurious assessments of the relationship between species occurrence and ecological predictor variables (see Sect. 5.2.1). However, advantages of using such data have been widely published (Graham et al. 2004; Kadmon et al. 2004; Elith et al. 2006; Craig and Huettmann 2009).

As with other data types, sub-sampling prior to modeling can create a more balanced sample across species and ecological conditions, but with the caveat that the inferences from the resulting model are limited to the range of resampled environmental conditions (Araújo and Guisan 2006). Moreover, one cannot design an appropriate sub-sampling scheme for unidentified biases in the data. Hirzel and Guisan (2002) recommend using a robust design for sub-sampling that is repeated multiple times to assess the sensitivity of model outcomes to the sample data. Sub-sampling reduces the sample size, so it is important to ensure sufficient data for model calibration.

Alternatively, with sample weighting all records are retained and weights adjust the influence of each record in the sample within the modeling process itself. Adding weights might result in the addition of a column to the database. With both sub-sampling and sample weighting, it is assumed that environmental space has been comprehensively sampled but in a very unbalanced way leading to over-representation of certain conditions (Araújo and Guisan 2006). When this assumption does not hold, additional targeted sampling may be employed to complete the data records in ecological space.

Another substantial, but considerably different, source of presence-only data comes from tracking the locations of individually marked animals over time. These data are unique in representing a time series of space-use by individuals, and thus having correlated records. Data may be collected in any number of ways, from mark-recapture techniques to direct observation as an individual hops, walks, swims, or flies; or affixing a radio-collar and remotely tracking the individual over larger space and longer time intervals. Such databases can become quite large due to the in-time collection of data, especially when GPS collars are used. Telemetry projects require major considerations regarding database setup, policy, management, sharing and delivery (see Tagging of Pelagic Predators; http://www.topp.org for an example that awaits implementation). For one, an additional column is required to identify multiple records belonging to a given marked individual. It is also important to consider the temporal spacing between geo-referencing events, in order to test for autocorrelation and when databases are merged. The effort and expense of capturing and tracking animals can be considerable, and datasets tend to contain relatively few individuals (on the order of 10 s for radio-collaring studies to 100 s for mark-recapture studies) within a restricted geographical space (anywhere from 1 to a few thousand km^2 depending upon the species and tracking method) (White and Garrott 1990; Rodgers 2001), in contrast to the more broad-scale surveys of animal occurrence described previously (Fig. 4.2). Collecting repeated location information on marked individuals provides unique opportunities to identify the behavioral or demographic mechanisms by which species occurrence changes over time, specifically by studying patterns of survival, movement and resource use within a study area (White and Garrott 1990; Turchin 1998; Marzluff et al. 2001).

A further contrast for radio-telemetry data to the data previously discussed is a particular focus on quantifying locational error. Regardless of their collection method, species occurrence locations are rarely fixed and telemetry records in particular usually include an estimate of error associated with each location in the form of an error ellipse (see BIOGEOMANCER; http://www.biogeomancer.org/ for more details and automatized software implementation). Collection locations are often recorded by location name (see BIOGEOMANCER for official location names) whereas telemetry locations are obtained either by bi- or triangulation methods, aerial reconnaissance, or on-board GPS devices. GPS-equipped collars provide by far the most spatially accurate (10 s of m for recent models, 100 s of m for older models) and temporally complete data, and extend species monitoring around the clock and into adverse weather conditions. This detail comes at the expense of a shortened monitoring duration due to battery limitations, with GPS collars typically deployed on animals for months to at most 2 years. Moreover, advanced GPS collars are expensive (few tend to be deployed in a given study) and heavy (so are suitable only for mid- to large-bodied species). As mentioned previously, telemetry location data usually are buffered to account for spatial inaccuracy, resulting into a new (polygon) database (i.e., shapefiles, used for the modeling project).

In contrast to GPS collars, traditional radio-telemetry or very-high frequency (VHF) collars generally result in fewer, considerably less accurate (100 s of m) (Withey et al. 2001), and temporally or weather-biased location-based data (White and Garrott 1990; Withey et al. 2001). Locations are solved by "homing in" on the source of the radio signal, by aerial tracking or triangulation via fixed or mobile stations (White and Garrott 1990), with each technique having different sources and magnitudes of error (Braun 2005). Many studies based on VHF data unfortunately neglect to report error in a meaningful way (White and Garrott 1990; Saltz 1994), possibly out of fear to invalidate the study, its publication, and reduce subsequent funding success. Although location error is increasingly focused upon, it is likely that the original data are stored without reference to the error estimate for each location. It is more likely that any published data comes with a single average error reported if any. Error estimates per location may be unattainable (such is the case for GPS collar data), in which case the expected distribution of error might be used to produce robust model predictions. We suggest data producers provide an honest reference on error, in the form of an error column when possible, or a description of the error distribution within the metadata.

A second type of error is implicit in both VHF and GPS studies, specifically missed fixes and failed location attempts. With VHF collars, missed fixes are usually a function of limited access to signals given that they may be heard only within a few kilometers of a hand-held receiving unit (White and Garrott 1990; Withey et al. 2001). In contrast, missing locations from GPS collars occur either randomly or as a predictable function of dense canopy and rugged terrain conditions (Moen et al. 1996, 1997; Rempel et al. 1995; Rempel and Rodgers 1997; D'Eon et al. 2002; Frair et al. 2004). Unlike missing locations in VHF studies, missed GPS locations represent biased sampling. Approaches to overcome habitat-bias in GPS collars include sample weighting or filling in missing data points in the time series using multiple imputation techniques (Frair et al. 2004).

4.2.2.4 Presence-only Database Management

Presence-only and occurrence databases contain many of the same requirements for data documentation, but the additional importance of location-based error is of greater importance in telemetry databases. One should also keep in mind that these data usually get merged with pseudo-absence, or random location data, and should allow for comparable formats (e.g., column names, column content and format). As such, telemetry data records should include: (1) a unique identification code for the individual animal, (2) spatial coordinates for the individual at time t, and (3) a time stamp for each location (Turchin 1998). When pseudo-absence data are involved, an additional column demarcating animal locations from pseudo-absence locations is also necessary. Another typical entry in telemetry databases is a technical estimate of the inaccuracy of the location record. Along with each GPS location record there should be some estimate of the positional dilution of precision (PDOP) based on satellite geometry, and whether the location was solved in two dimensions (2D; from <3 satellites) or three dimensions (3D; requiring >3+ satellites), with low DOP values and 3D locations considered most accurate (Moen et al. 1996, 1997; Rempel et al. 1995; Rempel and Rodgers 1997; D'Eon et al. 2002). These estimates have been used to effectively screen out locations having large spatial inaccuracies that might induce bias when relating species occurrence to habitat characteristics (D'Eon and Delparte 2005).

Raw telemetry records can be processed first using free or proprietary software (e.g., LOAS, http://gcmd.nasa.gov/records/ESS_LOAS.html or LOCATE III, http://www.locateiii.com/), which use compass bearings from known locations to the radio signal to estimate the animal's location with a measure of location precision. GPS collars process satellite bearings internally and require proprietary software from the manufacturer to communicate with the collars and download the data. After processing VHF bearings or downloading GPS collars one is left with a dataset that can be viewed and manipulated in any text editor, spreadsheet, or database package. The great majority of telemetry databases have not been made publicly available, even for studies receiving public funding. Ongoing studies need to protect monitored individuals from outside influence (to avoid introducing bias into the study), as well as to protect agreed upon intellectual property and publication priorities. However, once study objectives have been met according to the best professional practices and data policies, telemetry data should be made accessible for the wider good. For example, The Starkey Project led by the United State Forest Service has posted online access to 287,000 telemetry records collected from 1993 to 1996 for mule deer (*Odocoileus hemionus*), elk (*Cervus elaphus*), and cattle (*Bos taurus*), as well as ecological data in a GIS format, and complete metadata (http://www.fs.fed.us/pnw/starkey/). In other cases, data collected for educational or public outreach purposes may be published essentially "live" (e.g., Whalenet served by Wheelock College, http://whale.wheelock.edu/Welcome.html; Narragansett Bay Coyote Study, http://www.theconservationagency.org/coyote.htm). Raw data are still the key for providing synergies for a global audience. Data for endangered species may be time delayed, spatially degraded, or otherwise controlled specifically to protect the species from harm; metadata should specify these constraints further. Currently, there are only a few central "warehouses" for telemetry data itself, which hinders their broader use. Move Bank (http://www.movebank.org/), established in 2007, is intended to fill that void by serving as a standardized data repository for animal movement data, but its data policy is still restrictive and requires an approved membership registration, thus limiting full public access.

4.2.3 Data Availability and Dissemination

Depending on the type of data collected, a modeler needs to select a data model based on the considerations outlined above. In addition, the underlying data model needs to consider topics such as software stability, robustness, skill, staff, business model, infrastructure, and feasibility [see Data Information Service of the International Polar Year (IPY), http://ipydis.org/ or www.ecoinformatics.org for guidance]. A key consideration should be given to data import, export and delivery to the scientists, managers and the global public (we define global public here as audiences from China, India, Africa and Brazil for instance; all of which still have a major need for sound and available wildlife and biodiversity information for various sustainability questions). Although it might be tempting to go along with the latest computational trends and technology, robust and well-proven database models that include metadata should always be taken into account first. Flat database files have been considered as extremely robust while various software and hardware are subject to the forces, and the failures, of an international market that usually is beyond the ability of the user to control or handle. Care is necessary because (a) new technology is usually not well tested and thus unreliable and risky, (b) robust data models are relatively cheap, and (c) they can still be transferred into modern and better models at a later stage, if the modern models prove to be useful. Automation of any component of data handling will better facilitate modeling and data dissemination, but the user should always remain in control. The scientific content is the prime focus of any database. Indeed, the hiring of a permanent database manager, and assigning the budget accordingly, is currently a standard and critical role for the successful development of any digital database. Treating databases as a simple add-on to projects often results in high turn-over and inefficiency. Linking databases directly with the metadata and with an efficient workflow is necessary and will aid metadata creation greatly when data are edited and updated. It should be pointed out that any of these items are part of various funding agency rules, such as those of the National Science Foundation (NSF), National Institutes of Health (NIH) and US Geological Survey (USGS), as well as rules governing megascience projects [such as IPY, OBIS, National Centre for Ecological Analysis and Synthesis (NCEAS), Long Term Ecological Research (LTER) and GBIF]. Independent of granting agency rules, these items represent best professional practices (see Braun 2005 for wildlife applications).

4.2.3.1 Data in a Digital Form

In times of digitization, old-fashioned hardcopy data are becoming increasingly obsolete. However, there are three primary obstacles to the shift of hard copies to digital formats. First, although almost all data for purposes of predictive modeling are basically collected digitally, a huge volume of historical hardcopy data remain to be digitized (e.g., in archives and libraries). Hardcopies are still considered an important backup, and these data can present critical sources of information for predictive modeling (e.g., natural history collections). Second, although technological advances are making the collecting of digital data in the field more feasible (Travaini et al. 2007), many field data are still collected in field notebooks when difficult field conditions and remote regions are encountered. The advantages of using digital field data collection methods include immediate data availability (e.g., online), lack of labor-intense data key-in sessions afterward, and automated metadata and processing. Third, many digital datasets are still getting printed as hardcopy for cultural and logistical reasons. Although these obstacles can be difficult to overcome, there are various and convincing reasons why data should now all be digital (with necessary backup systems). Most importantly, only digital data will allow for progressive, state-of-the-art species distribution modelingusing the latest modeling algorithms available online. Despite the technological advantages, we believe that digital data currently are still not used to their full potential as many universities, governments, NGOs and funding agencies fail to embrace current technologies. Reasons are widely found in the lack of training, lack of skill, insufficient infrastructure and an outdated culture for comprehending digital data and information. These obstacles becomes painfully obvious when dealing with an appropriate online delivery, and ought to be overcome (Huettmann 2007; Karasti and Baker 2008).

4.2.3.2 Online Data Dissemination

The advent of online capabilities provides important opportunities for data dissemination, but also adds large complexities to data discovery, collection, maintenance, and delivery (Karasti and Baker 2008). Although we focus here on modeling aspects, the subject of online data dissemination is bigger and touches on the essence of science, job descriptions, and the funding and governance of nations (Esanu and Uhlir 2003). Here, we simply emphasize the importance of databases and data flow (from the field to digitization, online storage, modeling and online delivery), file format (e.g., .xls, .dbf), metadata standard (Federal Geographic Data Committee [FGDC], Ecological metadata language [EML], and Darwin Core), and map presentation (e.g., Web Feature Service, ArcIMS or Google Earth). Flat files, such as those in Excel or TXT/CSV format, are preferred and allow for uncomplicated Darwin Core formats and attributes. These files are either used directly in GIS, or delivered online through a single point access or application to distributed data sources (e.g., DiGIR, http://digir.sourceforge.net/) and formats, and then available to modelers through a free global download. Once an output from a statistical model is obtained, the data can be re-imported into a GIS, and traditionally, a final GRID layer is created presenting a relative index of occurrence. This GRID layer file is stored in GIS database formats (e.g., ESRI GRIDs) and can then be exported and offered in applications. For these purposes, and of increasing interest, are Google Earth (earth.google.com) applications based on KML (Keyhole Markup Language) files, an XML-based language schema for expressing geographic annotation and visualization (OpenSourceGIS Consortium). A key consideration in any database delivery is the intended audience. Indeed, global data dissemination must cater to a global clientele and many data downloads will be occurring in any other nations outside of the Western World.

4.2.3.3 Metadata and Re-usability of Datasets

Any of the topics outlined above require a sophisticated documentation of processing steps if the data are to be used properly by researchers not associated with their collection (Duke 2006, 2007; Peng et al. 2006; Hollister and Walker 2007). Specifically, if data and results are placed online and made widely available for peer-review and decision-making, the act of re-using data will occur. Asking researchers for their underlying data from publications is unfortunately still unusual, but it represents best professional practices and is highly encouraged by granting agencies such as the NSF and NIH, because it will allow one to assess findings, create synergies, further promote modeling, and ultimately obtain better decisions (Duke 2006, 2007; Peng et al. 2006; Hollister and Walker 2007). For these details, the concept of documenting *every relevant* processing step remains of great importance. Automated workflows and digital workbenches such as Open Modeler, Kepler and Taverna start to offer such concepts and help to make modeling more repeatable, more transparent and less prone to (human) error. It is unfortunate that the topic of data and database management has received so little attention, and that even well-respected journals, scientific academies, and high-quality educational institutions still do not formally support this subject. The North American FGDC Metadata standard is considered as one of the most thorough standards for achieving this goal. If biological data are involved, the NBII profile applies. The FGDC standard and its profiles are widely embraced by the US Federal Government as well as by international initiatives such as the IPY for instance. Several other metadata standards exist, and interest is amassing towards a more global standardization and its implementation. These ISO standards are implemented in XML, which also has ISO standards. Other Metadata standards contributing to a unified ISO standard include (a) DIF for a short telephone entry description and still widely used by British Antarctica Service (BAS), (b) EML for a rather detailed description of relational databases by LTER sites in the US and adopted by GBIF, and (c) SML (Sensor Metadata Language) for a very powerful and progressive description of high-performance Sensors Networks. There is also a wide array of metadata standards that have local relevance only, and are not compatible with global metadata standards. Such efforts impede global data availability. Alternatives to the FGDC NBII standard have been observed due to the belief by some decision-makers, agencies, and governments that a lack of documentation would be faster, cheaper, and sufficient, whereas all of these initiatives have simply resulted in large information loss, and have now reverted back to high-quality data descriptions (Michener et al. 1997). The concept of a techno-fix, where the simpler metadata concepts can simply be mapped, cross-walked, through automatic parsing software to other standards, such as FGDC NBII to satisfy delivery needs, proved fatal to data quality because once an information field is missing, its content basically can never again be filled. With over 50 collective years of database experience, we can state with certainty that a lack of metadata make databases and their resulting models entirely unusable. Thus, metadata and data management needs to occupy a major section of the modeling project budget (Huettmann 2009). This investment always pays off.

4.3 Past, Current, and Future Applications

Predictive species distribution modeling relies on ecological data that are inherently complex. Unfortunately, the management and curation of ecological data still has not received sufficient attention and budgetary allocations, yet. These tasks are often given to lower ranking members of the laboratory and research team, and are still considered "technician work" and thus not not part of science. If it cannot be achieved that way, databases are simply contracted out to others, and it is hoped the problem can be fixed that way. It is this "techno-fix" philosophy that holds us back from a more suitable and sophisticated way of resolving administrative problems, efficient data management, and eventually, science-based management for global sustainability. This is particularly discouraging when considering that data management has critical implications in the process of species distribution modeling, and must not be considered a distraction to ignore. To overcome this situation, management and leading institutions need to get the digital expertise and provide true leadership with vision.

In other disciplines relevant for globalization and human civilization (e.g., industry, biochemistry, banking, insurance, government, customs, immigration, and military) databases are more advanced and better maintained than in wildlife biology, landscape ecology, and natural resource management. Many ecological databases exist in raw or clumsy formats, are published as dead-end PDFs, are stored in older Excel-type worksheets, or have simply been lost or discarded. Transforming the results and efforts of ecology into a Digital Information Science (DIS) is of utmost relevance for global sustainability and decision-making. For research that is publicly funded, it is incumbent upon the researcher to provide public access to their data once funding and publishing obligations have been met. Editors of peer-reviewed journals are increasingly encouraging researchers to publish their analyzed and interpreted raw data along with their articles, which is a critical first step to making data and science more widely available. How to do so in a manner that ensures proper use of data is a critical consideration, and we encourage the following ISO data and metadata standards as a first step. Free public access may help increase rather than impede data accessibility for future research.

Species distribution modeling in landscape ecology is a complex endeavor. Ongoing digitization initiatives such as promoted by NSF, NIH, LTER, NCEAS, and IPY, and others will add to the (online) data jungle. Being able to deal with this information flood correctly will be the key issue for extracting valid information, tracking outliers, and for making the best possible decisions based on the data. Every project and modeling initiative requires a solid and robust data management model. With automated online deliveries of data, metadata, and visualizations we are moving into a new age of digitization; one that caters to all citizens of the world equally using the best available technology and science.

Acknowledgments Many colleagues have contributed to the ideas and concepts expressed in this chapter. This chapter benefited from the suggestions and reviews of B. McComb, W. Hochachka, M. Hooten, and an anonymous reviewer. We are very grateful for their input. We would also like to thank the editors for the invitation to contribute to this volume and we are grateful for their guidance.

References

Aldridge CL, Boyce MS (2007) Linking occurrence and fitness to persistence: habitat-based approach for endangered Greater Sage-Grouse. Ecol Appl 17:508–526.
Anderson DR (2008) Model based inference in the life sciences: a primer on evidence. Springer, New York, NY.
Anderson DR, Burnham KP, Gould WR, Cherry S (2001) Concerns about finding effects that are actually spurious. Wildl Soc Bull 29:311–316.
Araújo MB, Guisan A (2006) Five (or so) challenges for species distribution modelling. J Biogeogr 33:1677–1688.
Araújo MB, Luoto M (2007) The importance of biotic interactions for modelling species distributions under climate change. Global Ecol Biogeogr 16:743–753.
Araújo MB, Williams PH, Fuller RJ (2002) Dynamics of extinction and the selection of nature reserves. Proc R Soc Lond Ser B 269:1971–1980.
Austin MP (2002) Spatial prediction of species distribution: an interface between ecological theory and statistical modelling. Ecol Model 157:101–118.
Austin M (2006) Species distribution models and ecological theory: a critical assessment and some possible new approaches. Ecol Model 200:1–19.
Barry S, Elith J (2006) Error and uncertainty in habitat models. J Appl Ecol 43:413–423.
Bibby CJ, Burgess ND, Hill DA, Mustoe S (2000) Bird census techniques. Academic Press, San Diego, CA.
Bishop JA, Myers WL (2005) Associations between avian functional guild response and regional landscape properties for conservation planning. Ecol Indic 5:33–48.
Braun CE (2005) Techniques for wildlife investigations and management. The Wildlife Society, Bethesda, MD.
Breiman L (2001a) Random forests. Mach Learn 45:5–32.
Breiman L (2001b) Statistical modeling: the two cultures. Stat Sci 16:199–231.
Brennan JM, Bender DJ, Contreras TA, Fahrig L (2002) Focal patch landscape studies for wildlife management: optimizing sampling effort across scales. In Lui J, Taylor WW (eds) Integrating landscape ecology into natural resource management. Cambridge University Press, NY.
Brotons L, Thuiller W, Araújo MB, Hirzel AH (2004) Presence–absence versus presence-only modelling methods for predicting bird habitat suitability. Ecography 27:437–448.
Buckland ST (2001) Introduction to distance sampling: estimating abundance of biological populations. Oxford University Press, Oxford, UK.
Burnham KP, Anderson DR (2002) Model selection and inference: a practical information-theoretic approach. Springer-Verlag, New York.
Coudun C, Gégout JC (2006) The derivation of species response curves with Gaussian logistic regression is sensitive to sampling intensity and curve characteristics. Ecol Model 199: 164–175.
Craig E, Huettmann F (2009) Using "blackbox" algorithms such as TreeNET and Random Forests for data-mining and for finding meaningful patterns, relationships and outliers in complex ecological data: an overview, an example using golden eagle satellite data and an outlook for a promising future. In Wang HF (ed) Intelligent data analysis: developing new methodologies through pattern discovery and recovery. Information Science Reference, Hershey, PA.
D'Eon RG, Delparte D (2005) Effects of radio-collar position and orientation on GPS radio-collar performance, and the implications of PDOP in data screening. J Appl Ecol 42:383–388.
D'Eon RG, Serrouya R, Smith G, Kochanny C (2002) GPS radiotelemetry error and bias in mountainous terrain. Wildl Soc Bull 30:430–439.
Donald PF, Fuller RJ (1998) Ornithological atlas: a review of uses and limitations. Bird Study 45:129–145.
Duke CS (2006) Data: share and share alike. Front Ecol Environ 4:395–395.
Duke CS (2007) Beyond data: reproducible research in ecology and environmental sciences – the author replies. Front Ecol Environ 5:67.

Edwards TC, Cutler DR, Zimmermann NE, Geiser L, Moisen GG (2006) Effects of sample survey design on the accuracy of classification tree models in species distribution models. Ecol Model 199:132–141.

Elith J, Graham CH, Anderson RP, Dudik M, Ferrier S, Guisan A, Hijmans RJ, Huettmann F, Leathwick JR, Lehmann A, Li J, Lohmann LG, Loiselle BA, Manion G, Moritz C, Nakamura M, Nakazawa Y, Overton JM, Peterson AT, Phillips SJ, Richardson K, Scachetti-Pereira R, Schapire RE, Soberón J, Williams S, Wisz MS, Zimmermann NE (2006) Novel methods improve prediction of species' distributions from occurrence data. Ecography 29:129–151.

Elzinga CL (2001) Monitoring plant and animal populations. Blackwell Science, Malden, MA.

Esanu JM, Uhlir PF (2003) The role of scientific and technical data and information in the public domain: proceedings of a symposium. National Academies Press, Washington, DC.

Ferrier S, Guisan A (2006) Spatial modelling of biodiversity at the community level. J Appl Ecol 43:393–404.

Fortin MJ, Dale MRT (2005) Spatial analysis: a guide for ecologists. Cambridge University Press, Cambridge, UK.

Frair JL, Nielsen SE, Merrill EH, Lele SR, Boyce MS, Munro RHM, Stenhouse GB, Beyer HL (2004) Removing GPS collar bias in habitat selection studies. J Appl Ecol 41:201–212.

Frair JL, Merrill EH, Allen JR, Boyce MS (2007) Know thy enemy: experience affects elk translocation success in risky landscapes. J Wildl Manag 71:541–554.

Gibbons DW, Donald PF, Bauer HG, Fornasari L, Dawson IK (2007) Mapping avian distributions: the evolution of bird atlases. Bird Study 54:324–334.

Graham CH, Ferrier S, Huettman F, Mortiz C, Peterson AT (2004) New developments in museum-based informatics and applications in biodiversity analysis. Trends Ecol Evol 19:497–503.

Guisan A, Zimmermann NE (2000) Predictive habitat distribution models in ecology. Ecol Model 135:147–186.

Guisan A, Lehmann A, Ferrier S, Austin M, Overton JMcC, Aspinall R, Hastie T (2006) Making better biogeographical predictions of species' distributions. J Appl Ecol 43:386–392.

Guisan A, Graham CH, Elith J, Huettmann F, Dudik M, Ferrier S, Hijmans R, Lehmann A., Li J, Lohmann LG, Loiselle B, Manion G, Moritz C, Nakamura M, Nakawawa Y., Overton JMcC, Peterson AT, Phllips SJ, Richardson K, Scachetti-Pereira R, Schapire RE, Williams SE, Wisz MS, Zimmermann NE (2007) Sensitivity of predictive species distribution models to change in grain size. Divers Distrib 13:332–340.

Hames RS, Rosenberg KV, Lowe JD, Dhondt AA (2001) Site reoccupation in fragmented landscapes: testing predictions of metapopulation theory. J Anim Ecol 70:182–190.

Hastie AT, Tibshirani R, Friedman J (2001) The elements of statistical learning: data mining, inference, and prediction. Springer, New York.

Heikkinen RK, Luoto M, Virkkala R, Pearson RG, Körber JH (2007) Biotic interactions improve prediction of boreal bird distributions at macro-scales. Global Ecol Biogeogr 16:754–763.

Hernandez PA, Graham CH, Master LL, Albert DL (2006) The effect of sample size and species characteristics on performance of different species distribution modeling methods. Ecography 29:773–785.

Hirzel A, Guisan A (2002) Which is the optimal sampling strategy for habitat suitability modelling. Ecol Model 157:331–341.

Hochachka WM, Caruana R, Fink D, Munson A, Riedewald M, Sorokina D, Kelling S (2007) Data-mining discovery of pattern and process in ecological systems. J Wildl Manag 71:2427–2437.

Hollister JW, Walker HA (2007) Beyond data: reproducible research in ecology and environmental sciences. Front Ecol Environ 5:11–12.

Huettmann F (2005) Databases and science-based management in the context of wildlife and habitat: toward a certified ISO standard for objective decision-making for the global community by using the internet. J Wildl Manag 69:466–472.

Huettmann F (2007) The digital teaching legacy of the International Polar Year (IPY): details of a present to the global village for achieving sustainability. Proceedings 18th International Workshop on Database and Expert Systems Applications, DEXA: 673–677.

Huettmann F, Diamond AW (2006) Large-scale effects on the spatial distribution of seabirds in the Northwest Atlantic. Landsc Ecol 21:1089–1108.

Huettmann, F. (2009) *The Global Need for, and Appreciation of, High-Quality Metadata in Biodiversity work*. In: E. Spehn and C. Koerner (eds). Data Mining for Global Trends in Mountain Biodiversity. CRC Press, Taylor & Francis. pp 25–28.

Jan L (2006) Database model for taxonomic and observation data. In Sahni S (ed) Proceedings of the 2nd IASTED international conference on advances in computer science and technology. ACTA Press, Puerto Vallarta, Mexico.

Jochum K (2008) Benefits of using marginal opportunistic wildlife behavior data: constraints and applications across taxa – a dominance hierarchy example relevant for wildlife management. M.Sc. Thesis, University Hannover: Hannover, Germany.

Kadmon R, Farber O, Danin A (2004) Effect of roadside bias on the accuracy of predictive maps produced by bioclimatic models. Ecol Appl 14:401–413.

Karasti H, Baker KS (2008) Digital data practices and the long term ecological research program growing global. Int J Digit Curation 3:42–58.

Lutolf M, Kienast F, Guisan A (2006) The ghost of past species occurrence: improving species distribution models for presence-only data. J Appl Ecol 43:802–815.

MacKenzie DI (2005a) Was it there? Dealing with imperfect detection for species presence/absence data. Aust N-Z J Stat 47:65–74.

MacKenzie DI (2005b) What are the issues with presence–absence data for wildlife managers? J Wildl Manag 69:849–860.

MacKenzie DI, Nichols JD, Royle JA, Pollock KH, Bailey LL, HIines JE (2006) Occupancy estimation and modeling: inferring patterns and dynamics of species. Elsevier, Burlington, MA.

MacKenzie DI, Royle JA (2005). Designing occupancy studies: general advice and allocating survey effort. J Appl Ecol 42:1105–1114.

Magness DR, Huettmann F, and Morton JM (2008) Using Random Forests to provide predicted species distribution maps as a metric for ecological inventory & monitoring programs. Pages 209–229 in Smolinski TG, Milanova MG & Hassanien A-E (eds.). Applications of Computational Intelligence in Biology: Current Trends and Open Problems. Studies in Computational Intelligence, Vol. 122, Springer-Verlag Berlin Heidelberg. 428 pp.

Manel S, Williams HC, Ormerod SJ (2001) Evaluating presence–absence models in ecology: the need to account for prevalence. J Appl Ecol 38:921–931.

Manly BFJ, McDonald LL, Thomas DL, McDonald TL, Erickson WP (2002) Resource selection by animals: statistical design and analysis for field studies. Kluwer Academic Publishers, Boston, MA.

Marzluff JM, Knick ST, Millspaugh JJ (2001) High-tech behavioral ecology: modeling the distribution of animal activities to better understand wildlife space use and resource selection. In Marzluff JM, Millspaugh JJ (eds) Radio-tracking and animal populations. Academic Press, San Diego, CA.

McGowan K, Zuckerberg B (2008) Summary of results. In McGowan K, Corwin K (eds) The second atlas of breeding birds in New York State. Cornell University Press, Ithaca, NY.

Meyer CB (2007) Does scale matter in predicting species distributions? Case study with the Marbled Murrelet. Ecol Appl 17:1474–1483.

Michener WK, Brunt JW (eds) (2000) Ecological data: design, management, and processing. Blackwell Science, Malden, MA.

Michener WK, Brunt JW, Helly JJ, Kirchner TB, Stafford SG (1997) Nongeospatial metadata for the ecological sciences. Ecol Appl 7:330–342.

Moen R, Pastor J, Cohen Y, Schwartz CC (1996) Effects of moose movement and habitat use on GPS collar performance. J Wildl Manag 60:659–668.

Moen R, Pastor J, Cohen Y (1997) Accuracy of GPS telemetry collar locations with differential correction. J Wildl Manag 61:530–539.

Nemitz, D. 2008 An assessment of sampling detectability for global bioidversity monitoring: results from sampling GRIDs in different climatic regions. MINK program, University of Goettingen, Germany, unpublished Masters thesis.

Nielsen SE, Stenhouse GB, Boyce MS (2006) A habitat-based framework for grizzly bear conservation in Alberta. Biol Conserv 130:217–229.

Pearson RG (2007) Species' distribution modeling for conservation educators and practitioners – synthesis. American Museum of Natural History. http://ncep.amnh.org. Accessed 7 May 2008.

Pearson RG, Raxworthy CJ, Nakamura M, Peterson AT (2007) Predicting species distributions from small numbers of occurrence records: a test case using cryptic geckos in Madagascar. J Biogeogr 34:102–117.

Peng RD, Dominici F, Zeger SL (2006) Reproducible epidemiologic research. Am J Epidemiol 163:783–789.

Pulliam HR (2000) On the relationship between niche and distribution. Ecol Lett 3:349–361.

Rempel RS, Rodgers AR (1997) Effects of differential correction on accuracy of a GPS animal location system. J Wildl Manag 61:525–530.

Rempel RS, Rodgers AR, Abraham KF (1995) Performance of a GPS animal location system under boreal forest canopy. J Wildl Manag 59:543–551.

Rodgers AR (2001) Recent telemetry technology. In: Millspaugh JJ, Marzluff JM (eds) Radio-tracking and animal populations. Academic Press, San Diego, CA.

Royle JA, Dorazio RM (2008) Hiearchical modeling and inference in ecology: the analysis of data from populations, metapopulations, and communities. Academic Press, Boston, MA.

Saltz D (1994) Reporting error measures in radio location by triangulation – a review. J Wildl Manag 58:181–184.

Sauer JR, Hines JE, Fallon J (2007) The North American breeding bird survey, results and analysis 1966–2006. Version 10.13.2007. USGS Patuxent Wildlife Research Center, Laurel, MD.

Scott JM, Heglund PJ, Morrison ML (eds) (2002) Predicting species occurrences: issues of accuracy and scale. Island Press, Washington, DC.

Segurado P, Araújo MB (2004) An evaluation of methods for modelling species distributions. J Biogeogr 31:1555–1568.

Soberón JM, Llorente JB, Onate L (2000) The use of specimen-label databases for conservation purposes: an example using Mexican Papilionid and Pierid butterflies. Biodiv Conserv 9:1441–1466.

Stockwell DRB, Peterson AT (2002) Effects of sample size on accuracy of species distribution models. Ecol Model 148:1–13.

Sutherland WJ (2000) The conservation handbook: research, management and policy. Blackwell Science, Malden, MA.

Sutherland WJ (2006) Ecological census techniques: a handbook. Cambridge University Press, Cambridge, UK.

Thompson WL (2004) Sampling rare or elusive species: concepts, designs, and techniques for estimating population parameters. Island Press, Washington, DC.

Thompson WL, White GC, Gowan C (1998) Monitoring vertebrate populations. Academic Press, San Diego, CA.

Travaini A, Bustamante J, Rodríguez A, Zapata S, Procopio D, Pedrana J, Peck RM (2007) An integrated framework to map animal distributions in large and remote regions. Divers Distrib 13:289–298.

Trzcinski MK, Fahrig L, Merriam G (1999) Independent effects of forest cover and fragmentation on the distribution of forest breeding birds. Ecol Appl 9:586–593.

Turchin P (1998) Quantitative analysis of movement: measuring and modeling population redistribution in animals and plants. Sinauer Associates, Sunderland, MA.

Venier LA, McKenney DW, Wang Y, McKee J (1999) Models of large-scale breeding-bird distribution as a function of macro-climate in Ontario, Canada. J Biogeogr 26:315–328.

Venier LA, Pearce J, McKee JE, McKenney DW, Niemi GJ (2004) Climate and satellite-derived land cover for predicting breeding bird distribution in the Great Lakes Basin. J Biogeogr 31:315–331.

Vesley D, McComb BC, Vojta CD, Suring LH, Halaj J, Holthausen RS, Zuckerberg B, Manley PM (2006). Development of protocols to inventory or monitor wildlife, fish, or rare plants. General Technical Report WO-72. U.S. Department of Agriculture, Forest Service, Washington, DC.

White GC, Garrott RA (1990) Analysis of wildlife radio-tracking data. Academic Press, San Diego, CA.

Withey JC, Bloxton TD, Marzluff JM (2001) Effects of tagging and location error in wildlife radio-telemetry studies. In Millspaugh JJ, Marzluff JM (eds) Radio-tracking and animal populations. Academic Press, San Diego, CA.

Chapter 5
The Role of Assumptions in Predictions of Habitat Availability and Quality

Edward J. Laurent, C. Ashton Drew, and Wayne E. Thogmartin

5.1 Introduction

Abstracting a complex reality into ecological models composed of maps, diagrams, and mathematical equations forces modelers to organize information, distinguish essential from superfluous components, and define relationships among variables. Within this context, an assumption is a premise, stated or unstated, which characterizes model variables and relationships as essential or irrelevant to the model's setting and purpose. For example, assumptions about how species interact with their environment at a specific time and place can be used to justify the thematic, spatial, and temporal extent and grain of the input data, given a model's intended application. These assumptions also justify the use or rejection of specific model variables, parameters, and mathematical functions describing the relationship between focal species and their environment. Furthermore, models improve over time through incremental steps of testing assumptions as hypotheses to establish empirical knowledge. Hence, the utility of any habitat model is both empowered by and limited by its assumptions. Therefore it is critical that project objectives and ecological theory inform assumptions, rather than allowing these decisions to be driven by data availability and knowledge gaps.

The ecological modeling literature offers a growing wealth of publications that evaluate various statistical techniques that relate species and environmental data (Segurado and Araújo 2004; Elith et al. 2006) and assess model predictive performance (Pearce and Ferrier 2000; Hirzel et al. 2006). These comparisons provide guidance for the selection of appropriate modeling methods. What is generally missing, however, and what we have attempted to provoke here, is discussion of the assumptions underlying choices about the use of environmental and species data. Many decisions are made defining what Austin (2007) refers to as the data model prior to defining "habitat" by statistically associating environmental data with

---

E.J. Laurent (✉)
American Bird Conservancy, 4249 Loudoun Avenue, The Plains, VA 20198, USA
e-mail: elaurent@abcbirds.org

species data. These decisions include not only those regarding which variables and relationships to model, but also how to interpret and represent results in ecologically meaningful ways.

Given the increasingly important role of predictive species–habitat models in support of conservation planning (e.g., Amstrup et al. 2007), it is vital to recognize model assumptions and consider their appropriateness. We therefore focus here on assumptions specific to the representation of landscapes and species *prior* to generating statistical relationships between the two types of data. Such assumptions largely determine how applicable a given model will be for another species or in another location. As models become more easily accessible through widely dispersed and publicly available decision support tools, we hope this discussion will prompt model developers to take greater care by explicitly stating their assumptions. This, in turn, will empower model end-users (Table 5.1) to assess model utility in their unique conservation setting and will promote the testing and experimentation necessary to advance our ability to effectively model landscape pattern and process.

The following chapter is divided into three sections. The first section proposes a general semantic framework to aid both model developers and users in describing critical assumptions. The second section addresses two major areas where modeling decisions constrain how a model should be applied and what conclusions can be supported. These areas relate to assumptions about the choices made when characterizing landscapes as well as the distribution and abundance of species. Our treatment of this issue is not intended to be exhaustive but rather illustrative of the breadth of topics addressed by common assumptions. We also offer details, where applicable, about how these assumptions have evolved and what alternatives are suggested by recent ecological theory. Citations primarily reference either review articles that discuss progress in relevant ecological theory or research articles that experimentally test the validity of alternative assumptions. Finally, we conclude with a discussion of the critical role of validating model assumptions, and enhancing the timely integration of theoretical and experimental ecology results into predictive models.

Table 5.1 Who are model end-users?

Knowledge gained from models of species–habitat relationships is integral to many aspects of conservation science and planning, by enabling:	
Scientists	To summarize existing knowledge, formulate hypotheses, and test critical assumptions
Program administrators	To assess program effectiveness and allocate resources according to need and opportunity
Public and private landowners	To make informed management decisions
Public affairs specialists	To succinctly communicate urgent problems and conservation achievements
Land protection organizations	To identify properties of high value to wildlife
Agency directors	To justify support for conservation programs
Law makers	To craft public policies that reduce threats to wildlife
Regulators	To assess the risks of issuing development, emissions, and discharge permits

5.2 A Semantic Framework for Evaluating the Assumptions of Model Components and Data

In general, modelers are encouraged toward parsimony. At the same time, there is real danger in excluding key variables or ignoring critical processes. In this balancing act between simplicity and complexity, abstraction and realism, the objective is to produce a model that generates predictions at an acceptable level of accuracy and precision for how the model will be applied. The application drives predictive model development because it defines the model objectives, sets the spatial and temporal domain, quantifies what level of error is acceptable, and characterizes how the model will be evaluated. Only in light of this information can a modeler evaluate the assumptions underlying critical choices about the relative value and appropriate representation of environmental and species data sources.

A common vocabulary (i.e., semantic framework) for describing assumptions may assist in communicating the strengths and weaknesses of modeling approaches so that their idiosyncrasies may be evaluated and improved over time and their results applied appropriately. In the most general sense, assumptions may be classified according to whether they serve to reduce or expand the complexity of a model's representation of reality (*exclusion* vs. *inclusion*) and whether they are clearly stated (*explicit* vs. *implicit*) and justified. Exclusion assumptions lead to simpler models with fewer variables and fewer, or simpler, relationships among variables. The choice of a model's temporal and spatial scale is an exclusion assumption because it defines the domain of the model, thereby excluding all other possibilities outside that domain. Processes at coarser and finer scales are assumed irrelevant to the model method and application because coarser scale processes are assumed to be uniform over the model extent, whereas finer scale processes may be heterogeneous but their overall effect is assumed either inconsequential or "averaged out." Additional explicit exclusion assumptions may be justified based on available data; either there are insufficient data to characterize a variable's influence on the model system or there are sufficient data to characterize the variable as non-significant prior to modeling. Common examples of exclusion assumptions include conceptualizing a system as a closed box without interaction with the external environment (e.g., excluding immigration and emigration in a population model), treating a dynamic process as random or constant (e.g., excluding energetic constraints and behavioral choice by dispersing organisms), and grouping items based on similar traits (e.g., ignoring the diversity of non-habitat classes to treat them equally as non-habitat).

The use of exclusion assumptions, whether stated or unstated, is a very practical approach for simplifying models during development. However, they can also have unintended consequences. Exclusion assumptions treat influences external to the model as "noise," with the assumption that any trends that might occur in the noise is random or non-significant relative to the variability of the factors selected for model inclusion. If this assumption is false, models are under-parameterized (e.g., too simple) and will fail to explain key sources of variation in the response, which may in turn lead to inappropriate inferences. For example, failure to accommodate spatial correlation between response data often leads to biased parameter estimates and inappropriately narrow confidence estimates (Koenig 1999; Lennon 2000).

Inclusion assumptions justify the modeling approach (e.g., statistical method), data sources, and relationships among variables. Several inclusion assumptions are implicit in most species–habitat models, including the prerequisite assumptions that organisms are non-randomly distributed across the landscape, and that a close relationship exists between the occurrence, frequency or number of species observations and habitat quality. Generally, modelers assume, correctly or not, that such "facts" are "common knowledge," and therefore do not explicitly state them in the text. Therefore these assumptions can persist despite long-standing theoretical (Van Horne 1983; Pulliam 1988; Hobbs and Hanley 1990; Pulliam and Danielson 1991) and experimental (multiple examples reviewed in Van Horne 1983) evidence that species distribution and abundance patterns do not necessarily equate directly with habitat quality or carrying capacity. Of special concern when modeling within geographical information systems (GIS) are implicit inclusion assumptions about the suitability of the spatial datasets for characterizing the relationship of a species to its habitat. Ecologists employing these data and end users of the models are often unaware of the many technical constraints and processing decisions impacting the quality of the spatial datasets (Thogmartin et al. 2004a; Gallant 2009), such as errors of commission and omission that often vary among land cover classes or of smoothing algorithms used to "despeckle" maps so that they better represent the patterns that humans perceive across landscapes. Most data users implicitly use the limited data at hand and assume that their spatial, temporal, and thematic resolutions are relevant to the ecological patterns and processes under investigation, thus ignoring these quality issues.

Whether addressing inclusion or exclusion assumptions, explicit statements of assumptions promote better understanding of the modeled system through several mechanisms. Such statements identify knowledge gaps and potential weaknesses for future model development and testing. They provide modelers and decision makers with the information necessary to assess the relevance of models in both the original and novel settings. Also, when clearly stated, assumptions have value as working hypotheses. As new knowledge becomes available, explicit assumptions can be tested to determine whether they continue to offer a valid representation of the system. If assumptions are left unstated, there is a greater risk that models will be applied inappropriately or provide inappropriate and untested information.

5.3 Landscape Characterization

Environmental information represented as spatial data within a GIS are typically classified from remotely sensed imagery or are interpolated from field sampling locations. How these data are processed and used to delineate landscape features in species–habitat models requires careful consideration. The majority of spatially-explicit landscape ecological models classify the environment into discrete categorical units corresponding to habitat and non-habitat. However, several alternative approaches view landscapes as offering continuous, uncertain, and dynamic gradients in habitat quality.

5.3.1 Discrete Patches Versus Continuous Gradients

A patch-matrix model assumes that landscapes can be divided into discrete polygons (i.e., patches) that are delineated by hard boundaries and can be grouped into two or more categories on the basis of a variety of reasons (Wiens et al. 1993; Antrop 2007). Generally, environmental variation within a patch is ignored (assumed constant or inconsequential), as is variation among patches of the same category (Ovaskainen 2004; Dunn and Majer 2007). Patch characteristics, such as area, perimeter, and isolation, are then assumed to serve as direct proxies for habitat quality or suitability. Delineating habitat features with hard boundaries easily represents many developed landscapes. For example, in agrarian regions, transitions from agricultural farms, woodlots, and riparian corridors are clearly abrupt and the potential value of fields versus forest to woodland species is distinct. However, boundaries and transitions are not always so visibly defined. Even when clear to the human perspective, a patch-matrix model may not reflect the perceptions and use of the landscape by other species.

The patch-matrix data structure emphasizes land cover classes as the primary driver of and proxy for ecological processes. This can greatly simplify models but implies acceptance of many assumptions that may be inappropriate (Tang and Gustafson 1997). For example, if a species specializes in habitat with finer thematic precision than our classification system (e.g., only using oak stands within patches categorized as deciduous) then the assumption of species presence within homogeneous deciduous patches could result in unacceptably high commission errors (i.e., false predictions of species presence). Furthermore, many landscape metrics are known to be highly sensitive to the number of map classes (i.e., thematic resolution; Castilla et al. 2009); a single landscape will appear more fragmented as additional cover or use categories are defined and existing patches are split according to the finer classification system. If a model's purpose is to identify and prioritize locations for conservation action to ease fragmentation pressures, then assumptions made to justify thematic resolution have a very strong influence on the model predictions and resulting decisions.

Alternative methods used to represent and process spatial data incorporate hierarchical nesting of discrete patches (Dunn and Majer 2007) or simply avoid defining hard boundaries and categorical habitat classes by depicting landscapes as continuous environmental gradients (Manning et al. 2004; Fischer and Lindenmayer 2006; see Chap. 7). Gradient (also referred to as continuum) representations of environmental data claim to more directly focus on process, by considering gradients in food, shelter, and stress. Advocates of gradient approaches argue that there is no reason to *a priori* assume that environmental variability is categorical or that species or ecological processes respond categorically to it (McGarigal and Cushman 2005). Gradient models do not force arbitrary delineation of boundaries where transitions are gradual (e.g., fresh to saline gradients in coastal marshes), yet they still illustrate sharp environmental discontinuities where such juxtapositions exist (e.g., canopy closure at agriculture-forest edge). Critical consideration of a project's objectives in relation to available data may also lead to models that use a combination of categorical and continuous data for both predictor and response variables. For example, models of habitat quality for forest-dependent species existing within a dominantly agricultural landscape may use a categorical description of shelter based on land cover and land use (assuming use of fields as shelter is insignificant), but also include a description of stress quantified as a continuous variable (e.g., distance to field).

5.3.2 Static Versus Dynamic Landscapes

Land cover and land use patterns change through time, driven by natural processes and changing social, political, and economic values, resulting in an ever-shifting variety and combination of environmental conditions (Bürgi et al. 2004; Manning et al. 2009). Simple models, however, typically ignore historical events and temporal processes, despite their profound impact not only on the structure but also the function of present day landscapes (Bürgi et al. 2004; Rhemtulla et al. 2007; Gillson 2009). Incorporating knowledge of processes at historical, archeological, and palaeoecological scales can improve the precision and accuracy of predictive landscape ecological models (Graham et al. 2006; Von Holle and Motzkin 2007). However, when temporal processes are explicitly modeled to make predictions through time, a common claim is that data from the past and present informs the future – that there is a continuity of process and pattern. This assertion is increasingly challenged, as natural and social scientists recognize the possibility of a "no analog" future in light of unprecedented anthropogenic perturbation of global biogeochemical and energy cycles (Fox 2007; Williams and Jackson 2007). For example, the concept of landscape fluidity (Manning et al. 2009) describes an ebb and flow of species–habitat associations through time as species encounter and adapt to different environmental conditions. Whereas scientists may grasp the basic processes driving long-term temporal and spatial patterns, the accelerating pace and cascading effects of global change has the potential to produce non-equilibrium landscapes lagging in their response to key drivers (Harris 2007; Manning et al. 2009). Such short-term (decadal and sub-decadal) trends are poorly understood, yet represent the time horizon of most planning and monitoring activities (Willis et al. 2007). Assumptions regarding how to handle (or when to safely ignore) these trends within predictive models are extremely important, and more research is needed to guide model development and application.

As research into temporal processes advances, most predictive landscape ecological models continue to assume static conditions. In large part, this is because many available data to define environmental conditions at a given location currently only exist as a single value or summary statistic during a single snapshot in time; they typically do not represent historical conditions or dynamic processes. Even when models assume static conditions, it is still critical to consider the temporal validity of the data in relation to the project focus and scale (Table 5.2). For example, associating multiple data layers together requires the simplifying assumptions that (1) no significant change has occurred between the oldest and most recent data, and (2) this assumed-static landscape remains valid for the temporal extent over which the predictive model will apply (e.g., the period over which the model will be used to support conservation planning decisions). Ignoring temporal dynamics is likely to generate prediction errors if there is a long time lag between data acquisition, classification, and application; if the rates of change within modeled landscapes are rapid; or if there is an anomalously large disturbance (e.g., fire, hurricane) between the time of data collection and application. Different environmental data sources will have different "shelf-lives": snapshots of elevation or geologic substrate could remain valid for a long period, while a snapshot of land cover in regions experiencing strong economic

Table 5.2 An example of dynamic landscape processes.

Population connectivity among coral populations is maintained through dispersal of larvae carried by ocean currents. Dispersing larvae are often depicted as passive particles drifting on ocean surface currents, failing to recruit if these currents do not carry them by suitable reef habitat at the correct stage of development. The direction and strength of ocean currents are often summarized and incorporated into models as mean values over a season or year. However, ocean current patterns vary through the climate cycles of El Niño and La Niña, potentially altering the connectivity among coral populations. Models that consider only mean conditions, without the context of variation around that mean, can act to silence the signal in relationships among variables, especially when those relationships have evolved under dynamic, yet regularly repeating conditions.	
Example citation	Treml EA, Halpin PN, Urban DL, Pratson LF (2008) Modeling population connectivity by ocean currents, a graph-theoretic approach for marine conservation. Landsc Ecol 23: 19–36
Modeling decision	Incorporate dynamic influences of El Niño Southern Oscillation on ocean currents in the Tropical Pacific
Model objective	Predict dispersal pathways of coral larvae
Model application	Identify potential population sources, sinks, and stepping stones
Inclusion assumption	The authors increased model complexity by separately modeling dispersal pathways under three climate scenarios: strong La Niña year, strong El Niño year, and Neutral year
Justification	Working at the spatial scale of the Tropical Pacific allows for many simplifying assumptions about fine-scale processes (e.g., regarding individual larvae behavior) but increases the potential importance of large-scale, long-term spatial processes. Where gene flow, persistence, and metapopulation dynamics are driven by recruitment from external sources, it is especially important to accurately predict cycles in recruit supply over time. This information is necessary to distinguish population decline due to local conditions from population decline due to absence of recruitment. Sites that appear isolated under mean conditions could be critically important to the dispersal dynamics of coral during the occasionally strong El Niño or La Nina years when those locations offer the only stepping stone between source and sink populations

growth or a novel invasive pest species could quickly become outdated. The specific forces driving change are project specific, because they are determined by the spatial, temporal, and institutional scale of the system under study (Bürgi et al. 2004). Efforts to estimate rates, causes, and consequences of land cover change. (e.g., Loveland et al. 2002; Gallant et al. 2004), will assist modelers in determining the sensitivity of their project to dynamic processes affecting land cover and land use patterns.

5.3.3 Deterministic Versus Fuzzy Classification

Another area of model advancement has been the development of approaches to integrate and represent uncertainty of both input data and output population and habitat values. Here we refer not to the equally important methods of uncertainty analysis, sensitivity analysis, and other error assessment methods (Jager and King 2004; see Chaps. 6, 11, 13, and 14), but rather the explicit inclusion of uncertainty in the representation of environmental data. In the simplest models, each patch or pixel of a landscape is assigned a single value for each environmental data layer, and landscape pattern is a deterministic product of these environmental data. The use of maps when modeling species–habitat relationships therefore incorporates an implicit assumption that there is no spatial, temporal, or measurement uncertainty associated with these data. Deterministic classification systems assume that classes are mutually exclusive and exhaustive, and that each classified unit belongs to one class only with perfect certainty. The absence of confidence intervals on output products or assessment of model sensitivity to input errors unwittingly bolsters the assumption that data are precise and accurate. Ecological risk assessment models (also referred to as threats assessment) are a general exception, where various social, economic, and ecological threats are commonly evaluated within a probabilistic framework that explicitly accounts for uncertainty (Hunsaker et al. 1990; Harwood 2000).

The role of biocomplexity and uncertainty in predictive models is receiving increased attention as conservation planning shifts from a protected area focus to a whole-landscape focus and from short- to long-term planning horizons. As a consequence, modelers of ecological landscapes are increasingly turning from deterministic to fuzzy methods to represent uncertainty of both input data layers and output predictions (Adriaenssens et al. 2004; Malczewski 2006; Uusitalo 2007). Fuzzy set theory offers an alternative data model in that it incorporates classification uncertainty by defining an object's degree of belonging to each possible class, using either expert knowledge-based or rule-based algorithms (Rocchini and Ricotta 2007; Mouton et al. 2009).

A situation meriting consideration of fuzzy classification is the application of forest versus residential classes in rural regions, where a gradient of housing density often exists, such that neither class applies perfectly. In this type of situation, fuzzy classification systems can also be applied in a hierarchical fashion. For example, a landscape unit may first be coarsely classified as forest versus non-forest with high confidence. Then the forest units are assigned fuzzy membership values as deciduous versus evergreen forest. Finally, deciduous forests are assigned membership values as oak versus maple dominated canopy. Such an approach arguably better reflects our skill and confidence in the final thematic maps and allows hierarchical assessments of area estimates and thematic accuracy (Woodcock and Gopal 2000; Rocchini and Ricotta 2007). A fuzzy set approach provides transparency and flexibility when dealing with incomplete data, non-linear relationships between ecological variables, low precision data in highly heterogeneous regions, or situations where the spatial transition between land cover classes is gradual rather than abrupt.

5.4 Species Characterization Assumptions

Once landscapes have been defined, a second set of assumptions define how species will interact with the landscape as individuals, with conspecifics as populations, and in community with other species. Unlike the environmental data discussed above, available species data may or may not be spatially or temporally explicit (see Chap. 4). When species location data are available and overlap the available environmental data in space and time, correlative relationships between the two sets of data are frequently interpreted to characterize "habitat." When species location data are absent, habitat must be defined and extrapolated based on expert or literature defined hypotheses. In these cases, because empirical data are limiting, it is common to assume that results of a few studies or opinions of a few experts adequately represent the necessary knowledge to distinguish patterns of habitat use, selection, and importance (Garshelis 2000). Whether data come from field observations or expert opinion, common assumptions about the value and appropriate use of these data relate to concepts of species range, ecological niche, intra- and inter-specific competition, and carrying capacity. Further, species–habitat relationships are sometimes extrapolated to other areas or times, which adds additional assumptions such as stationarity (i.e., relationships among variables are constant in space and time; Dutilleul and Legendre 1993; Petitgas 2001). Again, as for landscapes, these assumptions may or may not be appropriate given the project objectives. Modelers must balance simplifying assumptions of homogeneity against needs or desires for incorporating complexity.

5.4.1 Structural Versus Functional Connectivity

Connectivity (and its inverse, isolation) is an issue that bridges the discussion of landscape and species characterizations, because estimates of connectivity will vary with the level of detail used to classify landscapes and describe species movements. Terms such as functional connectivity (Bélisle 2005) and functional heterogeneity (Wiens et al. 2002) are used to describe the ability of a species to assess and access patches within a landscape. The terms express a fusion of concepts that reflect appreciation of the complexity and importance of an individual organism's (often statistically summarized by population and metapopulation; Camus and Lima 2002) perceptual ranges, energetic balances, risk tolerances, and other behavioral responses to environmental and community information (Lima and Zollner 1996; Olden et al. 2004).

Although simulation models and experimental studies demonstrate the role of individual perceptual ability and behavioral choices in shaping population level distribution patterns (Zollner and Lima 2005), few predictive models have incorporated such levels of detail. Instead, structural connectivity assumptions – most simply quantified as Euclidean distance thresholds – have commonly substituted for functional connectivity. Using simple distance measures assumes equal dispersal rates and risks throughout the modeled space. Incorporating structural connectivity assumptions into models often constrains them to movements within aggregations of specified habitat types and emphasizes the role of corridors in connecting isolated habitat patches (Haddad et al. 2003).

Models attempting to describe the functional connectivity of landscapes with greater complexity typically do so by classifying non-habitat space in relation to its permeability or risk to species dispersal, and then calculating connectivity or isolation based upon least-cost pathways (Adriaensen et al. 2003). Graph theoretic models offer an approach that incorporates functional connectivity (Bunn et al. 2000; Urban and Keitt 2001) by emphasizing dispersal processes. In these models, habitat is simplified to binary habitat and non-habitat, but the "edges" that connect habitat patches can be weighted to identify optimal dispersal pathways using values representing energetic costs, predation risk, or species preferences. Models that avoid assumptions associated with simpler geographic distance cost-path values could be especially important in landscapes with large barriers or high risk features in the landscape (Bunn et al. 2000), where species avoid rather than move through these less permeable elements (e.g., mountains, lakes, roads).

5.4.2 Single Species Versus Species in Community

When species data are incorporated into species–habitat models, it is common to define the habitat association of individual species without direct reference to the distribution of other species. Many models ignore species interactions because of their dynamic nature, their poor representation in mapped form, or because the species interactions are not directly amenable to management action. This excludes the complexity of potential community level effects (e.g., competition, predation, commensalism) on the distribution and abundance patterns of the focal species. For example, many models designed for rapid assessments often describe only the habitat requirements (e.g., include only patches of adequate size), limitations (e.g., exclude patches above a certain elevation), and threats (e.g., exclude patches within proximity to urban areas) (Angelstam et al. 2004) of a single species. By excluding the possibility of species interactions, the reduced set of relationships among variables is useful to predict potential habitat or even relative suitability, in the broadest sense, for large scale conservation planning.

However, community ecology can have a strong influence on species distributions. Microcosm experiments have shown that the distribution of species in isolation deviates greatly from the distribution seen when in community (Pearson and Dawson 2003). Species interactions lead to resource partitioning, trophic dependencies, and other inter-species dynamics within this potentially suitable landscape (Soberón and Peterson 2004). These factors as well as limitations of carrying capacity and ability to move through the landscape, will generally bias single-species models toward over-predicting the amount of available and utilized habitat under present conditions (i.e., add commission error). If the present distribution of one species is constrained by another, and this constraint were removed in the future, a single-species model would fail to account for the resulting expansion of the realized niche of the species. Research on the phenomena of species introductions and ecological release (also competitive release, enemy release) provides abundant evidence (and debate) regarding how a species' realized niche breadth and abundance can increase within a given habitat as species diversity decreases (Cox and Ricklefs 1977; Vassallo and Rice 1982; Keane and Crawley 2002). Guidelines are available to assess the importance of species interactions in conservation applications, and they generally focus on potential community-wide effects of species removal (e.g., Soulé et al. 2005). Such guidelines could also inform decisions about whether to include community effects in models, by identifying situations when the distribution of a focal species is heavily dependent upon the distribution of one or more other species.

5.4.3 *Individual versus Population Habitat Associations*

Models accounting for how individual choice and fitness vary within a population and across the landscape are rare (but see Grimm and Railsback 2005; Goss-Custard and Stillman 2008). Instead, individual responses are often generalized as statistical population responses to environmental conditions. This approach excludes as irrelevant the variation among individuals with regard to physiological tolerances, habitat preferences, and overall fitness. It also excludes the possibility of range-wide or temporal variation in species–habitat associations. Often the choice to simplify models in this manner reflects data limitations, as expert knowledge or previous research results are either not available or are unequally distributed within the model's scope.

Imprecise regional predictions and inaccurate local predictions can result from models that assume constant range-wide species–habitat associations (Table 5.3). Across species ranges, latitudinal variation in species–habitat associations has been demonstrated in diverse taxa, reflecting possible differences in selection pressure, resource abundance, or habitat availability. For example, Cerulean Warblers occupy larger patches in the core of their range, but smaller patches along their periphery (Thogmartin et al. 2004b). Similarly, invasive species are known to expand their distribution into environments not present in their native range, such as fire ants

Table 5.3 An example of range-wide versus regional habitat associations.

Species–habitat relationships are rarely constant throughout their range. Geographic variation in the presence or strength of different limiting factors (e.g., resources, predators, disease) influences species' habitat selection and use patterns. Regional differences may be more pronounced for some parameters or some life-stages.	
Example citation	Thogmartin WE, Sauer JR, Knutson MJ (2009) Modeling and mapping abundance of American Woodcock across the Midwestern and Northeastern United States. J Wildl Manage 17:376–382
Modeling decision	American Woodcock (Scolopax minor) are managed separately as eastern and central populations because of differences in migration patterns and population dynamics. Despite these differences in management, the authors decided not to model species–habitat associations with respect to these management zones because of a lack of information regarding purported differences in habitat selection
Model objective	To model spatial patterns of relative abundance of American Woodcock across its breeding range in the United States
Model application	To focus management and monitoring on areas and habitat features of range-wide importance to breeding American Woodcock
Exclusion assumption	The authors reduce model complexity by assuming that American Woodcock's association with habitat and landscape features are consistent across the breeding range of the species
Justification	The model focused only on breeding American Woodcock because information on population dynamics for this species suggests that habitat associations do not differ significantly between regional populations during this life-stage

occupying colder and drier environments in their invasive North America range than their native South America range (Fitzpatrick et al. 2007). Simplifying assumptions of species–habitat association homogeneity and stationarity are most likely to be valid if (1) the entire landscape is available to all individuals (Hjermann 2000), (2) the individuals have *a priori* knowledge of the attributes of all locations (Heinrich 1979; Stephens and Krebs 1986; Orians and Wittenberger 1991; Pulliam and Danielson 1991; Badyaev et al. 1996), and (3) individuals act on this perfect knowledge and access by spending proportionally more time in locations that are "better" (Fretwell and Lucas 1969; Hjermann 2000). Closer examination of homogeneity and stationarity assumptions should occur if the modeled area is far from (or much larger than) the area contributing species data, if species would encounter different competitors or predators in different portions of the model extent, or if local adaptation has been shaped by different historical events.

5.4.4 Perfect versus Imperfect Habitat Use

Many prediction errors in distribution models can be traced back to implicit assumptions about the ability of species to efficiently locate and occupy all suitable habitat (Fielding and Bell 1997). Typically, species are implicitly assumed to always occupy only suitable habitat, to occupy the most suitable habitat prior to occupying less suitable habitat, and to fully saturate available habitat. Yet, theories of spatial population dynamics, such as metapopulation theory, clearly identify cases where suitable habitat can remain temporarily or permanently unoccupied due to reduced immigration, heightened emigration, or delay between localized extinction and recolonization (Thomas and Kunin 1999). Unsaturated habitat is a common condition for recovering populations or invasive species. In such cases, "absence" locations surveyed to characterize non-habitat environmental conditions are in fact similar or identical to "presence" locations. Similarly, predictions could falsely rank the suitability of locations for species that have been excluded from optimal habitat because of fragmentation, dispersal limitation, or pressure by competing species. In both these situations, presence and absence locations can be ecologically indistinguishable outside of the broader spatial and historical context.

Ideal distribution of individuals among the best habitat sites would require perfect knowledge of available resources and unlimited dispersal capability (Zimmerman et al. 2003). Neither exists outside of simulation models, but such simplifying assumptions may be more reasonable for some species than others. Discussions over the fundamental versus realized niche provide insight into factors influencing the suitability of assumptions that simplify models of species habitat associations. The greater the degree to which the realized niche (utilized habitat) is suspected to diverge from the fundamental niche (suitable habitat), the greater its importance in evaluating the assumptions that underlie predictions of species distribution and abundance patterns. Theoretical debate and field research investigating the prevalence of Ideal Free Distribution (i.e., wildlife density is an indication of habitat quality, because animals freely move to distribute themselves in proportion to available resources) and Ideal Despotic Distribution patterns (i.e., density can be lower in higher quality habitat because superior competitors monopolize resources) also highlights the need to critically consider such factors as the dispersal ability, life span, experience, and site fidelity of a species (Zimmerman et al. 2003) when predicting distributional patterns.

5.4.5 Demographically Closed versus Open Populations

In species–habitat modeling, it becomes more difficult (and less appropriate) to causally associate observed distribution and abundance patterns to local environmental conditions as the degree of demographic openness increases. All populations are demographically open at some spatial and temporal scales but closed at others (Hixon et al. 2002). Therefore model extent and resolution, and subsequent representation of habitat configuration, influence whether the simplifying assumption of closed populations is reasonable. If a modeler mistakenly assumes population closure where, for example, high immigration rates mask high death rates (e.g., population sinks), then observed high local abundance could mistakenly be interpreted as high habitat suitability. Similarly, population sources where high birth rates are masked by high emigration rates could be falsely modeled as low suitability habitat based on perceived low local abundance.

Many predictive landscape ecological models assume population closure, forecasting future recruitment solely as a function of present population size and local birth and death rates (Johnson 2005). In contrast, recruitment in demographically open populations can be less influenced by local birth and death rates if processes of immigration and emigration significantly affect population size. Modelers should pay special attention to assumptions of closure in situations where suitable habitat is patchily distributed but the inter-patch functional distances are less than the typical dispersal distance of a species (Thomas and Kunin 1999; Hixon et al. 2002). In addition, immigration and emigration processes will be especially important for species with life-histories that include at least one highly vagile stage, such plants or insects dispersed by wind and most marine organisms dispersed by ocean currents (Hixon et al. 2002).

5.4.6 Absence versus Non-detection

The vast majority of species data available for species–habitat modeling are presence-only (e.g., National Heritage Program Element Occurrences) or count data (e.g., Breeding Bird Survey). Data on true absence are difficult to obtain, because the survey site must be closed to individual movement and searched such that no individual escapes detection. Presence and count data, however, are subject to multiple sources of detection biases affecting the validity of the assumption that the data accurately represent the species distribution and abundance patterns (Gu and Swihart 2004). Each presence or absence record reflects the outcome of a certain level of effort under certain site conditions. Variability in effort (e.g., research and recreation surveys focused on public rather than private lands, timed point counts vs. transects) and variability in site conditions (e.g., dense vs. sparse forest understory, dry vs. wet years), among other sources of variability, influence the likelihood of detecting a given species. Thus, surveys of occupied habitat can often result in both presence or absence data, depending on chance and survey conditions, thus potentially confounding efforts to define species–habitat associations for predictive modeling. As even low levels of non-detection within occupied habitat can lead to spurious results (Gu and Swihart 2004), there is potential for significant predictive error in models assuming perfect correlation between presence data and habitat occupancy. The potential for error is greatest for species that are cryptic, rare, sparsely distributed within suitable habitat, or occupying a wide variety of habitat, as these situations present the greatest challenge to obtaining accurate presence–absence data. Incorporating assumptions of perfect knowledge of species occurrences can result in conservation network design recommendations that are smaller than required and less representative of the true habitat of a species (Rondinini et al. 2006; see also Table 5.4).

To reduce error and avoid assumptions of perfect presence (or absence) data, there is a strong push to establish relationships between species probability of detection and environmental covariates. Such relationships distinguish conditions

Table 5.4 An example of species occurrence as an indicator of habitat quality.

	Species absence from a given habitat is typically interpreted as indicating inadequate resources or other conditions hostile to the species' welfare. However, species occurrence patterns can be unrepresentative of habitat quality under certain conditions. When a species has not fully saturated a landscape, as is common for an invasive or recovering species, the absence of that species from a given region or habitat may not necessarily indicate poor habitat quality. True absence data are rare, and randomly generating absence locations in areas of non-detection (e.g., pseudo-absences) can increase omission errors and lead to inappropriate inferences if species were, in fact, present. Presence data, under these circumstances, can also be difficult to interpret, as a species may be observed in unsuitable habitat while dispersing to more favorable locations
Example citation	Mladenoff DL, Sickley TA, Haight RG, Wydeven AP (1995) A regional landscape analysis and prediction of favorable gray wolf habitat in the Northern Great Lakes Region. Conserv Biol 9:279–294
Modeling decision	The area containing 80% of a gray wolf (Canis lupus) pack's telemetry observations defines suitable habitat within the spatial extent of all observations, while similarly sized "territories" placed in non-use areas represent non-suitable habitat, even if the telemetry path of a dispersing wolf has occasionally entered this space
Model objective	Determine whether environmental differences between areas of pack use and non-use are accurate predictors of wolf distribution and recolonization patterns
Model application	Identify spatial population dynamics and management effects that promote or hinder gray wolf recolonization success. Estimate the potential size of regional wolf populations given available habitat in areas accessible to expanding populations
Exclusion assumption	The authors reduce model complexity by assuming perfect knowledge of suitable habitat, such that 80% isopleths of wolf telemetry data indicate suitable habitat and that random areas at least 10 km outside these isopleths were non-suitable habitat
Justification	The modeled wolf population had been monitored by telemetry for over 20 years. This long-term, continuous data set enabled the authors to identify certain non-use areas within the present range as truly non-preferred habitat (true absence rather than non-detection) with very high confidence

under which non-detection of a species represents true absence versus a failure to detect the species in occupied habitat (Royle and Nichols 2003; Royle et al. 2005; MacKenzie et al. 2006). These adjustments to raw detection data support more accurate predictions of species distribution patterns and estimates of abundance in heterogenous habitat. Research in this area is developing rapidly and examples of applying detection adjusted occupancy within predictive landscape ecology models are still few (but see De Wan et al. 2009).

5.5 Conclusions: Evaluating Assumptions

As abstractions of reality, all landscape ecological models imperfectly predict pattern and process in the real world, regardless of the strength of the underlying assumptions. Furthermore, because species–habitat relationships are fluid through time, the reality which our models represent will change as we progress into the future.

Therefore, the purpose of model validation is not to determine if a model is wrong, but to determine if it is "good enough" to serve the intended purpose (Shifley et al. 2009). Thus modeling is often an iterative process, as we must proceed in using model results while awaiting new data, and new theoretical and experimental insights, to compare against our assumptions (Starfield 1997; Shifley et al. 2009).

In some cases, the "right" or "best" choice during the modeling process is strongly supported by local, empirical evidence. In other cases, expert knowledge may provide useful insight and support for a given approach. However, there will always be cases where the evidence supporting or contradicting a given choice is equivocal and the modeler must depend on logic or default to the simplest option. Regardless, each choice entails assumptions about data quality, suitability, and appropriate use *to best meet model objectives*. The importance of considering model objectives during every step of the modeling process cannot be overemphasized, because given the same information but different objectives, the choices and their supporting assumptions could change.

Although the evaluation of assumptions is an important part of improving and adapting models, the assumptions themselves are typically not subject to validation. Instead, model evaluation aims to clarify the strengths, weaknesses, and utility of the assumptions in relation to the model objectives. The responsibility to evaluate assumptions is shared by the modeler, peer-reviewers, and the end-users of the model products. Explicit explanatory statements of data assumptions are perhaps the best protection against a model being applied in an inappropriate context or for an inappropriate purpose. They ensure that the modeler has not simply chosen the data because they are available (a common justification, as data for species–habitat models are extremely limited), but instead indicate that the modeler has given careful considerations to the data properties in relation to the project's scope and purpose. Implicit assumptions, however, are less likely to be acknowledged in model reviews and uses. Implicit assumptions are also less likely to result in iterative model improvements and overall knowledge of the relationships between species and their environments unless they are revealed through model validation.

Common approaches to testing assumptions of a predictive model include (1) seeking peer-review of model assumptions, (2) performing field or laboratory experiments, (3) simulation modeling, and (4) designing an effective monitoring program to test the significance of assumptions on model fit. Of these options, a well-designed monitoring program provides a practical opportunity to independently validate and improve predictions by advancing knowledge of the system. In an adaptive management context, the explicit assumptions of the models supporting management decisions are treated as hypotheses to be tested through both management action and monitoring outcomes of those actions (Williams et al. 2002).

A growing collection of publications offers helpful advice to modelers and model reviewers seeking to better integrate ecological and statistical theory into model development (e.g., Araújo and Guisan 2006; Austin 2007; Elith and Leathwick 2009). These sources offer background and practical guidance regarding modeling decisions in light of model objectives, ecological theory, data limitations, and statistical constraints. Given this rich history of theoretical debate and experimental insights, we suggest that the knowledge and theory to develop informative models is not lacking. The challenge for modelers is to build models that test objective-based hypotheses and then test the critical assumptions of those models that are most relevant to modeling objectives, given data availability, the abundance of theory, and the diversity of experimental conclusions.

Acknowledgments We thank N. Haddad, A. McKerrow, J. Collazo, M. Iglecia, and two anonymous reviewers who provided insightful comments at various stages of this manuscript's development.

References

Adriaensen F, Chardon JP, De Blust G, Swinnen E, Villalba S, Gulinck H, Matthysen E (2003) The application of "least-cost" modelling as a functional landscape model. Landsc Urban Plan 64:233–247
Adriaenssens V, De Baets B, Goethals PLM, De Pauw N (2004) Fuzzy rule-based models for decision support in ecosystem management. Sci Total Environ 319:1–12
Amstrup SC, Marcot BM, Douglas DC (2007) Forecasting the range-wide status of polar bears at selected times in the 21st Century. U.S. Geological Survey Administrative Report.
Angelstam P, Roberge J-M, Lõhmus A, Bergmanis M, Brazaitis G, Dönz-Beuss M, Edenius L, Koskinski Z, Kurlavicius P, Lārmanis V, Lūkins M, Mikusinski G, Račinskis E, Strazds M, Tryjanowski P (2004) Habitat modelling as a tool for landscape-scale conservation – a review of parameters for focal forest birds. Ecol Bull 51:427–453.
Antrop M (2007) The preoccupation of landscape research with land use and cover. In: Wu J, Hobbs R (eds) Key topics in landscape ecology. Cambridge University Press, Cambridge.
Araújo MB, Guisan A (2006) Five (or so) challenges for species distribution modelling. J Biogeogr 33:1677–1688.
Austin M (2007) Species distribution models and ecological theory: a critical assessment and some possible new approaches. Ecol Modell 200:1–19.
Badyaev AV, Martin TE, Etges WJ (1996) Habitat sampling and habitat selection by female wild Turkeys: ecological correlates and reproductive consequences. Auk 113:636–646.
Bélisle M (2005) Measuring landscape connectivity: the challenge of behavioral landscape ecology. Ecology 86:1988–1995.
Bunn AG, Urban DL, Keitt TH (2000) Landscape connectivity: a conservation application of graph theory. J Environ Manage 59:265–278.
Bürgi MA, Hersperger M, Schneeberger N (2004) Driving forces of landscape change – current and new directions. Landsc Ecol 19:857–868.
Camus PA, Lima M (2002) Populations, metapopulations, and the open-closed dilemma: the conflict between operational and natural population concepts. Oikos 97:433–438.
Castilla G, Larkin K, Linke J, Hay GJ (2009) The impact of thematic resolution on the patch-mosaic model of natural landscapes. Landsc Ecol 24:15–23.
Cox GW, Ricklefs RE (1977) Species diversity and ecological release in Caribbean land bird faunas. Oikos 28:113–122.
De Wan AA, Sullivan PJ, Lembo AJ, Smith CR, Maerz JC, Lassoie JP, Richmond ME (2009) Using occupancy models of forest breeding birds to prioritize conservation planning. Biol Conserv 142:982–991.
Dunn AG, Majer JD (2007) In response to the continuum model for fauna research: a hierarchical, patch-based model of spatial landscape patterns. Oikos 116:1413–1418.
Dutilleul P, Legendre P (1993) Spatial heterogeneity against heteroscedasticity: an ecological paradigm versus a statistical concept. Oikos 66:152–171.

Elith J, Leathwick JR (2009) Species distribution models: ecological explanation and prediction across space and time. Annu Rev Ecol Evol Syst 40:677–697.

Elith J, Graham CH, Anderson RP, Dudík M, Ferrier S, Guisan A, Hijmans RJ, Huettmann F, Leathwick JR, Lehmann A, Li J, Lohmann LG, Loiselle BA, Manion G, Moritz C, Nakamura M, Nakazawa Y, Overton JM, Peterson AT, Phillips SJ, Richardson KS, Scachetti-Pereira R, Schapire RE, Soberón J, Williams S, Wisz MS, Zimmermann NE (2006) Novel methods improve prediction of species' distributions from occurrence data. Ecography 29:129–151.

Fielding AH, Bell JF (1997) A review of methods for the assessment of prediction errors in conservation presence/absence models. Environ Conserv 24:38–49.

Fischer J, Lindenmayer DB (2006) Beyond fragmentation: the continuum model for fauna research and conservation in human-modified landscapes. Oikos 112:473–480.

Fitzpatrick MC, Weltzin JF, Sanders NJ, Dunn RR (2007) The biogeography of prediction error: why does the introduced range of fire ant over-predict its native range? Glob Ecol Biogeogr 16:24–33.

Fox D (2007) Back to the no-analog future? Science 316:823–825.

Fretwell SD, Lucas HL (1969) On territorial behavior and other factors influencing habitat distribution in birds I: theoretical development. Acta Biotheoretica 19:16–36.

Gallant AL (2009) What you should know about land-cover data. J Wildl Manage 73:796–805.

Gallant AL, Loveland TR, Sohl TL, Napton DE (2004) Using an ecoregional framework to analyze land-cover and land-use dynamics. Environ Manage 34:S89–S110.

Garshelis DL (2000) Delusions in habitat evaluation: measuring use, selection, and importance. In: Boitani L, Fuller TK (eds) Research techniques in animal ecology: controversies and consequences. Columbia Unievsrity Press, New York.

Gillson L (2009) Landscapes in time and space. Landsc Ecol 24:149–155.

Goss-Custard JD, Stillman RH (2008) Individual-based models and the management of shorebird populations. Nat Resour Model 21:3–71.

Graham CH, Moritz C, Williams SE (2006) Habitat history improves prediction of biodiversity in rainforest fauna. Proc Natl Acad Sci USA 103:632–636.

Grimm V, Railsback SF (2005) Individual-based modelling and ecology. Princeton University Press, Princeton.

Gu W, Swihart RK (2004) Absent or undetected? Effects of non-detection of species occurrence on wildlife-habitat models. Biol Conserv 116:195–203.

Haddad NM, Bowne DR, Cunningham A, Danielson BJ, Levey DJ, Sargent S, Spira T (2003) Corridor use by diverse taxa. Ecology 84:609–615.

Harris G (2007) Seeking sustainability in an age of complexity. Cambridge University Press, New York.

Harwood J (2000) Risk assessment and decision analysis in conservation. Biol Conserv 95:219-226.

Heinrich B (1979) Bumblebee economics. Harvard University Press, Boston.

Hirzel AH, Le Lay G, Helfer V, Randin C, Guisan A (2006) Evaluating the ability of habitat suitability models to predict species presences. Ecol Modell 199:142–152.

Hixon MA, Pacala SW, Sandin SA (2002) Population regulation: historical context and contemporary challenges of open vs. closed systems. Ecology 83:1490–1508.

Hjermann DØ (2000) Analyzing habitat selection in animals without well-defined home ranges. Ecology 81:1462–1468.

Hobbs NT, Hanley TA (1990) Habitat evaluation: do use/availability data reflect carrying capacity? J Wildl Manage 54:515–522.

Hunsaker CT, Graham RL, Suter GW, II, O'Neill RV, Barnthouse LW, Gardner RH (1990) Assessing ecological risk on a regional scale. Environ Manage 14:325–332.

Jager HI, King JW (2004) Spatial uncertainty and ecological models. Ecosystems 7:841–847.

Johnson MP (2005) Is there confusion over what is meant by "open population?" Hydrobiologia 544:333–338.

Keane RM, Crawley MJ (2002) Exotic plant invasions and the enemy release hypothesis. Trends Ecol Evol 17:164–170.

Koenig WD (1999) Spatial autocorrelation of ecological phenomena. Trends Ecol Evol 14:22–26.

Lennon JJ (2000) Red-shifts and red herrings in geographical ecology. Ecography 23:101–113.

Lima SL, Zollner PA (1996) Towards a behavioral ecology of ecological landscapes. Trends Ecol Evol 11:131–135.

Loveland TR, Sohl TL, Stehman SV, Gallant AL, Sayler KL, Napton DE (2002) A strategy for estimating rates of recent United States land cover changes. Photogram Eng Remote Sensing 68:1091–1099.

MacKenzie DI, Nichols JD, Royle JA, Pollock KH, Bailey LL, Hines JE (2006) Occupancy estimation and modeling. Elsevier, Oxford.

Malczewski J (2006) GIS-based multicriteria decision analysis: a survey of the literature. Int J Geogr Inf Sci 20:703–726.

Manning AD, Lindenmayer DB, Nix HA (2004) Continua and umwelt: novel perspectives on viewing landscapes. Oikos 104:621–628.

Manning AD, Fischer J, Felton A, Newell B, Steffen W, Lindenmayer DB (2009) Landscape fluidity – a unifying perspective for understanding and adapting to global change. J Biogeogr 36:193–199.

McGarigal K, Cushman SA (2005) The gradient concept of landscape structure. In: Wiens JA, Moss MR (eds) Issues and perspectives in landscape ecology. Cambridge University Press, Cambridge.

Mouton AM, De Baets B, Goethals PLM (2009) Knowledge-based versus data-driven fuzzy habitat suitability models for river management. Environ Model Software 24:982–993.

Olden JD, Schooley RL, Monroe JB, Poff NL (2004) Context-dependent perceptual ranges and their relevance to animal movements in landscapes. J Anim Ecol 73:1190–1194.

Orians GH, Wittenberger JF (1991) Spatial and temporal scales in habitat selection. Am Nat 137:S29–S49.

Ovaskainen O (2004) Habitat-specific movement parameters estimated using mark-recapture data and a diffusion model. Ecology 85:242–257.

Pearce J, Ferrier S (2000) Evaluating the predictive performance of habitat models developed using logistic regression. Ecol Modell 133:225–245.

Pearson RG, Dawson TP (2003) Predicting the impacts of climate change on the distribution of species: are bioclimate envelope models useful? Glob Ecol Biogeogr 12:361–371.

Petitgas P (2001) Geostatistics in fisheries survey design and stock assessment: models, variances and applications. Fish Fish 2:231–249.

Pulliam HR (1988) Sources, sinks, and population regulation. Am Nat 132:652–661.

Pulliam RH, Danielson BJ (1991) Sources, sinks, and habitat selection: a landscape perspective on population dynamics. Am Nat 137:S50–S66.

Rhemtulla JM, Mladenoff DJ, Clayton MK (2007) Regional land-cover conversion in the U.S. upper Midwest: magnitude of change and limited recovery (1850–1935–1993). Landsc Ecol 22:57–75.

Rocchini D, Ricotta C (2007) Are landscapes as crisp as we may think? Ecol Modell 204:535–539

Rondinini C, Wilson KA, Boitani L, Grantham H, Possingham HP (2006) Tradeoffs of different types of species occurrence data for use in systematic conservation planning. Ecol Lett 9:1136–1145.

Royle JA, Nichols JD (2003) Estimating abundance from repeated presence-absence data or point counts. Ecology 84:770–790.

Royle JA, Nichols JD, Kery M (2005) Modelling occurrence and abundance of species when detection is imperfect. Oikos 110:353–359.

Shifley SR, Rittenhouse CD, Millspaugh JJ (2009) Validation of landscape-scale decision support models that predict vegetation and wildlife dynamics. In: Millspaugh JJ, Thompson FR (eds) Models for planning wildlife conservation in large landscapes. Elsevier, New York.

Segurado P, Araújo MB (2004) An evaluation of methods for modelling species distributions. J Biogeogr 31:1555–1568.

Soberón J, Peterson AT (2004) Interpretation of models of fundamental ecological niches and species' distributional areas. Biodivers Inform 2:1–10.

Soulé ME, Estes JA, Miller B, Honnold DL (2005) Strongly interacting species: conservation policy, management, and ethics. BioSci 55:168–176.

Starfield AM (1997) A pragmatic approach to modeling for wildlife management. J Wildl Manage 61:261–270.

Stephens DW, Krebs JR (1986) Foraging theory. Princeton University Press, Princeton, New Jersey.

Tang SM, Gustafson EJ (1997) Perception of scale in forest management planning: challenges and implications. Landsc Urban Plan 39:1–9.

Thogmartin WE, Gallant A, Fox T, Knutson MG, Suárez M (2004a) Commentary: a cautionary tale regarding use of the 1992 National Land Cover Dataset. Wildl Soc Bull 32:960–968.

Thogmartin WE, Sauer JR, Knutson MG (2004b) A hierarchical spatial count model of avian abundance with application to Cerulean warblers. Ecol Appl 14:1766–1779.

Thomas CD, Kunin WE (1999) The spatial structure of populations. J Anim Ecol 68:647–657.

Uusitalo L (2007) Advantages and challenges of Bayesian networks in environmental modelling. Ecol Modell 203:312–318.

Urban D, Keitt T (2001) Landscape connectivity: a graph-theoretic perspective. Ecology 82:1205–1218.

Van Horne B (1983) Density as a misleading indicator of habitat quality. J Wildl Manage 47:893–901.

Vassallo MI, Rice JC (1982) Ecological release and ecological flexibility in habitat use and foraging of an insular avifauna. Wilson Bull 94:139–155.

Von Holle B, Motzkin G (2007) Historical land use and environmental determinants of nonnative plant distribution in coastal southern New England. Biol Conserv 136:33–43.

Wiens JA, Stenseth NC, Van Horne B, Ims RA (1993) Ecological mechanisms and landscape ecology. Oikos 66:369–380.

Wiens JA, Van Horne B, Noon BR (2002) Integrating landscape structure and scale into natural resources management. In: Liu J, Taylor WW (eds) Integrating landscape ecology into natural resources management. Cambridge University Press, New York.

Williams BK, Nichols JD, Conroy MJ (2002) Analysis and management of animal populations. Academic Press, San Diego, California.

Williams JW, Jackson ST (2007) Novel climates, no-analog communities, and ecological surprises. Front Ecol Environ 5:475–482.

Willis KJ, Araújo MB, Bennet KD, Figueroa-Rangel B, Froyd CA, Myers N (2007) How can a knowledge of the past help to conserve the future? Biodiversity conservation and the relevance of long-term ecological studies. Phil Trans R Soc B 362:175–187.

Woodcock CE, Gopal S (2000) Fuzzy set theory and thematic maps: accuracy assessment and area estimation. Int J Geogr Inf Sci 14:153–172.

Zimmerman GS, LaHaye WS, Gutiérrez RJ (2003) Empirical support for a despotic distribution in a California spotted owl population. Behav Ecol 14:433–437.

Zollner PA, Lima SL (2005) Behavioral trade-offs when dispersing across a patchy landscape. Oikos 108:219–230.

Chapter 6
Insights from Ecological Theory on Temporal Dynamics and Species Distribution Modeling

Robert J. Fletcher Jr., Jock S. Young, Richard L. Hutto, Anna Noson, and Christopher T. Rota

6.1 Introduction

> *Theory potentially has value far beyond the satisfaction of intellectual curiosity*
> Huston (2002)

Understanding species distributions in space and time is essential to ecology, evolution, and conservation biology. There is a growing need for robust habitat models that can adequately predict species distributions across broad spatial scales (Guisan and Thuiller 2005). An invaluable tool for conservation biologists (Norris 2004), species distribution models can be used to evaluate potential management actions, interpret the potential effects of climate change, and maximize biodiversity with reserve selection algorithms (Guisan and Thuiller 2005, see also Chap. 14). Yet the usefulness of such models is limited by a number of factors (Guisan and Thuiller 2005; Araújo and Guisan 2006), including poor incorporation of ecological theory in modeling approaches (Austin 2002; Huston 2002; Guisan et al. 2006). This is unfortunate because many theories in ecology can help guide the model building process, which may not only improve model predictions but may also provide greater inference regarding habitat quality in heterogeneous landscapes.

Our objectives are several-fold. First, we synthesize important, although often overlooked, assumptions and predictions from ecological theory, focusing on habitat selection and metapopulation theory, for variation in species occurrence over time. Second, we describe how these perspectives can be incorporated into models of species distribution with data on the detection/non-detection of species across environmental gradients. Third, as a case study, we apply these models to evaluate model fit and predictive ability relative to more conventional modeling approaches using a relatively large-scale and long-term dataset of bird populations inhabiting forests across Montana and northern Idaho, USA.

R.J. Fletcher Jr. (✉)
Department of Wildlife Ecology and Conservation, University of Florida, PO Box 110430, 110 Newins-Ziegler Hall, Gainesville, FL 32611, USA
e-mail: robert.fletcher@ufl.edu

6.2 Management Challenge, Ecological Theory, and Statistical Framework

Currently, a major limitation of species distribution models and their implementation for management is their lack of dynamic predictions (i.e., models are temporally static in the absence of perturbations of environmental variables) (Guisan and Zimmermann 2000; Castellón and Sieving 2006), even though we know that patterns of habitat use vary with population size and other factors (Gill et al. 2001; Sergio and Newton 2003). Whereas other, more mechanistic, modeling approaches exist that capture temporal dynamics (Lischke et al. 2007), such models typically contain numerous parameters that are difficult to parameterize for more than a few species across spatial scales relevant for management purposes. Incorporation of key dynamics into species distribution models could thus be very useful, since a major strength of species distribution models is the ability to fit models to many species across large spatial scales with relatively sparse data (Guisan and Thuiller 2005; Elith et al. 2006).

A second major limitation is that interpreting habitat quality from patterns of density or occurrence can be misleading (Van Horne 1983). Incorporating deterministic aspects of temporal dynamics into species distribution models may provide one way to address these issues by explicitly focusing on consistency of occupancy over time. To guide our perspective on temporal dynamics, we draw from both habitat selection and metapopulation theory.

6.2.1 Perspectives from Habitat Selection and Metapopulation Theory

Habitat selection and metapopulation theory both make general assumptions and predictions regarding occupancy and population persistence within habitats that are relevant to species distribution models. All common ecological theories on habitat selection to date – the ideal free, ideal despotic, and ideal pre-emptive distributions (Fretwell and Lucas 1970; Pulliam and Danielson 1991) – make two key assumptions relevant to developing species distribution models. First, animals are ideal. That is, animals can identify the best quality habitats available, in terms of fitness rewards. Even when animals behave less than ideally, as long as they can gain reliable information about habitat quality from sampling, settlement will approach what is expected of ideal behavior (Pulliam and Danielson 1991; Tyler and Hargrove 1997). This assumption is relevant to species distribution models because it suggests that occupancy can be a measure of habitat quality (Sergio and Newton 2003), as is often implicitly assumed, but it also leads to predictions relevant to temporal variation in occupancy (see below). Thus, while the ideal despotic and pre-emptive distributions assume territoriality/aggressive behaviors and pre-emption by early settlers can influence settlement of some individuals whereas the ideal free distribution assumes individuals are free to settle anywhere, each theory assumes that animals are generally ideal (Fretwell and Lucas 1970; Pulliam and Danielson 1991).

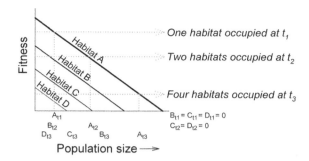

Fig. 6.1 Habitat selection theory generally assumes that as population size increases within a habitat, quality decreases. Organisms begin to use sub-prime habitat only after population size in the high-quality habitat grows to a point where the suitability therein is equal to the suitability in the next best habitat. Shown here is an example drawing from the ideal free distribution (Fretwell and Lucas 1970), where there are four different habitats that vary in quality with population size. *Lines* are arranged from *highest* (A) to *lowest* (D) quality. As population size increases over time, the number of habitats occupied increases. Thus, basic habitat selection theory predicts that observed species–environment relationships (specifically, the breadth of habitats occupied) should change with fluctuations in population size

Second, habitat selection theory assumes that the range of habitats used by any given species will vary with population density, where low-quality sites are more likely to be used as population density increases (Fig. 6.1). Because average fitness will decrease with increasing population size, this phenomenon can regulate populations (Brown 1969; cf. Rodenhouse et al. 1997). Density-dependent habitat use has been documented in a wide variety of taxa, from fish (Fraser and Sise 1980) to birds (Gill et al. 2001). Note that density-dependent habitat selection can occur at a variety of spatio-temporal scales, from arrival order of migrants within a breeding season (Shochat et al. 2005), to variation in habitat use among years where population size varies (Gill et al. 2001). Such density-dependent habitat use can have consequences for managing animal populations and can impose problems in diagnosing causes of population decline (Norris 2004). Variation in the range of habitats used as a function of population density makes conventional habitat use models limited (Wiens 1989). This limitation arises because habitat models built using information from a time of high population density will not be able to distinguish good- from poor-quality habitats, whereas models built using information from a time of low population density will fail to recognize the range of potentially suitable habitat types. Because the range of habitats used changes with population size (O'Connor 1981; Chamberlain and Fuller 1999; Gill et al. 2001; Sergio and Newton 2003) and land management strategies can influence population size (Newton 1998), accurate forecasting of the dynamics of animal communities will need to link density-dependent variation in habitat use with expected changes in population size that occur from landscape change.

Metapopulation and habitat selection theory also make a related prediction that populations will be more likely to persist in habitats of high suitability (Pulliam 1988; Hames et al. 2001), and thus be occupied more consistently through time.

This prediction follows from habitat selection and metapopulation theory for different reasons. Habitat selection theory suggests that low-quality habitat is occupied only at high population densities because animals behave ideally, and temporal variation in habitat use will be greater in low-quality habitats.

In metapopulation theory, persistence in areas of high habitat quality is predicted based on variation in local colonization/extinction dynamics. Much of metapopulation theory uses metrics that reflect variation in local population size (e.g., patch area) or surrounding population size (propagule pressure; e.g., isolation) to predict variation in extinction/colonization dynamics. Metapopulation theory predicts that local extinctions should be less likely, and local colonizations more likely, with increased habitat suitability, all else being equal (Moilanen and Hanski 1998; Thomas et al. 2001; Fleishman et al. 2002; Bonte et al. 2003). Extinction is less likely because higher density populations should occur in areas of higher quality. Accordingly, Hanski and Ovaskainen (2000) suggested that information on habitat quality might actually be used to adjust patch area metrics to reflect changes in local population size that occur via variation in habitat quality (see also Moilanen and Hanski 1998). Colonizations could also be more likely in high-quality, but unoccupied, habitat when animals use some component of habitat quality in their searching (e.g., patch detection) or settlement decisions (Tyler and Hargrove 1997; Fletcher 2006). The net result is that temporal variation in occupancy should lead to high-quality habitat being more frequently occupied through time than low-quality habitat, all else being equal, and thus create a higher likelihood of persistence, than low-quality habitat.

6.2.2 Statistical Framework

Incorporating these perspectives may not only improve predictions of species distribution models by providing dynamic predictions, but may also provide a more refined perspective on habitat quality by identifying areas that are consistently occupied (Sergio and Newton 2003). Yet, to date, there has been little attempt to incorporate these perspectives into model building on a wide suite of species at scales relevant to management strategies. Here, we incorporate these ideas into species distribution models that accommodate detection/non-detection (presence–absence) data. Specifically, we (1) incorporate temporal variation in occurrence via changes in population size, and (2) model local indices of persistence rather than patterns of occurrence alone. To do so, we use a generalized linear modeling (GLM) framework because GLMs can address spatio-temporal complexities (e.g., repeated measures, spatial autocorrelation) in datasets, which is essential for appropriately addressing temporal dynamics.

Conventional GLMs for detection/non-detection data come in the form of logistic regression models, where we specify a logit link function and a binomial error distribution. Expanding these models to accommodate multi-year data requires addressing the potential for non-independence (i.e., repeated measures) across years, which can be accomplished either using generalized estimating equations (GEE) or by specifying random effects terms in GLMs. We refer to this model structure as a "conventional model" throughout.

Extending conventional models to incorporate temporal variation in occurrence can be solved by either developing separate models for each time period considered and subsequently identifying if coefficients change consistently with changes in population size, or by including temporal indices of population size directly into multi-year models. Here we focus on the latter approach, because it provides a more parsimonious modeling solution that easily allows for translation into species distribution maps. By incorporating a temporal index of population size and interactions of this index with relevant environmental variables, models can test directly for variation in occurrence and whether observed habitat-relationships change with fluctuating population size.

Modeling indices of persistence can be accomplished using numerous approaches. Markov models that estimate local colonization/extinction events have seen widespread use in metapopulation ecology, but to link such models to predicting the occurrence of species across landscapes and/or regions requires the strong assumption of equilibrium (Moilanen 2000), and predictions represent an "equilibrium probability of occupancy." Because species distribution models are often used to interpret changes in land use or to understand reasons for species declines (Guisan and Thuiller 2005), the assumption of equilibrium may limit the use of metapopulation models in many situations. Alternatively, estimates of the frequency of occurrence through time at a given location can provide another measure of persistence, where locations that are more frequently (consistently) occupied are more likely to persist. With this approach, the response variable is the proportion of time a sampling unit is occupied (e.g., 3 out of 5 years a species is present), and the explanatory variables are the average values through time. Note that this approach assumes the environment is relatively constant through time, which may not be appropriate in some situations. These kinds of data have been analyzed with logistic regression using binomial responses rather than binary responses (Dunford et al. 2002), linear regression (Johnson and Igl 2001) or ordinal logistic regression techniques (Hames et al. 2001). Using binomial responses in logistic regression assumes counts across years are independent Bernoulli trials, which may limit that approach in many cases where species show site fidelity or when there are other reasons to assumed non-independence across years. Linear regression is limited because it will not provide appropriate variance estimates (constrained to the 0–1 interval) and can make predictions beyond observable values (Guisan and Harrell 2000). Consequently, we focus here on ordinal logistic regression as a potential approach for directly modeling indices of persistence.

Ordinal logistic regression, in our case employing the cumulative logit model (also known as the proportion odds model), is an extension of logistic regression where the response variable is a series of ranked values (Allison 1999). For modeling occupancy through time, the dependent variables are categories of the frequency of occurrence. Note that the model makes no assumptions regarding the distances between observed categories (Allison 1999), such that it can accommodate datasets where time between sampling periods varies. The cumulative logit model structure is formalized as:

$$\log\left(\frac{P(Y \leq j)}{1 - P(Y \leq j)}\right) = \alpha_j + \beta_1 x_1 + \ldots \beta_k x_k, \quad j = 1, \ldots J-1, \quad (6.1)$$

where the response variable represents the log odds of a point being occupied for Y out of j years, J is the total number of categories (time periods sampled), α_j is the intercept, β is the habitat coefficient, and k is the number of explanatory variables. In this model, the coefficients of predictor variables are the same for the ordered categories, but the intercepts for each category are different, resulting in J equations.

This approach may be useful for understanding habitat occupancy and habitat quality based on the way that this model is interpreted. As the name implies, a cumulative logit analysis provides the probability of being in an ordered category or lower. Consider a dataset where sites were sampled on an annual basis for 5 years across an environmental gradient. A cumulative logit analysis would provide predicted estimates of the frequency of occurrence through time as a function of the environmental gradient; however, a 60% probability of occupancy at a location 4 out of 5 years would actually be interpreted as a 60% chance that the species would occupy the location 4 or fewer years out of a total of 5 years. The model can be rearranged to give the likelihood of a minimum probability of occurrence, which could be useful for conservation purposes by providing a close analog to a measure of persistence. Here, a 60% probability of occupancy at a location 4 out of 5 years would be interpreted as a 60% chance that the species would occupy the location a minimum of 4 years out of a total of 5 years. A model with this structure is:

$$\log\left(\frac{P(Y \geq j)}{1 - P(Y \geq j)}\right) = \alpha_j + \beta_1 x_1 + \ldots \beta_k x_k, \quad j = 1, \ldots J - 1 \quad (6.2)$$

6.2.3 Comparing Models

To determine whether the incorporation of these aspects of theory improves model performance, we need ways to compare results among models. Models that incorporate temporal indices of population size can be compared with conventional model selection criteria and validation statistics (Fielding and Bell 1997); however, cumulative logit models have a different currency. Because there are a variety of statistics developed for estimating the robustness of binary models, an intuitive approach would be to convert putative persistence models to a similar scale as conventional models. There are at least two ways to do this for frequency of occurrence data. First, ordinal responses can be recoded to presence–absence predictions, using some sort of cutoff (Guisan and Harrell 2000), similar to cutoffs used for binary models. Second, ordinal responses can be back-transformed to estimate the average probability of occurrence, p_{avg}, at a location for any given year:

$$f_0 = (1 - p_{avg})(1 - p_{avg}) = (1 - p_{avg})^t \quad (\ldots)$$

$$p_{avg} = 1 - (f_0)^{1/t}, \quad (6.3)$$

where t is the number of years surveyed and f_0 is the probability that a location is predicted to be occupied in none of the years (from (6.1)). Back-transformation allows for direct comparisons with evaluation datasets using the same approaches as for binary data.

6.3 Model and Model Validation Techniques

6.3.1 Modeling Database

As a case study, we apply the above modeling approaches to a large database that incorporates multiple years of data across broad spatial scales. The USFS Northern Region Landbird Monitoring Progam (NRLMP) involves monitoring all diurnal landbird species that can be detected through a single methodology throughout Montana and northern Idaho (Hutto and Young 2002). The NRLMP surveys birds on a series of transects located using a geographically stratified random sampling design on Forest Service lands across the Northern Region and on lands of several cooperating agencies (Fig. 6.2). The NRLMP includes approximately 350 transects on USFS lands (30 on each National Forest unit) and 200 transects on other lands. Each transect consists of ten permanently marked points, at which 10-min bird counts are conducted according to a standard point-count protocol (Hutto et al. 1986; Ralph et al. 1995). We truncated detections to include only those ≤100 m from the center points.

For the purposes of incorporating predictions from ecological theory, we focus on transects that occurred over a subset of the region with adequate GIS-based information (see Fig. 6.2). In this area, we divided the dataset into transects that had been surveyed for at least 6 years (1995, 1996, 1998, 2000, 2002, 2004) to develop models (6,768 point × year combinations; Fig. 6.2), and used the remaining transects that had been surveyed for 1–5 years to test models (6,256 point × year combinations). We chose this partitioning approach for model development and testing to ensure that we captured potential temporal dynamics in model development and to test whether the inclusion of dynamics improved predictions of occurrence.

The NRLMP generates long-term data on bird distribution in association with detailed local-level and GIS-based landscape-level vegetation information. Here we focus solely on GIS-based measures that allow for predictive modeling across the study area. However, the local field measures were used to identify the points with substantial changes in habitat during the monitoring time period. We used this record of habitat changes over time to remove survey points ($n = 296$) that changed substantially in vegetation structure (usually due to timber harvesting activity), because persistence models are less interpretable if environmental changes occur over the time period used for model development.

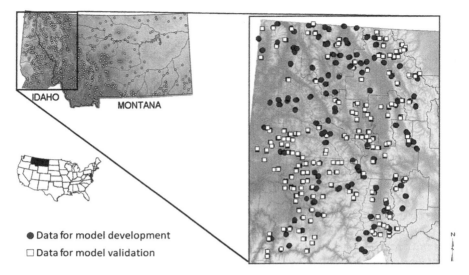

Fig. 6.2 The database used for modeling species distribution was generated from the USFS Northern Region Landbird Monitoring Program, which spans Montana and northern Idaho (inset). We focused modeling efforts on the western part of this area, where relatively accurate GIS-based environmental information was available. In this region, we divided the dataset into model-building and test sets, with model-building data being drawn from all transects that had 6 years of temporal replication. The remaining data were used for model testing

6.3.2 Focal Species

We contrasted models developed for bird species that vary in their use of different habitats and in their variation in occurrence across years. We expected that explicitly incorporating temporal variation in population size into models should improve model performance for species known to use multiple habitats and that show greater temporal variation in occurrence, but should be less useful for those species that are rigid in their associations and less variable in occurrence over time. Similarly, we expected that using measures of persistence would be more useful for species that varied substantially across years in their occurrence than for species with less temporal variation.

To choose focal species, we considered those species that occurred on at least 5% of points throughout the study period. From this list, we calculated the coefficient of variation (CV) in occurrence across the 6 years available in the database, and we calculated vegetation breadth (Colwell and Futuyma 1971), based on local habitat categories identified in Hutto and Young (1999). We then chose species that represent extremes in their variation in occupancy through time and in their variation in habitats used. Specifically, we ranked the upper and lower quartiles for CV and habitat breadth and factorially selected the four species that represent these extremes (Table 6.1): ruby-crowned kinglet (*Regulus calendula*), hermit thrush (*Catharus guttatus*), dark-eyed junco (*Junco hyemalis*), and dusky flycatcher (*Empidonax oberholseri*). Thus, we

Table 6.1 Differences in temporal variability in occurrence across years and vegetation breadth (CV, vegetation breadth) for focal species used in species distribution modeling

Vegetation breadth	Coefficient of variation across years	
	Low	High
Low	Dusky flycatcher (0.11, 5.96)	Hermit thrush (0.42, 6.66)
High	Dark-eyed junco (0.11, 11.75)	Ruby-crowned kinglet (0.38, 8.56)

expected that persistence models should most improve predictions for the thrush and kinglet, which vary considerably across years in occurrence, with less improvement for the junco and flycatcher, and that temporal models would most improve models for the kinglet and provide least improvement for the flycatcher.

6.3.3 Environmental Covariates

GIS-based vegetation measures were derived from a 15 m resolution, digital landcover map developed by the USFS (Northern Region Vegetation Mapping Project, R1-VMP) based on Landsat TM imagery, aerial photo interpretation, and field data collected in 2001 (Brewer et al. 2004). This is the most accurate GIS database currently available for this region. We used three separate data layers from this database, general cover type (conifer forest, deciduous forest, shrub), successional stage (four tree-diameter categories [dbh]), and percent canopy cover (three categories). We also calculated elevation using a 30-m resolution digital elevation model (Gesch et al. 2002) and derived the mean annual precipitation from a Parameter Elevation Regressions on Independent Slopes Model (PRISM), which was based on climate data from 1961 to 1990 (PRISM Climate Group, Oregon State University, http://www.prismclimate.org, 2004). We used this GIS-based information to measure both local- (within 100 m of each point) and landscape-scale (within 1 km of each point) environmental variables. The 1 km landscape scale was chosen on the basis of other investigations in this region that showed strong correlations at this scale (Tewksbury et al. 2006). These variables capture the major biotic and abiotic gradients in the region relevant to avian distribution. For the four successional stage and three percent canopy cover categorical variables, we initially subjected variables to a principal component analysis (PCA), to derive new, independent variables that represent variation in dbh and percent cover information. For both PCAs, two new components were derived (explaining 92% of the variation in each analysis), one of which described a linear gradient in dbh/canopy cover, and the other described non-linearity of each variable (Table 6.2).

6.3.4 Model Development

We compared conventional modeling approaches involving binary response data with approaches that explicitly incorporate temporal dynamics. To develop models, we

Table 6.2 Explanatory variables used in development of models

Variable	Description
Physical/abiotic environment	
Elevation (km)[a]	Elevation taken from digital elevation model
Precipitation[a]	Mean annual precipitation (based on data from 1961 to 1990)
Latitude[a]	North-south gradient, in decimal degrees (dd.dddd°)
Longitude[a]	East-west gradient, in decimal degrees (dd.dddd°)
Local vegetation (100 m)	
CanopyPC1	Principal component reflecting moderate canopy cover
CanopyPC2	Principal component reflecting linear gradient in canopy
DbhPC1	Principal component reflecting linear gradient in dbh
DbhPC2	Principal component reflecting moderate dbh
Conifer[a]	Percent of conifer
Deciduous	Presence of broadleaf deciduous forest
Shrub	Percent of shrub-dominated areas; often clearcuts
Landscape structure (1 km)	
Conifer forest[a]	Percent of conifer
Deciduous forest	Percent of deciduous forest
Shrub	Percent of shrub-dominated areas; often clearcuts
Road density	Kilometers of roads within 1 km
Population	
Annual index	Annual index of population size (frequency of points occupied)
Other factors	
Time of year[a]	Julian date to accommodate potential time of year bias

[a]Nonlinear effects considered

used GLMs, considering repeated samples (points) over time as repeated measures using GEE in conventional models. While other approaches exist and are useful for modeling species distributions (e.g., Elith et al. 2006), dealing with potential biases (e.g., detection probability; MacKenzie et al. 2002), and uncertainty in model predictions (e.g., Karanth et al. 2004), we chose this framework to allow for standardized and seamless relative comparisons among the model types described above. Overall, we considered three general types of GLMs: conventional logistic regression (hereafter referred to as conventional models), conventional models with temporal indices of population size (hereafter referred to as temporal models), and cumulative logit models (hereafter referred to as persistence models). Each model type considered the same explanatory variables, except that temporal models also considered a temporal index in population size (i.e., frequency of points occupied each year across the region) and its pair-wise interactions with environmental variables.

We initially screened environmental explanatory variables for strong correlations ($r > 0.7$) and subsequently removed highly correlated variables. Then, we used univariate screening of linear and non-linear (quadratic) relationships of species occurrence and environmental variables (Fletcher and Hutto 2008). Based on this screening, for further model analyses we included only explanatory variables that improved model fit over an intercept-only model, as judged by QAICc (Pan 2001), a model selection criterion for GEE models analogous to Akaike's Information Criteria (AIC; Burnham and Anderson 2002). Given this initial screening, we used a manual backward elimination procedure (Thogmartin et al. 2004), where we determined variables for removal based on the 95% confidence limits of the parameter estimates; we stopped removing variables when QAICc (or AICc for cumulative logit models) was not lowered with the exclusion of variables (i.e., removal of more variables did not improve model fit).

Using these reduced models, we then tested for predictive accuracy using the independent dataset mentioned above. Predictive accuracy was evaluated by calculating the area under the ROC curve (AUC), the Kappa statistic, and the correct classification rate (Fielding and Bell 1997). AUC is a frequently used index that ranges between 0 and 1 and does not require threshold cutoffs to be determined. Kappa measures the proportion of correctly predicted points after the probability of chance agreement has been removed. The correct classification rate, also known as "overall prediction success" (OPS; Liu et al. 2005) is the proportion of detections/ non-detections correctly classified in the test data. For the Kappa statistic and the correct classification rate, we used a threshold cutoff value based on the prevalence of each species in the model building dataset (Liu et al. 2005).

6.3.5 *Model Results*

Overall, in the dataset used for model building, kinglets occurred on 19.4% of points, flycatchers on 10.3%, thrushes on 6.4%, and juncos on 48.3% of points. Models that incorporated temporal indices of population size fit the data for all species much better than conventional models that did not include temporal indices (Table 6.3). The relative model improvement, as evidenced by change in QAICc, was greatest for the kinglet and least for the flycatcher, consistent with our

Table 6.3 Model fit, based on QAICc, of the best model that includes and excludes temporal variation in population size

Species/model	Kappa statistic	Quasi-likelihood	QICc	ΔQICc	Model weight
Dark-eyed junco					
Best conventional model	10	−4,527.9	9,075.8	52.9	0.000
Best temporal model	14	−4,497.5	9,022.9	0.0	1.000
Dusky flycatcher					
Best conventional model	12	−1,923.2	3,870.4	10.9	0.004
Best temporal model	19	−1,910.8	3,859.5	0.0	0.996
Hermit thrush					
Best conventional model	11	−1,279.9	2,581.7	72.9	0.000
Best temporal model	12	−1,242.4	2,508.8	0.0	1.000
Ruby-crowned kinglet					
Best conventional model	12	−2,767.3	5,558.6	223.8	0.000
Best temporal model	16	−2,651.4	5,334.8	0.0	1.000

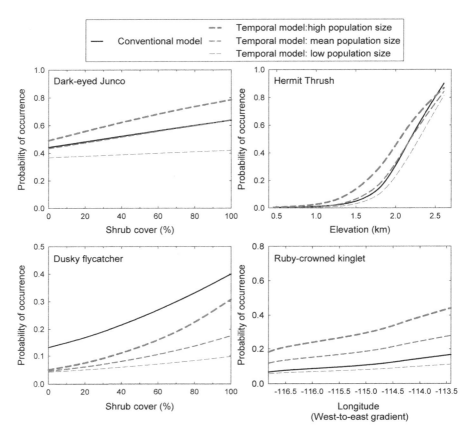

Fig. 6.3 Some examples of how temporal variation alters model predictions for each species relative to conventional models that do not incorporate such variation. Shown are partial relationships that exhibited significant interactions with population size

expectations (Table 6.3). In general, density-dependent relationships identified from temporal models suggested that relationships in low population size years tended to be weak, whereas relationships were stronger in high population size years (Fig. 6.3). Note that because persistence models were based on a different type of response variable, model selection criteria could not be compared directly with criteria for conventional and temporal models. Nonetheless, significant predictor variables explaining distributions for each model type tended to be the same, with the main exception being that the addition of the temporal index in population size altered some relationships with occurrence (Fig. 6.3).

Predictive accuracy measures suggested that, overall, temporal models tended to provide modest improvements over conventional models for all species (Fig. 6.4). As we expected on the basis of differences in the temporal variation in their occupancy rates and habitats used among species (Table 6.1), improvements based on AUC were greatest for the kinglet and least for the flycatcher. Persistence models

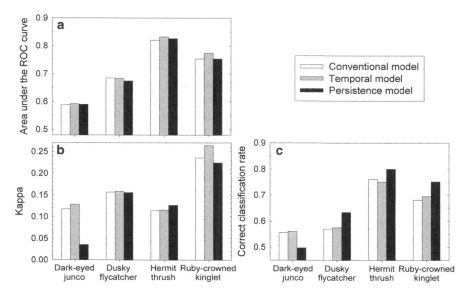

Fig. 6.4 Assessment indices of model performance against test data, using (**a**) area under the ROC curve, (**b**) the Kappa statistic, and (**c**) the correct classification rate (or overall predictive success; OPS). For Kappa and the correct classification rate, we used prevalence values of each species to determine appropriate threshold cutoff values

improved accuracy predictions for hermit thrush only, based on the Kappa statistic and OPS statistic. However, OPS was also greater for persistence models for the kinglet and flycatcher relative to conventional and temporal models, even though Kappa and AUC statistics did not indicate improvement of persistence models for those species.

6.3.5.1 Past, Current, and Future Applications

Aspects of ecological theory are relevant to, though often ignored in, the development of species distribution models. In the past development of models, when theory has been appreciated, it is primarily perspectives from niche theory that have been emphasized in modeling approaches (Austin 2002). Niche theory reminds us, for example, that species–environment relationships can often be non-linear (Austin 2002, 2007). Here we show that additional perspectives gleaned from habitat selection and metapopulation theory can further improve species distribution modeling efforts, by emphasizing that species–environment relationships should vary in consistent ways as populations vary in size over time. Spatio-temporal variation in population size might help to explain past discrepancies in environmental relationships, such as variation in animal sensitivity to patch size (Johnson and Igl 2001).

The current application of these ideas provides an important step in linking perspectives from theory to species distribution models and evaluating whether such perspectives matter. These perspectives greatly improved model fit and temporal models consistently improved predictive ability, although predictive ability was only moderately increased for all species. There are some notable limitations of the NRLMP database that likely limit predictive ability at the point-count level, regardless of modeling method. For instance, counts were conducted only once over a 1.5 month window each year, and consequently Julian date was a strong predictor of occurrence for all species considered. Nonetheless, the fact that these theoretical perspectives generally increased model fit and predictive ability in the face of data limitations suggests that such perspectives should be considered in other investigations. In addition, Austin (2007) recently argued that tests of predictive ability of models are insufficient and that models must be grounded in ecological realism as well. We believe that by incorporating temporal perspectives in models, and thus grounding models in theoretical predictions that have been documented in a wide range of taxa (e.g., Fraser and Sise 1980; Gill et al. 2001; Sergio and Newton 2003), ecological realism may be greatly improved. Furthermore, whereas our focus has been on vagile organisms, these perspectives are also relevant for species distribution modeling of more sessile organisms, such as plants. Numerous investigations on plant metapopulation dynamics highlight the need to incorporate temporal measures of persistence (e.g., Verheyen et al. 2004), and spillover effects that can occur in plants (Brudvig et al. 2009) are analogous to density-dependent habitat selection described above.

Future applications of these perspectives are potentially diverse. Appreciation of the variation in species–environment relationships with population size could be incorporated into other species distribution modeling approaches, including Bayesian approaches, regression trees, and various machine-learning techniques. Future applications should consider these perspectives during model development, and perhaps other approaches to modeling persistence and temporal variation in population size could further refine models. For example, while we focused on ordinal logistic regression for modeling persistence (cf. Hames et al. 2001), other approaches – such as Markov-based models – may be more flexible and provide greater insight for interpreting variation in temporal dynamics (MacKenzie et al. 2003). Finally, our approach allows for models to become more dynamic, which could be very useful in future applications of species distribution models applied to landscape and global change. Models could be used, for instance, to directly predict how species distribution changes with ongoing habitat loss that invariably reduces population size, which could alter predictions for occupancy in remaining habitats relative to predictions from conventional models.

6.4 Data Availability and Suitability

Many investigations that sample species distribution have at least limited temporal replication that could capture some relevant dynamics, yet temporal variability is often treated as a nuisance rather than as an opportunity (but see Bissonette and Storch 2007). Our modeling efforts suggest that temporal replication could provide opportunities for developing better models of species distribution, such that investigations of species distribution might be better off repeating samples over time rather than sampling new locations each time period. This potential space-for-time tradeoff in sampling for species distribution models is an ongoing issue in monitoring designs and model development.

The utility of repeating samples over time at the expense of spatial replication will undoubtedly be greater for species that have greater temporal variation in occupancy, such as the ruby-crowned kinglet modeled here. Temporal replication may also be relatively more useful than spatial replication when models are being developed to interpret ongoing landscape change. However, if spatial extents of data collection do not capture relevant spatial variation in environmental and land use gradients, temporally adequate data may not be sufficient for developing useful models in the first place. Further analysis of the temporal dynamics described here and the tradeoff of capturing such dynamics relative to spatial sampling procedures would be informative.

Indices of temporal variation in population size can be gleaned from survey protocols used in developing models or through use of independent datasets. Here, we chose to derive indices using the NRLMP database, because of the broad spatial and temporal extent of the data base. However, independent datasets, such as the ongoing Breeding Bird Survey in the US and in parts of Canada, could also be integrated into modeling efforts by providing temporal indices of population size. In doing so, even when an investigation is completed, models could be useful into some point in the future by altering predictions based on population size indices obtained from ongoing monitoring programs.

As species distribution modeling continues to advance as a field, explicit consideration of ecological theory in sampling designs and in the modeling process should prove helpful for improving accuracy of predictions, increasing insights regarding resource quality, and interpreting landscape change. We contend that while ecological theories might not always be "right," they can still be useful (cf. Box and Draper 1987).

Acknowledgments This work was supported by the National Research Initiative of the USDA Cooperative State Research, Education and Extension Service, grant #2006-55101-17158. The landbird database was created through support from USFS Northern Region (03-CR-11015600-019). We thank two anonymous reviewers who provided valuable comments on a previous version of this manuscript, which improved the ideas presented here.

References

Allison PD (1999) Logistic regression using the SAS system. SAS Institute, Inc., Cary, NC.
Araújo MB, Guisan A (2006) Five (or so) challenges for species distribution modelling. J Biogeogr 33:1677–1688.
Austin MP (2002) Spatial prediction of species distribution: an interface between ecological theory and statistical modelling. Ecol Model 157:101–118.
Austin M (2007) Species distribution models and ecological theory: a critical assessment and some possible new approaches. Ecol Model 200:1–19.

Bissonette JA, Storch I (eds) (2007) Temporal dimensions of landscape ecology: wildlife responses to variable resources. Springer, New York.

Bonte D, Lens L, Maelfait JP, Hoffmann M, Kuijken E (2003) Patch quality and connectivity influence spatial dynamics in a dune wolfspider. Oecologia 135:227–233.

Box GEP, Draper NR (1987) Empirical model-building and response surfaces. Wiley, New York.

Brewer CK, Berglund D, Barber JA, Bush R (2004) Northern region vegetation mapping project summary report and spatial datasets, version 42. Northern Region USFS.

Brown JL (1969) The buffer effect and productivity in tit populations. Am Nat 103:347–354.

Brudvig LA, Damschen EI, Tewksbury JJ, Haddad NM, Levey DJ (2009) Landscape connectivity promotes plant biodiversity spillover into non-target habitats. Proc Natl Acad Sci 106:9328–9332.

Burnham KP, Anderson DR (2002) Model selection and inference: a practical information-theoretic approach. Springer-Verlag, New York.

Castellón TD, Sieving KE (2006) Landscape history, fragmentation, and patch occupancy: models for a forest bird with limited dispersal. Ecol Appl 16:2223–2234.

Chamberlain DE, Fuller RJ (1999) Density-dependent habitat distribution in birds: issues of scale, habitat definition and habitat availability. J Avian Biol 30:427–436.

Colwell RK, Futuyma DJ (1971) On the measurement of niche breadth and overlap. Ecology 52:567–576.

Dunford W, Burke DM, Nol E (2002) Assessing edge avoidance and area sensitivity of Red-eyed Vireos in southcentral Ontario. Wilson Bull 114:79–86.

Elith J, Graham CH, Anderson RP, Dudik M, Ferrier S, Guisan A, Hijmans RJ, Huettmann F, Leathwick JR, Lehmann A, Li J, Lohmann LG, Loiselle BA, Manion G, Moritz C, Nakamura M, Nakazawa Y, Overton JM, Peterson AT, Phillips SJ, Richardson K, Scachetti-Pereira R, Schapire RE, Soberón J, Williams S, Wisz MS, Zimmermann NE (2006) Novel methods improve prediction of species' distributions from occurrence data. Ecography 29:129–151.

Fielding AH, Bell JF (1997) A review of methods for the assessment of prediction errors in conservation presence/absence models. Environ Conserv 24:38–49.

Fleishman E, Ray C, Sjögren-Gulve P, Boggs CL, Murphy DD (2002) Assessing the roles of patch quality, area, and isolation in predicting metapopulation dynamics. Conserv Biol 16:706–716.

Fletcher RJ, Jr. (2006) Emergent properties of conspecific attraction in fragmented landscapes. Am Nat 168:207–219.

Fletcher RJ, Jr., Hutto RL (2008) Partitioning the multi-scale effects of human activity on the occurrence of riparian forest birds. Landsc Ecol 23:727–739.

Fraser DF, Sise TE (1980) Observations on stream minnows in a patchy environment: a test of a theory of habitat distribution. Ecology 61:790–797.

Fretwell SD, Lucas HL, Jr (1970) On territorial behavior and other factors influencing habitat distribution in birds. I. Theoretical development. Acta Biotheo 19:16–36.

Gesch D, Oimoen M, Greenlee S, Nelson C, Steuck M, Tyler D (2002) The National Elevation Dataset. Photogram Eng Remote Sensing, 68(1):5–11.

Gill JA, Norris K, Potts PM, Gunnarsson TG, Atkinson PW, Sutherland WJ (2001) The buffer effect and large-scale population regulation in migratory birds. Nature 412:436–438.

Guisan A, Harrell FE (2000) Ordinal response regression models in ecology. J Veg Sci 11:617–626.

Guisan A, Thuiller W (2005) Predicting species distribution: offering more than simple habitat models. Ecol Lett 8:993–1009.

Guisan A, Zimmermann NE (2000) Predictive habitat distribution models in ecology. Ecol Model 135:147–186.

Guisan A, Lehmann A, Ferrier S, Austin M, Overton JMC, Aspinal R, Hatie T (2006) Making better biogeographical predictions of species' distributions. J Appl Ecol 43:386–392.

Hames RS, Rosenberg KV, Lowe JD, Dhondt AA (2001) Site reoccupation in fragmented landscapes: testing predictions of metapopulation theory. J Anim Ecol 70:182–190.

Hanski I, Ovaskainen O (2000) The metapopulation capacity of a fragmented landscape. Nature 404:755–758.

Huston MA (2002) Introductory essay: critical issues for improving predictions. In: Scott JM, Heglund PJ, Morrison ML (eds) Predicting species occurrences: issues of accuracy and scale. Island Press, Washington DC.

Hutto RL, Young JS (1999) Habitat relationships of landbirds in the Northern Region, USDA Forest Service. General Technical Report RMRS-GTR-32. U.S. Department of Agriculture, Forest Service, Rocky Mountain Research Station, Ogden, UT.

Hutto RL, Young JS (2002) Regional landbird monitoring: perspectives from the Northern Rocky Mountains. Wildl Soc Bull 30:738–750.

Hutto RL, Pletschet SM, Hendricks P (1986) A fixed-radius point count method for nonbreeding and breeding-season use. Auk 103:593–602.

Johnson DH, Igl LD (2001) Area requirements of grassland birds: a regional perspective. Auk 118:24–34.

Karanth KU, Nichols JD, Kumar NS, Link WA, Hines JE (2004) Tigers and their prey: predicting carnivore densities from prey abundance. Proc Natl Acad Sci 101:4854–4858.

Lischke H, Bolliger J, Seppelt R (2007) Dynamic spatio-temporal landscape models. In: Kienast F, Wildi O, Ghosh S (eds) A changing world: challenges for landscape research. Springer, Dordrecht, Netherlands.

Liu CR, Berry PM, Dawson TP, Pearson RG (2005) Selecting thresholds of occurrence in the prediction of species distributions. Ecography 28:385–393.

MacKenzie DI, Nichols JD, Lachman GB, Droege S, Royle JA, Langtimm CA (2002) Estimating site occupancy rates when detection probabilities are less than one. Ecology 83:2248–2255.

MacKenzie DI, Nichols JD, Hines JE, Knutson MG, Franklin AB (2003) Estimating site occupancy, colonization, and local extinction when a species is detected imperfectly. Ecology 84:2200–2207.

Moilanen A (2000) The equilibrium assumption in estimating the parameters of metapopulation models. J Anim Ecol 69:143–153.

Moilanen A, Hanski I (1998) Metapopulation dynamics: effects of habitat quality and landscape structure. Ecology 79:2503–2515.

Newton I (1998) Population limitation in birds. Academic Press, Inc., San Diego, CA.

Norris K (2004) Managing threatened species: the ecological toolbox, evolutionary theory and declining-population paradigm. J Appl Ecol 41:413–426.

O'Connor RJ (1981) Habitat correlates of bird distribution in British census plots. Stud Avian Biol 6:533–537.

Pan W (2001) Model selection in estimating equations. Biometrics 57:529–534.

Pulliam HR (1988) Sources, sinks, and population regulation. Am Nat 132:652–661.

Pulliam HR, Danielson BJ (1991) Sources, sinks, and habitat selection: a landscape perspective on population dynamics. Am Nat 137:S50–S66.

Ralph CJ, Sauer JR, Droege S (1995) Monitoring bird populations by point counts. USDA Forest Service General Technical Report PSW-GTR 149.

Rodenhouse NL, Sherry TW, Holmes RT (1997) Site-dependent regulation of population size: a new synthesis. Ecology 78:2025–2042.

Sergio F, Newton I (2003) Occupancy as a measure of territory quality. J Anim Ecol 72:857–865.

Shochat E, Patten MA, Morris DW, Reinking DL, Wolfe DH, Sherrod SK (2005) Ecological traps in isodars: effects of tallgrass prairie management on bird nest success. Oikos 111:159–169.

Tewksbury JJ, Garner L, Garner S, Lloyd JD, Saab V, Martin TE (2006) Tests of landscape influence: nest predation and brood parasitism in fragmented ecosystems. Ecology 87:759–768.

Thogmartin WE, Sauer JR, Knutson MG (2004) A hierarchical spatial model of avian abundance with application to Cerulean Warblers. Ecol Appl 14:1766–1779.

Thomas JA, Bourn NAD, Clarke RT, Steward KE, Simcox DJ, Pearman GS, Curtis R, Goodger B (2001) The quality and isolation of habitat patches both determine where butterflies persist in fragmented landscapes. Proc R Soc Lond Ser B 268:1791–1796.

Tyler JA, Hargrove WW (1997) Predicting spatial distribution of foragers over large resource landscapes: a modeling analysis of the ideal free distribution. Oikos 79:376–386.

Van Horne B (1983) Density as a misleading indicator of habitat quality. J Wildl Manag 47:893–901.

Verheyen K, Vellend M, Van Calster H, Peterken G, Hermy M (2004) Metapopulation dynamics in changing landscapes: a new spatially realistic model for forest plants. Ecology 85:3302–3312.

Wiens JA (1989) The ecology of bird communities: Volume 1, Foundations and Patterns. Cambridge University Press.

Part III
Simplicity, Complexity, and Uncertainty in Applied Models

Chapter 7
Focused Assessment of Scale-Dependent Vegetation Pattern

Todd R. Lookingbill, Monique E. Rocca, and Dean L. Urban

7.1 Introduction

Ecological processes frequently occur at multiple spatial scales simultaneously. For example, fires imprint the landscape at a variety of spatial scales, from small areas of high burn intensity due to patchy surface fuels, to large stands within fires that escape conflagration entirely (Fig. 7.1). These types of complex disturbances can increase environmental heterogeneity and thus species diversity by creating a variety of microhabitats and by increasing patch diversity (Romme and Knight 1982; Christensen 1985; Denslow 1985; Pickett and White 1985; Turner et al. 1998). The flow of organisms, genes, and populations provides another excellent example of scale-dependent ecological processes (see Chap. 8).

Because ecological data are scale-specific, any model based on these data will provide inferences at a specific spatial scale. The scale of inference of statistical models is a critical consideration in predictive modeling (see Chap. 4). Some confusion can arise if the variables used to build a model are themselves scaled differently. For example, soils variables and terrain features have different characteristic scales (Urban et al. 2000). As a consequence a regression based on these data would be multi-scaled (Lookingbill and Urban 2004). Peters et al. (2004) refer to this approach to landscape modeling as spatially *implicit*, a nonspatial model built from geospatial data. By far, most models of species distribution are spatially implicit.

By contrast, a spatially *explicit* model would include predictive variables that account for spatial processes explicitly. From a modeling perspective, few approaches are currently available to ecologists to account for scale explicitly. Multi-level modeling (Gelman and Hill 2007) offers a framework for multi-scale

---

T.R. Lookingbill (✉)
Department of Geography and the Environment, University of Richmond, Richmond, VA 23173, USA
e-mail: tlooking@richmond.edu

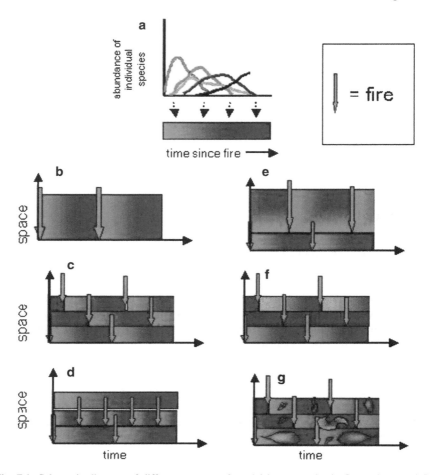

Fig. 7.1 Schematic diagram of different sources of spatial heterogeneity in fire regimes and fire effects. Color gradient represents turnover of species with space (vertical axis) and time (horizontal axis). (**a**) Red represents an early successional species/post-fire colonizers, blue represents late successional species/competitors, and orange represents the most severely burned areas; (**b**) uniform landscape with synchronous disturbance; (**c**) heterogeneity in time-since-fire with uniform fire regime; (**d**) heterogeneity in fire regime (fire frequency) across the landscape; (**e**) variable fire size (note dispersal gradients); (**f**) heterogeneity in season-of burn (top row burned under driest, hottest conditions resulting in orange/high severity; second row burned under mild conditions resulting in little setback of successional clock); (**g**) within-fire variability in fire intensity (fine-scale patchiness in intensity indicated by small patches with colors ranging from orange to red to blue)

models, if the levels of the nested ANOVA design are explicitly scaled by the user; that is, the levels specified in terms of variables with known characteristic scaling. Thus far, such applications tend to be nested logically (i.e., as treatments nested within blocks) rather than in a spatially explicit manner, but there is no reason why these models could not be framed across spatial scales. Similarly, hierarchical Bayesian models (Clark 2005; Gelman and Hill 2007) might be constructed to explicitly nest spatial scales, but are a more general construct.

We suggest that the multi-scale nature of ecological processes presents at least two challenges to landscape-scale species distribution modeling: choice of sample design and choice of inferential model. First, to capture landscape variability, fine-grained data need to be collected over a large spatial extent – a task that appears, at first glance, to be difficult if not impossible (Urban et al. 2002). In this chapter, we present a response to this data challenge that identifies and focuses sampling in locations of high resource heterogeneity. Ecologists have tended to focus on homogeneous environments to understand ecological processes (e.g., Whittaker 1956; Peet 1981), deliberately avoiding sampling locations of high local variability. However, locations of fine-scale heterogeneity can be extremely informative of broader-scale vegetation patterns. Ecotone studies provide the opportunity to efficiently collect detailed data on species-environment relationships at the competitive limits of the ecological tolerances of a species (Neilson 1991; Hansen and di Castri 1992; Risser 1995; but see Gosz 1993 for a warning of the dangers of not strictly defining the scale of transition in ecotone studies). Because these regions may be especially sensitive to environmental change, an increasing number of studies have been gathering data at treeline and other potentially sensitive vegetation ecotones (Camarero et al. 2000; Camill and Clark 2000; Fortin et al. 2000; Bunn et al. 2005).

Next, incorporating the data into predictive models presents a second major challenge. Just as it is logistically prohibitive to sample exhaustively at all spatial scales, no model can represent all details of an environmental system simultaneously. Models, by their nature, must simplify. Identification of dominant scales of ecological patterns and processes can be used by modelers to help determine which details are important to retain and which can be safely ignored (Levin 1992; Denny et al. 2004). Borrowing from fields of mathematics and computer science, much attention has been paid to identifying dominant scales of ecological patterns (Greig-Smith 1961; Dale 1999; Fortin and Dale 2005). Relating these patterns to their formative processes requires careful attention to the scale of inference of the statistical model. We provide a review of published approaches for linking pattern to process within a scale-specific framework. The direct assessment of these relationships provides opportunities for improved prediction and management of species responses to environmental change.

In the sections below, we describe methods for collecting landscape data at multiple spatial scales, the models that can be used to interpret these observations of pattern, and example applications of our approach to modeling scale-dependent vegetation pattern for the Sierra Nevada and Cascade Mountains of the western United States. The two examples demonstrate how accounting for spatial autocorrelation, capturing multi-scale pattern, and exploring the causes of pattern at each scale can be used to better predict vegetation shifts in response to changing climate and to improve fire management.

7.2 Management Challenge, Ecological Theory, and Empirical Framework

7.2.1 Management Challenge

The decision not to take a spatially explicit, cross-scale approach to environmental assessment can have profound management implications. Managers dealing with rapidly changing systems may not understand how these changes propagate at specific scales and locations on the landscape. A classic example is provided in the area of global climate change. Habitat models using the broad-scale predictions from global climate models as a basis for their predictions have predicted the upslope and northward migration of species in response to increasing temperatures (Peters and Darling 1985). The acquisition and protection of habitat corridors parallel to these gradients has been advised as a potential mitigation strategy (Noss 2001). However, as demonstrated by Halpin (1997) and others, basing management decisions on these coarse-scale models without regard to specific spatial characteristics of a site can be risky and ill-advised. Predictions based on regional variables such as temperature do not capture local-scale factors that drive the water balance (Stephenson 1990, 1998) and therefore may badly predict species physiological responses to future climates. Recent efforts to model species distributions in the context of global change have taken more sophisticated approaches and include more physiologically relevant data. These include the work of Iverson et al. (2004) to estimate potential migration of tree species into suitable habitat using a suite of data on climate, soils, land use, and landscape pattern. Biogeography models such as MAPPS (Neilson 1995) also use multiple soil, climate, and landscape variables to predict vegetation distribution at coarse scales. These models are increasingly linked to biogeochemistry models to provide estimates of major shifts in biomes under projected climate change (Nielson and Drapek 1998; Bachelet et al. 2001). New predictive models that rely heavily on detailed physiological data incorporate factors such as budburst phenology, frost hardiness, and drought tolerance (Morin et al. 2007, 2008). However, these modeling efforts remain largely coarse-scaled relative to the fine-scale variability in soils and climate that are critical to species migration patterns.

Wildland fire provides another example of an area in which managers are faced with making difficult decisions about a complex, multi-scale process. Despite the potential ecological importance of fire-generated heterogeneity at multiple spatial scales to ecosystem health (Davis et al. 1989; Moreno and Oechel 1992; Odion and Davis 2000; Brooks 2002), fire management rarely adopts a spatially explicit perspective that incorporates fine-scale variability. Fire suppression has dominated US natural resource policy and management for a century, and only recently has fire been reintroduced to some fire-adapted landscapes in the form of controlled burning. In many western forest types, especially those that historically experienced frequent low- or moderate-severity fire regimes, fire suppression has led to heavy fuel accumulations and shifts in plant community composition and structure, and has increased the risk of catastrophic wildfire (Agee 1998; Stephenson 1999). In addition to increasing overall loads, the legacy of fire suppression has also acted to generally homogenize fuel loads in these ecosystems. This homogenization can cause continuous, uniform burn patterns at the fine scale and a larger burn patch or fire size at the landscape scale. Whereas there is considerable variability in the effectiveness of recent fuel reduction efforts across the western US (Schoennagel et al. 2004), many fire-adapted landscapes cannot be restored without active management, nor can fires be entirely controlled (Miller and Urban 1999).

Replicating the full range of natural variability in fire behavior may be critical to maintaining biodiversity and ecosystem functioning. However, this goal has not been fully embraced by practitioners in the field, perhaps because of the difficulty associated with describing multi-scale fire patterns and their ecological effects. As a consequence, fire and mechanical thinning restoration treatments have been widely used without a clear understanding of the influence of fire on species distribution patterns. These applications tend to focus on the temporal domain in attempting to mimic historical disturbance frequency and to restore pre-suppression forest structure, but fire also has an important spatial component that creates heterogeneity in the physical and competitive environment. Spatial considerations have been limited to the among-patch scale (i.e., maintaining a mosaic of patch ages). The importance of reproducing the within-patch fire heterogeneity created by natural burns deserves further consideration in management prescriptions (Gill et al. 2002; Knapp and Keeley 2006; Rocca 2009).

7.2.2 *Ecological Theory*

Ecology has a rich tradition of studying species response to the variability along environmental gradients (Shreve 1922; Austin 1987). In these studies, species or communities of co-occurring species often are projected into a parameter space derived from environmental "proxy" variables. Some of the seminal examples of gradient work are provided by Whittaker (1956, 1960, 1967), who arrayed species abundance as distinct domains along elevation gradients in a variety of montane systems. However, correlations between species abundance and indirect environmental variables, such as elevation, are at best indicative of the underlying relationships driving the physiological responses of species. The importance of replacing indirect gradients such as elevation in vegetation models with direct and resource gradients such as temperature and moisture has been strongly emphasized by Austin and Smith (1989) among others. Vegetation analysis should be more than just correlation analysis, and studies should be geared more towards developing and testing vegetation theory (Austin 1987). This more mechanistic approach to modeling vegetation community pattern is reflected in a number of more recent gradient analyses (Ohmann and Spies 1998; Lookingbill and Urban 2005; Littell et al. 2008).

However, few empirical studies have attempted to link pattern and process in a predictive sense within a framework that explicitly considers the scale of interaction. The importance of scale in ecology has been well documented and is central to the discipline of landscape ecology (Wiens 1989; Wu and Hobbs 2007). According to hierarchy theory, lower levels of organization provide the mechanisms for patterns observed at higher levels (Allen and Starr 1982). Thus, ecological systems can be viewed as spatially nested hierarchies and understanding (or managing) pattern at the landscape level requires a rather comprehensive understanding of influential processes at finer resolutions. For example, disturbance processes of multiple types and at multiple times create a mosaic of patches of various ages for a given landscape (Pickett et al. 1989). The overlay of these many patch-scale disturbances may be expressed as complex landscape patterns.

Wu and Loucks (1995) argue that spatially explicit *simulation* modeling is the only practical approach to deal effectively with these problems of spatial patchiness, scale and hierarchical structure. Certainly, a new approach is required that develops a more mechanistic understanding of species patterns at landscape extents. We argue that traditional analytic techniques can be adapted to these challenges, but they require data inputs that are carefully selected to describe the multi-scale variability in direct environmental gradients. Conventional empirical methods are poorly suited to gather fine-resolution information on the distribution of these covariates at extents of kilometers to tens of kilometers. Refining landscape models to reflect the relative importance of fine-scale ecological processes requires a daunting field effort. One way to constrain this logistical challenge is to focus field campaigns on areas where we can expect high rather than low plot variability.

7.2.3 Empirical Framework

In his classic paper on pattern and scale in ecology, Levin (1992) provides one approach to developing predictive models in light of these issues: (1) describe pattern, (2) look for correlations with pattern to suggest potential mechanisms, and (3) improve understanding of pattern through careful examination of relationships with new 'mechanistic' variables. Though a true mechanistic understanding of ecological pattern may not be possible without *in situ* experimentation, well-designed modeling studies can go a long way towards disentangling the complex environmental gradients that are often invoked to explain ecological patterns.

Our approach focuses on the iterative relationship between data and models. The model-guided approach to sampling takes advantage of our ability to model at larger scales than we would reasonably be able to sample (Urban 2000). Preliminary hypotheses can be used to design initial sampling efforts. Field samples are then used to build models, which guide future sampling to answer new hypotheses and build better models. As an example, coarse-scale data available from land cover and digital elevation models can be used to identify areas of high species and environmental turnover on the landscape. Once areas of high variability in pattern over a relatively small space have been identified, those locations can be sampled intensively to attempt to uncover the process that underlies the pattern. The coarse and fine-scale datasets can then be compared to assess whether similar mechanisms are acting at the different spatial scales.

7.3 Data Availability and Suitability

The foundation of any predictive model is the structure of the underlying data, which are often gathered using a simple random sampling design. Unfortunately, random placement of plots can have significant drawbacks for species distribution modeling. There is no guarantee when using a simple random design that sample locations will be spatially balanced; part of the area being studied may be over-sampled and part may have few or no plots at all. This can be particularly troublesome for developing models for rare species, species with highly constrained distributions, or when some parts of a species range are inherently more informative than others for deciphering the ecological mechanisms behind the species distribution.

A systematic sampling design can increase the spatial balance of the sample locations. An even distribution of plots over the entire landscape ensures that no area is over- or under-represented, but is limited by logistical trade-offs in sample grain and extent. Very few studies have the luxury of collecting information using a fine-resolution grid over a large geographic area. Particular sites also may not be suitable for monitoring due to the absence of attributes of interest or due to issues such as safety or accessibility. This can leave a "hole" in the uniform sample. Stratification can be used to spread samples across a full range of conditions and to guarantee minimum sample sizes for different subpopulations. Stratification of samples can also make a field campaign easier to operate. For example, roads can be used as a stratum to improve ease of access for data gathering (Theobald et al. 2007).

We argue that species are rarely distributed randomly or uniformly and these sample designs are inefficient for many modeling efforts. Instead, hybrid designs such as Generalized Random-Tessellation Stratified (Stevens and Olsen 2004) and adaptive clustering (Philippi 2005) have recently become popular among managers interested in gathering spatially sensitive data. Multi-scale designs, in which a subset of plots is sampled at fine resolution, can be highly efficient at collecting data over large areas for intensive spatial analysis (Stohlgren et al. 1995, 1997; Urban et al. 2002). Multi-stage sampling involves completely different kinds of sampling at each stage. Nusser et al. (1998) strongly advocate multi-stage sampling designs for detecting broad-scale ecological patterns and for understanding the dynamics that produce observed changes in pattern. They provide a useful example using the National Resources Inventory (NRI) data.

The NRI sample design was developed to assess natural resource attributes for a broad geographic coverage (nonfederal rural lands of the United States) while acquiring a sufficient density of sample units for local subpopulation management. To accomplish these objectives, a stratified two-stage sample design was used. Stratification was based on small political or geographic areas and one to four primary sample units were located in each stratum during stage one sampling. These typically cover a total of between 2 and 6% of the entire land area. In stage two, a small number of sample points are placed in each primary sample unit. The exact number of sample points depends upon the various demands and constraints at the specific locale.

In addition to sample arrangement, the types of variables sampled at each stage determine the kinds of predictive models that can be used. Early stages typically rely upon indirect, coarse-resolution variables. For example, basic climate, land use, and ownership information is collected in stage one of an NRI sampling. These data are used to determine broad-scale trends and to develop initial weighting for stage two sampling. For our purposes in modeling complex vegetation pattern, these types of coarse variables are less useful than the fine-resolution, resource variables that have a more direct bearing on ecological processes. These types of more mechanistic but logistically more demanding variables can be sampled during later, "focused" stages of a multi-stage design.

7.4 Model and Model Validation Techniques

Spatially explicit, multi-scale data require spatially explicit, scale-specific models. These models can be found sprinkled throughout the literature in community, multivariate, and spatial ecology. We argue that developing species distribution models from spatially focused sampling schemes proceeds most powerfully from modeling and validation techniques that meet three criteria. First, they should recognize and account for spatial autocorrelation in the data that are used to fit the model. Second, they should identify the dominant spatial scales of species distributions and relate these to the dominant spatial scales associated with the predictor variables. Finally, we propose that techniques that explore quantitative relationships between species occurrences and predictor variables across a range of characteristic spatial scales provide the most mechanistic and insightful understanding of the drivers of species distributions. The first two criteria are well described elsewhere (Cressie 1993; Wagner 2001) and we review them only briefly. The final criterion about exploring scale-specific relationships between predictors and species distributions requires more novel methods that have only recently been explored by ecologists.

Because intensive focus plot sampling, by definition, collects spatially explicit information at fine spatial scales, statistical techniques used to examine species-environment relationships must account for the spatial autocorrelation present in these data as well as in any spatial processes affecting the response (Wagner 2001). In an intensive sampling scheme, data points are located close together so that closer points are likely to be similar to each other (i.e., each data point is not statistically independent). Explanatory variables that are autocorrelated inflate the effective sample size and may bias parameter estimates (Keitt et al. 2002; Dormann 2007). In working with these data, statistical techniques must either (a) remove spatial autocorrelation by averaging or removing data points that are within the spatial range of autocorrelation, or (b) explicitly account for spatial autocorrelation in explanatory variables, perhaps most simply by including a spatial blocking variable in the model design. Habitat modeling approaches include geostatistical methods such as kriging (Chong et al. 2001; Miller 2005), generalized additive models (Hastie and Tibshirani 1990), general estimating equations (Zeger and Liang 1986;

Gumpertz et al. 2000; Underwood et al. 2007), and autoregressive models (Keitt et al. 2002). Bayesian approaches have also been used (Hoeting et al. 2000; Lynch et al. 2006). Several excellent reviews of these methods are available (e.g., Dormann et al. 2007; Miller et al. 2007).

Modeling methods should also account for autocorrelation of residuals caused by spatially contagious processes that affect species distributions. Ecologically important yet many times unmeasured processes, such as propagule dispersal, drive species distributions and result in spatially correlated regression residuals (Legendre 1993; Wagner 2001). It may be difficult to distinguish whether spatial dependence in a response from a species is caused by autocorrelation of environmental variables or by a spatial biological process. Statistical approaches that go beyond consideration of spatial autocorrelation simply as a "nuisance" and instead describe the scales of spatial autocorrelation of both the predictor variables and the response may provide insight. Testing carefully crafted hypotheses and investigating spatial scales in detail can help to address these issues (Fraterrigo and Rusak 2008). In a compelling illustration of this approach, McIntire and Fajardo (2009) use six ecological examples to describe the benefits of using the scales of regression residuals to make statistical inferences about spatial processes.

A useful approach to testing for spatial dependence in explanatory variables takes advantage of the Mantel statistic (Mantel 1967). The statistic is simply calculated as the Pearson correlation between the elements in one distance matrix with the corresponding elements in a second distance matrix (Manly 1991). For a partial Mantel test, control variables are factored out, and the residuals are subsequently correlated with the variables of interest in a manner analogous to partial regression (Smouse et al. 1986). Mantel tests are ideal for testing relationships between species composition and the environment for several reasons. First, the effects of spatial autocorrelation can be tested for explicitly, or partialed out of analyses to detect relationships between variables after controlling for space. Second, the significance of a Mantel statistic is calculated using permutation procedures that eliminate problems associated with independence in parametric regression (Legendre and Fortin 1989). Third, because the correlation is calculated between distance matrices, the multivariate effects of multiple predictor variables (environmental factors) and multiple response variables (species) can be tested for simultaneously, as long as an appropriate distance metric is chosen (McCune and Grace 2002). Mantel tests average over all distances when calculating correlation coefficients. To identify the scales at which the environment and species are spatially autocorrelated, the Mantel correlation between a variable and space can be calculated separately within discrete distance classes (Goslee and Urban 2007). The results are displayed as a multivariate analog to a correlogram (Rossi et al. 1992), with distance class on the x-axis and Mantel correlation plotted on the y-axis.

Recently, Peres-Neto et al. (2006) have advocated principal coordinates analysis of truncated distance (neighbor) matrices (PCNM) (Legendre et al. 2008) as a means of exploring species turnover at multiple scales. PCNM captures spatial structure in terms of sine waves of varying wavelength, and summarizes species compositional patterns in terms of the compositional variance (dissimilarity) accounted at different wavelengths (scales). Legendre et al. (2008) argue that this new approach has more statistical power than the Mantel tests they had previously championed for this type of modeling.

Ecologists have also begun to use wavelet analysis to identify spatial scaling in variables. Like a Fourier transform, wavelet analysis decomposes a spatial or temporal data series into components at different scales. Wavelets are ideal for ecological data, because they can detect scale-specific patterns without the assumptions of stationarity required by semi-variance analysis or spectral analysis. Wavelet energies represent the proportion of variability in a data series expressed at each spatial scale, for scales in powers of two. Ogden (1997) provides a computational treatment of wavelets, though several attempts have been made to make wavelets more accessible to ecologists (e.g., Dale and Mah 1998; Torrence and Compo 1998; Cazelles et al. 2008; Dong et al. 2008). Most of these ecological treatments focus on using wavelets for time-series analysis, but data from a regularly spaced spatial grid are also appropriate for wavelet analysis (Bradshaw and Spies 1992; Mi et al. 2005; He et al. 2007).

We expect that relationships between environmental variables and species habitat often will change with spatial scale of analysis (Levin 1992; He et al. 2007). Ideally, we would be able not only to describe the scales of spatial variability of predictors and response but also to investigate relationships between variables at each spatial scale (Keitt and Urban 2005; Blanchet et al. 2008). Wavelet analysis can once again prove helpful for meeting this challenge. Wavelet covariance is the covariance between the wavelet coefficients for two variables at a defined spatial scale (Keitt and Urban 2005). Several ecological studies have demonstrated that scale-specific relationships between variables are revealed through analysis of wavelet covariance (Keitt and Fischer 2006; termed "wavelet cross-spectrum" in Cazelles et al. 2008). In a particularly innovative example, Mi et al. (2005) apply a statistical test to compare relationships between two transects.

Keitt and Urban (2005) take scale-specific inference a step further by introducing the concept of wavelet regression. They demonstrate that a linear regression between wavelet coefficients, extracted separately by spatial scale, can identify scale-specific relationships between several predictor variables and a response. Scale-specific regressions using wavelets have not yet caught on in the ecological community, yet we propose that they offer a promising approach for species distribution modeling. Carl and Kuhn (2008) analyze spatial ecological data using wavelet regression, but their objective appears to be to remove the effects of autocorrelation in an attempt to predict habitat for a plant species, rather than using it to extract scale-specific relationships between environmental predictors of habitat and plant occurrence. However, their analysis expands upon the methods of Keitt and Urban (2005) in two ways that are of interest to species distribution modelers: (1) they demonstrate that wavelet regression can proceed in a logistic regression setting, and (2) they take advantage of a two-dimensional (grid) dataset. More work is needed to determine how scale-specific regression analysis can be applied in a predictive mode, as often desired by species distribution modelers.

7.5 Case Studies in Western US Forests

The best species distribution models will incorporate the influence of all the important variables affecting species distributions, from coarse-scale environmental tolerances to fine-scale dispersal and competition processes. We have described the data and modeling challenges presented by the need to consider the influences of variables whose relationship to species may vary depending on spatial scale. In this section, we demonstrate our iterative approach to linking models and data in scale-dependent assessments of vegetation pattern with two examples. For the first, we emphasize the data aspect of developing multi-scale, mechanistic models. We describe how a multi-stage sampling design can be used to create a hierarchical set of models and demonstrate the utility of ecotone focus plots for modeling species distributions in old-growth forest habitats of the Pacific Northwest United States. The second example emphasizes the analytic side of our approach. We use scale-specific modeling approaches – such as wavelet analysis and the Mantel correlogram – to determine whether planned fires in Sequoia National Park appreciably homogenize the environment by burning through heavy, continuous fuel beds, thus leading to altered distributions of herbaceous species within the park.

7.5.1 *Predicting Spatial Shifts in Old-Growth Forest Habitat*

In an effort to develop a predictive model of forest community spatial pattern for the Western Cascades, we conducted intensive field sampling at areas of spatial transition between the *Tsuga heterophylla* (western hemlock) vegetation zone and the *Abies amabilis* (Pacific silver fir) vegetation zone. An improved understanding of this ecotone would better inform the potential impacts of changes in climate or management within these ecologically and economically important forests. For example, an ecotone formed primarily by differences in growth rates associated with temperature may respond linearly to changes in temperature, whereas an ecotone maintained primarily by winter precipitation may not simply migrate up slope in response to increasing temperature. In this study, we considered directly the effect of temperature, snowpack, radiation and moisture on seedling establishment and relative growth rates of trees in an effort to extend our knowledge base beyond the simple correlation of plant communities with the elevation gradient complex.

The multi-stage approach that we followed relies upon a process of successive refinement of the modeled species–environment relationships. The preliminary model (Lookingbill and Urban 2005) was based upon the correlation between species abundance and terrain proxy variables (e.g., elevation) for 164 (20×20 m) vegetation samples stratified across the H.J. Andrews Experimental Forest (HJA). The HJA is a Forest Service Experimental Watershed and a Long-Term Ecological Research site representative of the soils, geology, and climate of the Western Cascades (McKee 1998). Elevation varies from 425 to 1,620 m, and the sampling for this initial stage extended across the entire range. Variables included measurements of soil nutrients and chemistry, though only terrain variables were significant in the model. The geographic realization of this model identified areas of short but steep gradients in which discrete changes in community type were predicted (Fig. 7.2). A follow-up stage of new field studies at these key locations provided an efficient new set of data to help refine the model and provide a deeper understanding of the fine-scale processes associated with observed vegetation pattern.

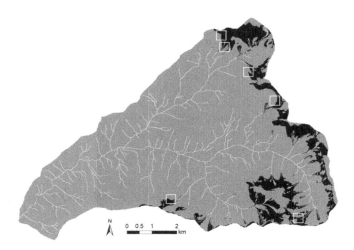

Fig. 7.2 Ecotone plot locations (boxes) within the H.J. Andrews Experimental Forest. Red area represents transition zone as identified by a forest community CART model using temperature, soil moisture and radiation as explanatory variables (described in Lookingbill and Urban 2005)

Six "ecotone focus plots" were designed to explicitly consider fine-scale environmental constraints along this region of active forest community transition. Whereas the preliminary model confirmed the well-documented shift from the *T. heterophylla* vegetation zone to the *A. amabilis* vegetation zone along an elevation gradient (Franklin and Dyrness 1988), the second stage of sampling addressed the relative importance of temperature, moisture, radiation, and snowpack as potential drivers of species distributions within the ecotone. Efforts to identify the physiological mechanisms responsible for this transition are surprisingly few, dated, and somewhat contradictory (Krajina 1969; Thornburgh 1969). In this study, we addressed the relationship between regeneration and physical drivers through logistic regression of point measurement and kriged data. We also considered dispersal constraints through bivariate Ripley's K-analysis of seedling and potential seed tree data. The Ripley's K-function differs from conventional nearest neighbor analyses in that it considers distances between all observed points and not just the first or second nearest neighbor (Moeur 1997; Haase 1995). An advantage of preserving all spatial relationships in the data is that Ripley's K-tests can assess pattern at multiple scales, and can thus be used to evaluate spatial scales of clustering, in a univariate sense, or attraction/repulsion, in a bivariate sense (Haase et al. 1996; Lookingbill and Zavala 2000).

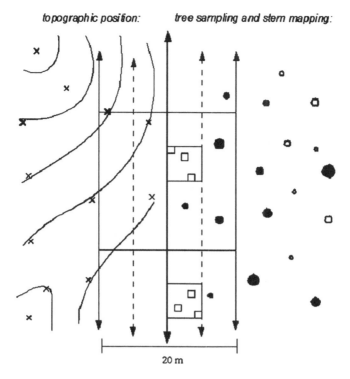

Fig. 7.3 Transect based sampling layout for microtopograpy (drawn on left side) and vegetation (*right side*) on ecotone focus plots. Transects were 20 m wide by 100–180 m in length. Centerline was paralleled by 5-m bands used for randomly locating seedling (1 × 1 m) and sapling (5 × 5 m) quadrats. All seedlings within 1 m of the centerline were mapped. All trees within the transect were mapped. In addition, trees sighted as "in" with a basal area prism were mapped (filled symbols on *right*). Topographic points were surveyed with sufficient density (drawn as x's) to generate a high-resolution DEM. Direct measures of temperature, soil moisture, snow and radiation also were taken

Our focus plots were 20 m in width and between 100 and 180 m in length. Within each of the plots, all dead and live trees were measured at breast height, cored for age and growth rate analysis, and mapped using a laser surveying system (Fig. 7.3). In addition, potential seed trees outside the transects were identified according to a plotless sampling design using a 2.5-factor basal area prism along the transect centerline. Nested within each transect were 3 (1 × 1 m) quadrats per every 20 m in which all seedlings were tallied by size class (young of the year, 0–10 cm in height, 10–50 cm in height, and 50–137 cm in height). We also mapped all seedlings within 1 m and all saplings within 5 m of the transect centerline.

Using the laser surveying system, we recorded critical points of topographic change and used these points to interpolate a high-resolution digital elevation model (DEM) of the plots. An average of nearly 40 measurements were taken for every 20 m of transect. Surface soil moisture (0–20 cm in depth) was recorded synoptically at these locations and at all seedling quadrats using a handheld volumetric moisture sampling device (Lookingbill et al. 2004). Three soil depth measurements also were taken at each of the moisture locations using a 1 m tile probe. Canopy closure was estimated at the seedling quadrats using a concave spherical densiometer. Temperature sensors (Lookingbill and Urban 2003) were located at several key locations along each transect, which recorded temperature at 30-min increments. Several complementary approaches were used to quantify snow levels and melt on the plots. First, we synoptically measured snow depth (up to a maximum depth of 100 cm) in the spring of 2002 at 1 m intervals along the centerline of three of the plots. Lichen height acts as a reasonable proxy for the average maximum snow depth (Winkler and Schultz 2000), and we recorded the average height at which lichens began growing on tree boles for each 20 m plot segment. Finally, we distributed additional temperature sensors at ground level across the plots. These sampling devices allowed remote monitoring of the beginning and end of winter snowcover for specific locations on the plots. When covered with snow, these sensors would consistently record a temperature of 0°C.

For each focus plot, we first looked for geospatial patterns in tree regeneration, growth, and mortality. No significant spatial patterns emerged in the distribution of snags on the plots. The density of dead trees of *A. amabilis* and *A. procera* were highly variable from plot to plot, but consistently greater than for *T. heterophylla*. These findings are consistent with those of Acker et al. (1996) who found low *T. heterophylla* mortality relative to *A.* spp mortality in a 27-year study of a forest stand within the ecotone zone at the HJA. We also found that growth rates were not significantly associated with elevation, temperature, or any of the other environmental variables gathered at this scale (Fig. 7.4). Earlier models relied heavily on these relationships in predicting how this ecotone would respond to changes in climate (Urban et al. 1993). Our models, instead, focus on the importance of regeneration in maintaining observed community patterns.

To evaluate whether dispersal limitations may be constraining the distribution of species, bivariate Ripley's K-analysis was conducted using a bivariate label permutation test of seedlings and trees in each of the plots. All tree and seedling locations were held constant, while we randomly reassigned the species labels of the seedlings. The distances from seedlings to conspecific adults for 99 of these randomized trials were then compared with the distances for the actual data ($P<0.01$). Observations higher than the randomized data were considered to be positively associated. Observations lower than the randomized data were considered to be negatively associated. The scale of positive association should be reflective of species' dispersal capabilities. The point pattern analysis confirmed that the heavier seeded *A. amabilis* and *A. procera* may be more prone to dispersal limitations than *T. heterophylla*, but none of the species were likely constrained by dispersal within the extents of the ecotone plots.

Point measurements of the environmental variables were then kriged to 1 m resolution grids and logistic regression was used to model the presence or absence of seedlings by species as a function of radiation, temperature, soil moisture and snow cover along the 1×1 m seedling sample quadrats running up the middle of the plots. Regressions were also conducted using the point measurements of seedling presence/absence, light, soil moisture, and soil depth at each of the 1×1 m seedling quadrats. Results of the logistic regression analyses ($n=653$ *T. heterophylla* seedlings, $n=603$ *A. amabilis* seedlings, and $n=232$ *A. procera* seedlings) indicated the

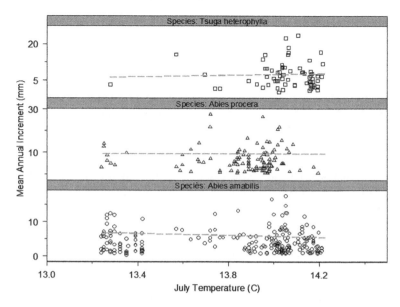

Fig. 7.4 Trends in relative growth rates. In contrast to prior modeling assumptions emphasizing the importance of differences in growth rates to species distributions in this system, growth was not significantly associated with temperature or any of the environmental variables considered at the ecotone scale. Species shown are *A. amabilis*, *A. procera*, and *T. heterophylla*

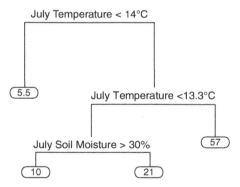

Fig. 7.5 Regression tree model of *T. heterophylla* seedling density on seedling plots. Circles provide mean number of seedlings for the plots described by that end-node. Length of branch corresponds to the amount of variance explained by that variable

importance of temperature and moisture as explanatory variables, but emphasized the high plot-to-plot variability. Temperature differences were highlighted as the strongest predictor of *T. heterophylla* seedling presence/absence across all plots; a finding also supported by regression tree analysis of seedling density in subplots (Fig. 7.5). Relationships were not consistent, however. For example, January temperatures rather than July temperatures were significantly different on Plot 3, for which soil moisture was the strongest predictor variable of seedling abundance.

Radiation was consistently the least important of the potential explanatory variables in the logistic regression and regression tree analyses.

Our results generally suggest that regeneration rather than growth or mortality is likely to be limiting the range of *T. heterophylla* through a combination of snow, temperature, and moisture limitation. Therefore, changes in climate that would alter the competitive dynamics between species would be most apparent in the regeneration niche. Landscape-level management activities such as timber harvesting within the shifting ecotone could result in unattended consequences if these new dynamics were ignored. An interesting observation from the bivariate Ripley's K-analyses was that *T. heterophylla* seedlings were more common under canopies of large trees (of any species) than canopy gaps. One explanation for this finding lies in the shade-tolerance of *T. heterophylla*. However, *A. amabilis*, an equally or even more tolerant competitor in this vegetation zone (Fonda and Bliss 1969; Mitchell et al. 2007), did not show the same spatial patterning. An alternative explanation is suggested by examining the patterns of snowpack on the plots. Snowfall interception by branches and needles can substantially decrease the amount of accumulation under tree crowns. The rate of snowmelt also is modified considerably in the vicinity of large stems that can re-radiate longwave radiation (Anderson 1963). It is possible that *T. heterophylla* establishment in the transition zone, and by extension *T. heterophylla* migration upslope, is aided by these melt cones. The single year of snow sampling that we conducted for this analysis is insufficient to definitively address this issue; however, the model results using these exploratory data do serve as a guide for future work. Targeted data collection to test this hypothesis provides the next round of study in our data gathering → modeling → data framework. In light of the potential interactions between climate and disturbance, experimental studies of montane conifer regeneration under alternative silviculture systems (Mitchell et al. 2007) may be the most efficient way of improving model predictions in this landscape.

7.5.2 Predicting Herbaceous Response to Prescribed Fire

In many ecosystems, fire-adapted landscapes cannot be restored without active management. However, how plant communities respond to prescribed fire and how these responses may differ from responses under natural fire regimes are poorly documented. Some scientists have raised the concern that prescribed fires may be too homogeneous to restore pre-suppression forest structure (Bonnicksen and Stone 1982; Allen et al. 2002). A handful of field-based studies have explored the biodiversity consequences of large-scale, fire-generated, environmental heterogeneity in Yellowstone National Park (Romme and Knight 1982; Turner et al. 1997, 1999). Schoennagel et al. (2008) documented spatial variation in post-fire structure, composition, and ecosystem function at multiple scales following the 1988 Yellowstone fires, but fine-scale heterogeneity under moderate-severity fire regimes have not been thoroughly explored. In the few systems where fine-scale heterogeneity in fire effects and species responses have been reported (Davis et al. 1989; Moreno and

Oechel 1992; Odion and Davis 2000; Brooks 2002), fires have been shown to reinforce already existing patterns, neither increasing nor decreasing spatial pattern and scale of variability. There are few well-studied examples of the effects of fire intensity and spatial patterning in determining post-fire spatial distributions of species.

We tested the role of within-fire variability in fire severity in structuring understory plant communities in a fire-adapted, mixed-conifer forest of the Sierra Nevada, California, USA. We examined the effects of six prescribed fires on plant community structure and spatial distributions, and compared the effects of two management alternatives: early-season prescribed fire (June burn) and late-season prescribed fire (October burn). We collected high-resolution floristic, fuels, and environmental data at every meter along 256-m transects through six prescribed fires (three in each season) and asked whether fire changes the scales of species distributions. If fire creates heterogeneity by subdividing previously homogeneous forest floor patches, we would expect plants to sort along finer-scale environmental gradients after fire and exhibit smaller-scale spatial autocorrelation than they did prior to fire. An increased spatial scale of species turnover, by contrast, would support a model in which fire homogenizes the environment.

Mantel correlograms on species composition show reduced spatial autocorrelation in species distributions after fire with a more pronounced effect observed in the June burns (Fig. 7.6). Before fire, species distributions along all transects exhibited spatial structure at scales up to approximately 150 m. After fire, species distributions on the October burn transects showed autocorrelation at scales similar to pre-fire patterns, whereas June burn transects had a shortened range of approximately 100 m. Perhaps more importantly, after fire the June burns showed a small ($p<0.05$) but significantly reduced autocorrelation at scales between 5 m and 100 m (the October burns also significantly reduced autocorrelation, but only to a modest degree in a few short portions of the range). These results support a model in which patchy June burns create heterogeneity in the environment, which in turn increases the variability in species distributions at fine-spatial scales. October burns, on the other hand, do not appear to appreciably change the scaling of the environment in a matter that affects understory herbaceous plants.

To further investigate the fire behavior responsible for these results, we used wavelet analysis to examine the spatial scales of burn pattern (for the June and October burns) and fire temperature (for the June burns). Then, we asked whether fire temperature and burn pattern can be predicted with pre-fire site data on fuels and topography, or whether they are affected more by less predictable and difficult to measure features such as ignition, moisture, and local wind patterns. In addition to the fine-scale measurements of fuel and topography taken at every meter along each transect, we recorded maximum fire temperature in the June prescribed burns by installing a "pyrometer" (containing streaks of temperature sensitive paints, which permanently change appearance once their particular melting temperatures are reached) at every meter along the transect.

Our results show that prescribed fires can be remarkably heterogeneous in burn pattern (Fig. 7.7) and fire temperature (Fig. 7.8). It appears possible to create a patchy burn pattern even within methodically ignited prescribed burns through

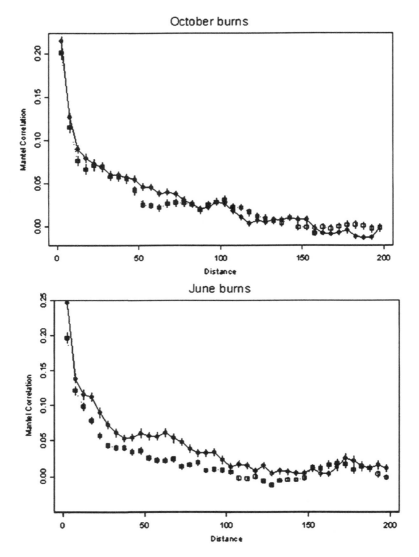

Fig. 7.6 Mantel correlograms of pre-fire (*circles* connected by *solid lines*) and post-fire (*squares*) species distributions, calculated using the Bray and Curtis (1957) index of dissimilarity. Filled symbols represent correlation significantly different than zero, and error bars represent 95% confidence intervals

fire-suppressed forest understories. In this experiment, four out of six management burns displayed significantly more variability in burn pattern and fire temperature than might have been expected, given high fuel continuity. The four heterogeneous fires included all three of the early season (June) burns and one October burn (plot 2). Plot 2 spans a relatively wet, level area, and was burned under rather humid weather conditions for the time of year. Plots 5 and 6, the other two October burns were

7 Focused Assessment of Scale-Dependent Vegetation Pattern

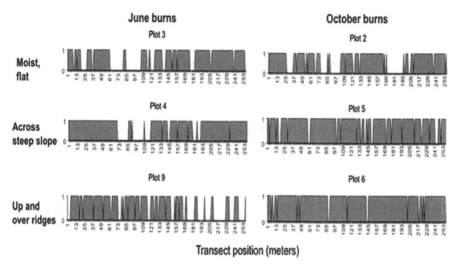

Fig. 7.7 Meters burned along each transect for spring and fall burns (1 = burned, 0 = unburned)

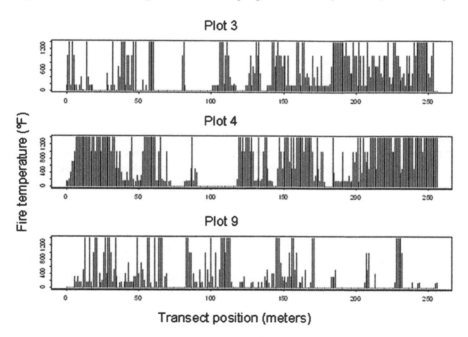

Fig. 7.8 Maximum fire temperatures along the three June transects

considerably more homogeneous in their fire effects. In comparison with the fuels in plot 2, fuels in plots 5 and 6 units were uniformly dry, with no areas of moisture accumulation encountered along the transect.

The wavelet energies for burn pattern showed a complex relationship in which the dominant spatial scales of variability are not easily categorized based on season

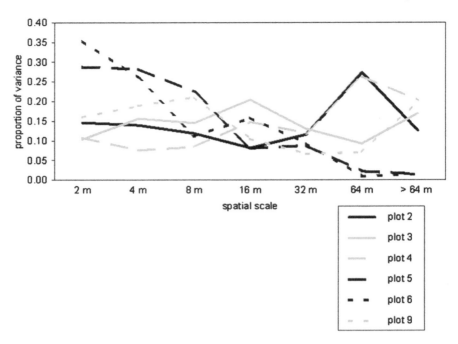

Fig. 7.9 Wavelet energies for burn pattern in the three June burn transects (*gray*) and the three October burn transects (*black*)

(Fig. 7.9). Plots 5 and 6, October burns that burned almost completely, expressed most of their (nominal) variability at the finest spatial scales. Plots 2 and 4 showed maximum variability at the 64 m scale. Plots 3 and 9, both highly patchy June burns, expressed their spatial variability in burn pattern across a wide range of scales.

The causes of these patterns in burn pattern were explored using a logistic regression on pre-fire variables related to fuels and topography. Because many of the explanatory variables taken prior to the burns are highly correlated, we first conducted a factor analysis on the pre-fire variables to create orthogonal "factors" that represent the major components of variability along the transect prior to fire, then used the resulting factors as the explanatory variables in the logistic regression models to predict burn pattern. The six factors identified through factor analysis together account for 61% of the original variability in the environmental dataset (Table 7.1).

Surprisingly, the important explanatory variables for predicting burn pattern depended on transect orientation more than burn season (Table 7.2). Plots 4 and 5, the two transects crossing across hillslopes, had burn patterns driven by fuel load (e.g., litter/canopy, woody fuel, and CWD). Burn patterns in the gently sloping, moist transects (plots 2 and 3), were also affected by fuel loads, with moisture and elevation decreasing the probability of burning. Plots 6 and 9, whose transects go over small ridges, depend on fuel to explain their burn pattern. The topographic factors insolation and elevation affected probability of burning, but only for the June burn (plot 9). Pre-burn factors account for between 14 and 36% of the variation in burn pattern.

Table 7.1 Variable loadings from factor analysis

Original variables	Factors					
	"Insolation"	"Woody fuel"	"Moisture"	"CWD"	"Litter/canopy"	"Elevation"
Elevation			0.51			0.65
Slope			−0.75			
T-aspect	0.96					
Radiation	0.94					
Soil moisture			0.56			
Light					−0.57	
Litter cover					0.39	
Litter depth					0.56	
1-h fuel		0.67				
10-h fuel		0.98				
100-h fuel		0.26				
CWD				1.00		
% variability	16.0	13.0	10.2	9.6	7.6	4.5
Cum % variability	16.0	29.0	39.2	48.8	56.4	60.9

Factors are combinations of variables that tend to correlate with each other, and loadings are the correlation coefficients between the original variable and the factors. We gave factors names to make the results easier to interpret. The percent variability in the original dataset that is explained by each factor is also provided

Table 7.2 Results of stepwise logistic regression on burn pattern

	Factors						% Variance in burn pattern explained	n
	"Insolation"	"Woody fuel"	"Moisture"	"CWD"	"Litter/canopy"	"Elevation"		
June burns								
Plot 3		(+)	−	+	+	−	17	242
Plot 4				+	+	(+)	36	254
Plot 9	+	+		+	+	+	33	242
October burns								
Plot 2		(+)	−	+	+	−	14	236
Plot 5	+			+	+		21	192
Plot 6	+		(+)	+	(+)		23	242

Positive factors increase probability of burning while negative factors reduce probability of burning. Signs in parentheses indicate factors that were included in the optimal model, but were not statistically significant at the 0.05 level. Percent of variance in burn pattern explained by each model (pseudo R^2) and sample size (n) are also provided

This case study demonstrates that prescribed fires can generate remarkable variability in fire temperature when conducted under particular conditions, and this variability can, in turn, play a pivotal role in shaping the distribution of flora. The factors that affect burn pattern depend on topographic context more than season. In particular, moisture patterns driven by topographic variability had a stronger influence on burn pattern than did seasonal differences (i.e., early summer vs. late-season burns). Despite the large amounts of data collected for this analysis, burn pattern and fire temperature were approximately 25% (±15%) predictable, suggesting that random or unmeasured factors (e.g., wind, ignition pattern) have an important influence on fire behavior and fire effects. Even so, this improved predictive ability translates directly into management recommendations about how prescriptions might be tailored to create within-burn heterogeneity. If fuel loads are continuous and homogeneous, as they typically are for first-entry burns in much of the Sierra Nevada, heterogeneity in fuel moisture due to topographic drainage patterns (which influence seasonal soil moisture patterns) or aspect and elevation (which influence diurnal fuel drying) can create heterogeneity. Heterogeneity associated with either drainage or diurnal drying occurs more often during the early season, before fuels are uniformly dry, but may extend later into the summer on moist sites. If fuels are distributed heterogeneously, either because of large tree gaps or because of previous fuel reduction burns, heterogeneous fuels will probably simulate a more natural burn pattern, even in uniformly dry conditions. In areas where both fuel and topography are homogeneous, then the possibility of altering ignition pattern should be explored in an effort to ensure heterogeneity.

7.6 Conclusions

The multi-scale nature of many ecological processes presents a challenge to modelers projecting species distributions across large geographic extents. Our approach of coupling focused sample plots with analytic models that explicitly account for variability in space and scale is appropriate when any one of the following is true:

1. The ecological process in question occurs at a fine spatial scale, but the manifestations of that process influence large-scale patterning. In these situations, a more traditional sampling scheme may capture the pattern, but it will not distinguish between alternative mechanisms for the fine-scale process that creates the pattern.
2. Data are expensive in terms of time or money. Focusing on areas of high heterogeneity may be necessary when the costs of travel across a research area are high or when field equipment – such as dataloggers or sensors – are in limited supply relative to the size of the study area. Given these restrictions, the most statistically powerful placement of samples may be in areas of high contrast in one variable, controlling for as many other influences as possible.
3. The ecology of areas of contrast, such as ecotones or edges, is the focus of the project. Examples might include tree-line studies of climate change effects or studies of the spread of invasive species across land-use boundaries in mixed-use landscapes.
4. Describing the scales and/or magnitudes of spatial variability of a pattern or process is an explicit goal of the study. In many cases, spatially averaged results are not adequate for answering a research question. Alternate hypotheses generated around ideas of scale can only be tested when spatial scale is explicitly explored.

Our second case study provides an example of this latter criterion. A better understanding of the scale of variability in burn pattern and intensity imposed by human prescriptions of fire can provide important insights into the influence of disturbance on species distributions. To capture this variability, fine-scale data need to be collected over a large sample area. This can be a daunting task that is often most efficiently accomplished by selecting sample locations of high, rather than low, plot heterogeneity as demonstrated in our initial case study. Whereas data collected at these sites are noisy and may be somewhat limited by small sample sizes, they offer the advantage that the environmental variables and demographic processes can be measured directly and thoroughly rather than inferred from larger-scale correlations.

Once these data are in hand, they can be analyzed using emerging statistical techniques to create powerful predictive models. The challenges of spatial patchiness, scale, and hierarchical structure need not demand a spatially explicit *simulation* modeling approach. Purely *statistical* approaches may suffice given that they (1) recognize and account for spatial autocorrelation in the data, (2) identify the dominant spatial scales of species distributions and of potential predictor variables, and (3) explore quantitative relationships between species distributions and predictor variables across the range of characteristic spatial scales.

In general, more landscape studies are needed that attempt to link pattern and process in a predictive sense within a framework that explicitly considers the scale of interaction (i.e., providing a more mechanistic understanding of species patterns and informing specific scales of management). We attempt to address this challenge in our work by maintaining a tight link between the data, hypotheses and models. The results are often fine-scale field studies that can be scaled up to their landscape-scale management implications.

Acknowledgments We would like to thank Y. Wiersma and three anonymous reviewers for valuable comments on earlier drafts of this manuscript. Field data were collected in the HJA as part of NSF awards IBN-9652656 and DEB-0108191. We thank the small army of field assistants that helped gather data for both of the case studies. The prescribed fire research would not have been possible without colleagues at USGS and NPS: N. Stephenson, J. Keeley, E. Knapp, E. Ballenger, T. Caprio, M. Keifer, and B. Jacobs.

References

Acker SA, Harmon ME, Spies TA, McKee WA (1996) Spatial patterns of tree mortality in an old-growth *Abies pseudotsuga* stand. Northwest Sci 70:132–138.
Allen CD, Savage M, Falk DA, Suckling KF, Swetnam TW, Schulke T, Stacey PB, Morgan P, Hoffman M, Klingel JT (2002) Ecological restoration of Southwestern ponderosa pine ecosystems: a broad perspective. Ecol Appl 12:1418–1433.
Allen TFH, Starr TB (1982) Hierarchy: perspectives for ecological complexity. The University of Chicago Press, Chicago.
Anderson HW (1963) Managing California's snow zone lands for water. U.S. Department of Agriculture, Forest Service, Pacific Southwest Research Station, Sacramento CA.
Agee JK (1998) The landscape ecology of western forest fire regimes. Northwest Sci 72:24–33.

Austin MP (1987) Models for the analysis of species' response to environmental gradients. Vegetatio 69:35–45.

Austin MP, Smith TM (1989) A new model for the continuum concept. Vegetatio 83:35–47.

Bachelet D, Neilson RP, Lenihan JM, Drapek RJ (2001) Climate change effects on vegetation distribution and carbon budget in the United States. Ecosystems 4:164–185.

Blanchet FG, Legendre P, Borcard D (2008) Modelling directional spatial processes in ecological data. Ecol Modell 215:325–336.

Bonnicksen TM, Stone EC (1982) Managing vegetation within U.S. national parks: a policy analysis. Environ Manage 5:101–102 and 109–122.

Bradshaw GA, Spies TA (1992) Characterizing canopy gap structure in forests using wavelet analysis. J Ecol 80:205–215.

Bray JR, Curtis JT (1957) An ordination of the upland forest communities of southern Wisconsin. Ecol Monogr 27:325–349.

Brooks ML (2002) Peak fire temperatures and effects on annual plants in the Mojave Desert. Ecol Appl 12:1088–1102.

Bunn AG, Graumlich LJ, Urban DL (2005) Trends in twentieth-century tree growth at high elevations in the Sierra Nevada and White Mountains, USA. Holocene 15:481–488.

Camarero JJ, Gutierrez E, Fortin M-J (2000) Spatial pattern of subalpine forest-alpine grassland ecotone in the Spanish Central Pyrenees. For Ecol Manage 134:1–6.

Camill P, Clark JS (2000) Long-term perspectives on lagged ecosystem responses to climate change: permafrost in boreal peatlands and the grassland/woodland boundary. Ecosystems 3:534–544.

Carl G, Kuhn I (2008) Analyzing spatial ecological data using linear regression and wavelet analysis. Stoch Environ Res Risk Assess 22:315–324.

Cazelles B, Chavez M, Berteaux D, Menard F, Vik JO, Jenouvrier S, Stenseth NC (2008) Wavelet analysis of ecological time series. Oecologia 156:287–304.

Chong GW, Reich RM, Kalkhan MA, Stohlgren TJ (2001) New approaches for sampling and modeling native and exotic plant species richness. West N Am Nat 61:328–335.

Christensen NL (1985) Shrubland fire regimes and their evolutionary consequences. In: Pickett STA, White P (eds) The ecology of natural disturbance and patch dynamics. Academic Press, Orlando, pp. 85–100.

Clark JS (2005) Why environmental scientists are becoming Bayesians. Ecol Lett 8:2–14.

Cressie NAC (1993) Statistics for spatial data. Wiley, New York.

Dale MRT (1999) Spatial pattern analysis in plant ecology. Cambridge University Press, Cambridge.

Dale MRT, Mah M (1998) The use of wavelets for spatial pattern analysis in ecology. J Veg Sci 9:805–814.

Davis FW, Borchert MI, Odion DC (1989) Establishment of microscale vegetation pattern in maritime chaparral after fire. Vegetatio 84:53–67.

Denny MW, Helmuth B, Leonard GH, Harley CDG, Hunt LJH, Nelson EK (2004) Quantifying scale in ecology: lessons from a wave-swept shore. Ecol Monogr 74:513–532.

Denslow JS (1985) Disturbance-mediated coexistence of species. In: Pickett STA, White P (eds) The ecology of natural disturbance and patch dynamics. Academic Press, Orlando, pp. 307–323.

Dong XJ, Nyren P, Patton B, Nyren A, Richardson J, Maresca T (2008) Wavelets for agriculture and biology: a tutorial with applications and outlook. Bioscience 58:445–453.

Dormann CF (2007) Effects of incorporating spatial autocorrelation into the analysis of species distribution data. Glob Ecol Biogeogr 16:129–138.

Dormann CF, McPherson JM, Araújo MB, Bivand R, Bolliger J, Carl G, Davies RG, Hirzel A, Jetz W, Kissling WD, Kuhn I, Ohlemuller R, Peres-Neto PR, Reineking B, Schroder B, Schurr FM, Wilson R (2007) Methods to account for spatial autocorrelation in the analysis of species distributional data: a review. Ecography 30:609–628.

Fonda RW, Bliss LC (1969) Forest vegetation of the montane and subalpine zones, Olympic Mountains, Washington. Ecol Monogr 39:271–301.

Fortin M-J, Dale MRT (2005) Spatial analysis: a guide for ecologists. Cambridge University Press, Cambridge.

Fortin M-J, Olson RJ, Ferson S, Iverson L, Hunsaker C, Edwards G, Levine D, Butera K, Klemas V (2000) Issues related to the detection of boundaries. Landsc Ecol 15:453–466.

Franklin JF, Dyrness CT (1988) Natural vegetation of Oregon and Washington. Oregon State University Press, Corvallis.

Fraterrigo JM, Rusak JA (2008) Disturbance-driven changes in the variability of ecological patterns and processes. Ecol Lett 11:756–770.

Gelman A, Hill J (2007) Data analysis using regression and multilevel/hierarchical models. Cambridge University Press, New York.

Gill AM, Bradstock RA, Williams JE (2002) Fire regimes and biodiversity: legacy and vision. In: Bradstock RA, Williams JE, Gill AM (eds) Flammable Australia: the fire regimes and biodiversity of a continent. Cambridge University Press, Cambridge, pp. 429–446.

Goslee SC, Urban DL (2007) The ecodist package for dissimilarity-based analysis of ecological data. J Stat Softw 22:1–19.

Gosz JR (1993) Ecotone hierarchies. Ecol Appl 3:369–376.

Greig-Smith P (1961) Data on pattern within plant communities: I. The analysis of pattern. J Ecol 49:695–702.

Gumpertz ML, Wu C, Pye JM (2000) Logistic regression for southern pine beetle outbreaks with spatial and temporal autocorrelation. For Sci 46:95–107.

Haase P (1995) Spatial pattern analysis in ecology based on Ripley's K-function: introduction and methods of edge correction. J Veg Sci 6:575–582.

Haase P, Pugnaire FI, Clark SC, Incoll LD (1996) Spatial patterns in a two-tiered semi-arid shrubland in southeastern Spain. J Veg Sci 7:527–534.

Halpin P (1997) Global climate change and natural-area protection: management responses and research directions. Ecol Appl 7:828–843.

Hansen AJ, di Castri F (eds) (1992) Landscape boundaries: consequences for biotic diversity and ecological flows. Springer-Verlag, New York.

Hastie T, Tibshirani R (1990) Exploring the nature of covariate effects in the proportional hazards model. Biometrics 46:1005–1016.

He YH, Guo XL, Si BC (2007) Detecting grassland spatial variation by a wavelet approach. Int J Remote Sens 28:1527–1545.

Hoeting JA, Leecaster M, Bowden D (2000) An improved model for spatially correlated binary responses. J Agric Biol Environ Stat 5:102–114.

Iverson LR, Schwartz MW, Prasad AM (2004) Potential colonization of newly available tree-species habitat under climate change: an analysis for five eastern US species. Landsc Ecol 19:787–799.

Keitt TH, Fischer J (2006) Detection of scale-specific community dynamics using wavelets. Ecology 87:2895–2904.

Keitt TH, Urban DL (2005) Scale-specific inference using wavelets. Ecology 86:2497–2504.

Keitt TH, Bjornstad ON, Dixon PM, Citron-Pousty S (2002) Accounting for spatial pattern when modeling organism-environment interactions. Ecography 25:616–625.

Knapp EE, Keeley JE (2006) Heterogeneity in fire severity within early season and late season prescribed burns in a mixed-conifer forest. Int J Wildland Fire 15:37–45.

Krajina VJ (1969) Ecology of forest trees in British Columbia. Ecol West N Am 2:1–146.

Legendre P (1993) Spatial autocorrelation – trouble or new paradigm. Ecology 74:1659–1673.

Legendre P, Fortin M (1989) Spatial pattern and ecological analysis. Vegetatio 80:107–138.

Legendre P, Borcard D, Peres-Neto PR (2008) Analyzing or explaining beta diversity? Comment. Ecology 89:3238–3244.

Levin SA (1992) The problem of pattern and scale in ecology. Ecology 73:1943–1967.

Littell JS, Peterson DL, Tjoelker M (2008) Douglas-fir growth in mountain ecosystems: water limits tree growth from stand to region. Ecol Monogr 78:349–368.

Lookingbill TR, Urban DL (2003) Spatial estimation of air temperature differences for landscape-scale studies in montane environments. Agric For Meteorol 114:141–151.

Lookingbill TR, Urban DL (2004) An empirical approach towards improved spatial estimates of soil moisture for vegetation analysis. Landsc Ecol 19:417–433.

Lookingbill TR, Urban DL (2005) Gradient analysis, the next generation: towards more plant-relevant explanatory variables. Can J For Res 35:1744–1753.

Lookingbill TR, Zavala MA (2000) Spatial pattern of *Quercus ilex* and *Quercus pubescens* recruitment in *Pinus halepensis* dominated woodlands. J Veg Sci 11:607–612.

Lookingbill TR, Goldenberg NE, Williams BH (2004) Understory species as soil moisture indicators in Oregon's Western Cascades old-growth forests. Northwest Sci 78:214–224.

Lynch HJ, Renkin RA, Crabtree RL, Moorcroft PR (2006) The influence of previous mountain pine beetle (*Dendroctonus ponderosae*) activity on the 1988 Yellowstone fires. Ecosystems 9:1318–1327.

Manly BFJ (1991) Randomization and Monte Carlo methods in biology. Chapman and Hall, London.

Mantel N (1967) The detection of disease clustering and a generalized regression approach. Cancer Res 27:209–220.

McCune B, Grace JB (2002) Analysis of ecological communities. MjM Software Design, Gleneden Beach, OR.

McKee A (1998) Focus on field stations: H.J. Andrews Experimental Forest. Bull Ecol Soc Am 79:241–246.

McIntire EJB, Fajardo A (2009) Beyond description: the active and effective way to infer processes from spatial patterns. Ecology 90:46–56.

Mi XC, Ren HB, Ouyang ZS, Wei W, Ma KP (2005) The use of the Mexican Hat and the Morlet wavelets for detection of ecological patterns. Plant Ecol 179:1–19.

Miller C, Urban DL (1999) A model of surface fire, climate and forest pattern in Sierra Nevada, California. Ecol Modell 114:113–135.

Miller J (2005) Incorporating spatial dependence in predictive vegetation models: residual interpolation methods. Prof Geogr 57:169–184.

Miller J, Franklin J, Aspinall R (2007) Incorporating spatial dependence in predictive vegetation models. Ecol Model 202:225–242.

Mitchell AK, Koppenaal R, Goodmanson G, Benton R, Bown T (2007) Regenerating montane conifers with variable retention systems in a coastal British Columbia forest: 10-year results. For Ecol Manage 246:240–250.

Moeur M (1997) Spatial models of competition and gap dynamics in old-growth *Tsuga heterophylla* and *Thuja plicata* forests. For Ecol Manage 94: 175–186.

Moreno JM, Oechel WC (1992) Factors controlling postfire seedling establishment in southern California chaparral. Oecologia 90:50–60.

Morin X, Augspurger C, Chuine I (2007) Process-based modeling of tree species' distributions: what limits temperate tree species' range boundaries? Ecology 88:2280–2291.

Morin X, Viner D, Chuine I (2008) Tree species range shifts at a continental scale: new predictive insights from a process-based model. J Ecol 96:784–794.

Neilson RP (1991) Climatic constraints and issues of scale controlling regional biomes. In: Holland MM, Risser PG, Naiman RJ (eds) Ecotones: the role of landscape boundaries in the management and restoration of changing environments. Chapman and Hall, New York, pp. 31–51.

Neilson RP (1995) A model for predicting continental-scale vegetation distribution and water balance. Ecol Appl 5:362–385.

Neilson RP, Drapek RJ (1998) Potentially complex biosphere response to transient global warming. Glob Change Biol 4:505–521.

Noss RF (2001) Beyond Kyoto: forest management in a time of rapid climate change. Conserv Biol 15:578–590.

Nusser SM, Breidt EJ, Fuller WA (1998) Design and estimation for investigating the dynamics of natural resources. Ecol Appl 8:234–245.

Odion DC, Davis FW (2000) Fire, soil heating, and the formation of vegetation patterns in chaparral. Ecol Monogr 70:149–169.

Ogden RT (1997) Essential wavelets for statistical applications and data analysis. Birkhauser, Boston.

Ohmann JL, Spies TA (1998) Regional gradient analysis and spatial pattern of woody plant communities of Oregon forests. Ecol Monogr 68:151–182.

Peet RK (1981) Forest vegetation of the Colorado Front Range – composition and dynamics. Vegetatio 45:3–75.

Peres-Neto PR, Legendre P, Dray S, Borcard D (2006) Variation partitioning of species data matrices: estimation and comparison of fractions. Ecology 87:2614–2625.

Peters DP, Herrick JE, Urban DL, Gardner RH, Breshears DD (2004) Strategies for ecological extrapolation. Oikos 106:627–636.

Peters RL, Darling JD (1985) The greenhouse effect and nature reserve design. Bioscience 35:707–717.

Philippi T (2005) Adaptive cluster sampling for estimation of abundances within local populations of low-abundance plants. Ecology 86:1091–1100.

Pickett STA, White PS (1985) The ecology of natural disturbance and patch dynamics. Academic press, Orlando.

Pickett STA, Kolasa J, Armesto JJ, Collins SL (1989) The ecological concept of disturbance and its expression at various hierarchical levels. Oikos 54:129–136.

Risser PG (1995) The status of the science examining ecotones. Bioscience 45:318–325.

Rocca ME (2009) Fine-scale patchiness in fuel load can influence initial post-fire understory composition in a mixed conifer forest, Sequoia National Park, California. Nat Areas J 29:126–132.

Romme WH, Knight DH (1982) Landscape diversity – the concept applied to Yellowstone Park. Bioscience 32:664–670.

Rossi RE, Mulla DJ, Journel AG, Fran EH. (1992) Geostatistical tools for modeling and interpreting ecological spatial dependence. Ecol Monogr 62:277–314.

Schoennagel T, Veblen TT, Romme WH (2004) The interaction of fire, fuels, and climate across rocky mountain forests. Bioscience 54:661–676.

Schoennagel T, Smithwick AHE, Turner MG (2008) Landscape heterogeneity following large fires: insights from Yellowstone National Park, USA. Int J Wildland Fire 17:742–753.

Shreve F (1922) Conditions indirectly affecting vertical distribution on desert mountains. Ecology 3:269–274.

Smouse PE, Long JC, Sokal RR (1986) Multiple regression and correlation extensions of the Mantel test of matrix correspondence. Syst Zool 35:627–632.

Stephenson NL (1990) Climatic control of vegetation distribution: the role of the water balance. Am Nat 135:649–670.

Stephenson NL (1998) Actual evapotranspiration and deficit: biologically meaningful correlates of vegetation distribution across spatial scales. J Biogeogr 25:855–870.

Stephenson NL (1999) Reference conditions for giant sequoia forest restoration: structure, process, and precision. Ecol Appl 9:1253–1265.

Stevens Jr DL, Olsen AN (2004) Spatially balanced sampling of natural resources. J Am Stat Assoc 99:262–278.

Stohlgren TJ, Falkner MB, Schell LD (1995) A modified-Whittaker nested vegetation sampling method. Vegetatio 117:113–121.

Stohlgren TJ, Chong GW, Kalkhan MA, Schell LD (1997) Multiscale sampling of plant diversity: effects of minimum mapping unit size. Ecol Appl 7:1064–1074.

Theobald DM, Stevens DL, White D, Urquhart NS, Olsen AR, Norman JB (2007) Using GIS to generate spatially balanced random survey designs for natural resource applications. Environ Manage 40:134–146.

Thornburgh DA (1969) Dynamics of the true fir-hemlock forests of the west slope of the Washington Cascade Range. University of Washington, Seattle.

Torrence C, Compo GP (1998) A practical guide to wavelet analysis. Bull Am Meteorol Soc 79:61–78.

Turner MG, Romme WH, Gardner RH, Hargrove WW (1997) Effects of fire size and pattern on early succession in Yellowstone National Park. Ecol Monogr 67:411–433.

Turner MG, Baker WL, Peterson CJ, Peet RK (1998) Factors influencing succession: lessons from large, infrequent natural disturbances. Ecosystems 1:511–523.

Turner MG, Romme WH, Gardner RH (1999) Prefire heterogeneity, fire severity, and early post-fire plant reestablishment in subalpine forests of Yellowstone National Park, Wyoming. Int J Wildland Fire 9:21–36.

Underwood EC, Ustin SL, Ramirez CM (2007) A comparison of spatial and spectral image resolution for mapping invasive plants in coastal California. Environ Manage 39:63–83.
Urban DL (2000) Using model analysis to design monitoring programs for landscape management and impact assessment. Ecol Appl 10:1820–1832.
Urban DL, Harmon ME, Halpern CB (1993) Potential response in Pacific Northwestern forests to climate changes: effects of stand age and initial composition. Clim Change 23:247–266.
Urban DL, Miller C, Halpin PN, Stephenson NL (2000) Forest gradient response in Sierran landscapes: the physical template. Landsc Ecol 15:603–620.
Urban D, Goslee S, Pierce K, Lookingbill T (2002) Extending community ecology to landscapes. Ecoscience 9:200–212.
Wagner HH (2001) Spatial covariance in plant communities: integrating ordination, geostatistics, and variance testing. Ecology 84:1045–1057.
Wiens JA (1989) Spatial scaling in ecology. Funct Ecol 3:385–397.
Whittaker RH (1956) Vegetation of the Great Smoky Mountains. Ecol Monogr 26:1–80.
Whittaker RH (1960) Vegetation of the Siskiyou Mountains, Oregon and California. Ecol Monogr 30:279–338.
Whittaker RH (1967) Gradient analysis of vegetation. Biol Rev Camb Philos Soc 42:207–264.
Winkler JB, Schultz F (2000) Seasonal variation of snowcover: inexpensive method for automatically measuring snow depth. In: Schroeter B, Schlensog M, Green TGA (eds) New aspects in cryptogamic research: contributions in honour of Ludger Kappen. Biblitheca Lichenologica 75. Cramer, Berlin, pp. 381–338.
Wu J, Hobbs RJ (eds) (2007) Key topics in landscape ecology. Cambridge University Press, Cambridge.
Wu J, Loucks OL (1995) From balance of nature to hierarchical patch dynamics: a paradigm shift in ecology. Q Rev Biol 70:439–466.
Zeger SL, Liang KY (1986) Longitudinal data-analysis for discrete and continuous outcomes. Biometrics 42:121–130.

Chapter 8
Modeling Species Distribution and Change Using Random Forest

Jeffrey S. Evans, Melanie A. Murphy, Zachary A. Holden, and Samuel A. Cushman

8.1 Introduction

Although inference is a critical component in ecological modeling, the balance between accurate predictions and inference is the ultimate goal in ecological studies (Peters 1991; De'ath 2007). Practical applications of ecology in conservation planning, ecosystem assessment, and bio-diversity are highly dependent on very accurate spatial predictions of ecological process and spatial patterns (Millar et al. 2007). However, the complex nature of ecological systems hinders our ability to generate accurate models using the traditional frequentist data model (Breiman 2001a; Austin 2007). Well-defined issues in ecological modeling, such as complex non-linear interactions, spatial autocorrelation, high-dimensionality, non-stationary, historic signal, anisotropy, and scale contribute to problems that the frequentist data model has difficulty addressing (Olden et al. 2008). When one critically evaluates data used in ecological models, rarely do the data meet assumptions of independence, homoscedasticity, and multivariate normality (Breiman 2001a). This has caused constant reevaluation of modeling approaches and the effects of reoccurring issues such as spatial autocorrelation. Model misspecification problems such as the modifiable aerial unit (MAUP) (Cressie 1996; Dungan et al. 2002) and ecological fallacy (Robinson 1950) have also arisen as clearly defined challenges to ecological modeling and inference.

Expert knowledge and well-formulated hypotheses can lend considerable insight into ecological relationships. However, given the complexities in ecological systems, it may be difficult to develop hypotheses or to select variables without first knowing if there is a correlative relationship. These correlative relationships may be highly non-linear, exhibit autocorrelation, be scale dependent, or function as an interaction with another variable. To uncover these relationships non-parametric data mining approaches provide obvious advantages (Olden et al. 2008). Scale can

---

J.S. Evans (✉)
The Nature Conservancy, North America Science, Fort Collins, CO 80524, USA
e-mail: jeffrey_evans@tnc.org

generate complex interactions in space and time that are inherently unobservable, given standard sample designs and modeling approaches (Wiens 1989; Dungan et al. 2002). Machine learning provides a framework for identifying these variables, building accurate predictions, and exploring mechanistic relationships identified in the model. We advocate performing a critical evaluation of variables used in a model, with careful *a priori* selection of those variables believed to directly relate a proposed explanatory hypothesis linking mechanisms to responses. Developing theories on mechanistic relationships that can account for non-linear variable interaction and process across scale is a particular challenge. Machine learning can provide a starting point for these investigations. The unique advantage of machine learning is that complex relationships and spatial patterns can be discovered more readily than in the traditional probability data model that assumes normality. A scientist should not stop at the identification of key predictor variables using a machine learning approach, but treat this as a starting point to develop and test new theory in an experimental framework.

The issue of machine learning in ecology is as contentious as the frequentists/ Bayesian debate (Cressie et al. 2009; Lele and Dennis 2009). We do not wish to inflame this argument, but unfortunately due to the nature of proposing a fundamental paradigm shift in ecological modeling, it is difficult not to offend certain sensibilities. We are not recommending the abandonment of well established methods; quite the contrary. If the goal of an analysis is prediction rather than formal explanation of hypotheses, machine learning provides a set of tools that can dramatically improve results. We hope to provide insight into ways that machine learning can augment our current toolbox and hope that broader application of algorithmic modeling will lead to increased understanding of ecological system and the development of new ecological theory. The goal of this chapter is to illustrate the emerging and potential role that machine learning approaches can assume in ecological studies and to introduce a powerful new model, Random Forest (Breiman 2001b; Cutler et al. 2007; Rogan et al. 2008), that is becoming an important addition to ecological studies. We provide a case study of species distribution modeling using the Random Forest model. As well, we illustrate the utility of Random Forest for exploring the impact of climate change by projecting the model into new climate space.

8.2 Ecological Theory and Statistical Framework

An emerging consensus in quantitative ecology is that spatial complexity across scale fundamentally alters pattern–process relationships (Wiens 1989; Dungan et al. 2002). This has major implications for both research and management. Effective and informed management decisions depend on accurate and precise estimates of current ecological conditions and reliable predictions of future changes (Austin 2007). Given the complexity across space and time inherent in high-dimensional ecological data, there is a critical need for an expanded statistical framework for ecological analysis based on ecological informatics (Park and Chon 2007).

The fusion of individualistic community ecology (Gleason 1926; Curtis and McIntosh 1951; Whittaker1967) with the Hutchinsonian niche concept (Hutchinson 1957) and its extension to spatially complex and temporally dynamic systems provides an excellent example of the application of powerful non-parametric analytical methods to this new paradigm. Each species responds to local environmental and biotic conditions and the biotic community is an emergent collective of species that are occurring together at a particular place and a particular time due to overlapping tolerances of environmental conditions and vagaries of history, rather than an integrated and deterministic mixture (Whitaker 1967; McGarigal and Cushman 2005). The natural level of focus of such analyses is the species, not community type, assemblage, or patch type; the natural focal scale for such analyses is the location, rather than the stand or patch (McGarigal and Cushman 2005; Cushman et al. 2008).

Adopting a species-level, gradient paradigm based on application of powerful and flexible algorithmic models greatly improves predictions for management, thereby allowing for improved decision making (Evans and Cushman 2009). For example, accurate prediction of current species distributions is a foundation for many management decisions (see Chap. 14). Decisions based on these highly accurate predictions are facilitated by flexible, algorithmic modeling approaches based on combinations of topographic and climatic limiting factors (Evans and Cushman 2009). In contrast with traditional approaches that classify data into categorical community assemblages (i.e., remote sensing) or presence/absence (niche models), predictions that are continuous in nature (i.e., proportion, probability) provide considerable more sensitivity to inference. The very nature of many classification schemas ignores scale, convolves the results, and potentially introduces aggregation errors, reducing the flexibility of data use for meaningful ecological inference (Chesson 1981; Cushman et al. 2008). Given the rapid changes in climate, disturbance regimes, and human impact on ecosystems, reliable prediction of future ecological change is equally important as an understanding of current conditions. The prediction of future changes faces the additional complexity of predicting the decoupling of species responses as responding on an individual level to alterations in limiting factors and disturbance regimes. Species-level predictions are necessary to address this non-equilibrium (Cushman et al. 2008; Evans and Cushman 2009). Whereas reliably predicting complex interactions between species and environmental change over space and time is extremely difficult (Dungan et al. 2002; Austin 2007), algorithmic modeling approaches applied to multi-scale gradient databases provide an effective tool, in that they can address complex interactions, non-linearity, and are ideally suited to modeling individualistic species responses in dynamic, high-dimensional systems (Olden et al. 2008).

A great strength of algorithmic methods, such as Random Forest, is their ability to identify and explore non-intuitive relationships. In traditional inferential approaches the scientist is trained to start by proposing an *a priori* hypothesis relating mechanisms to responses and then using inferential statistics to determine if the hypothesis can be rejected given the observed patterns in data. Whereas this has certain advantages, it does assume that the scientist is able to formulate the correct hypothesis prior to exploring the data (Cook and Campbell 1979). However, nature is full of surprises and complex ecological systems often behave in non-intuitive ways that defy our *a priori* expectations. Algorithmic methods provide a tool that facilitates development of ecological theory through an iterative process of exploration and discovery, followed by hypothesis generation and testing. Specifically, the flexibility of algorithmic approaches – such as Random Forest – to handle complex, high-dimensional interactions allows them to discover relationships that are hidden in traditional parametric analysis and are unlikely to be proposed *a priori* by a non-omniscient observer. The patterns and relationships identified then provide fertile material that the scientist may use to fashion new explanations and develop new theories.

8.3 Random Forest

8.3.1 Classification and Regression Trees

Classification and Regression Trees (CARTs) (Breiman et al. 1983) have gained prominence in both ecology and remote sensing due to their easy of interpretaion and ability to address data that interacts in a non-linear or hierarchical manner (Breiman et al. 1983; De'ath and Fabricius 2000; Rogan et al. 2008). CARTs are a binary recursive partitioning approach where the response is iteratively partitioned into nodes based on a measure of impurity (e.g., Gini index, sum of squares, entropy information). Each partition represents exclusive portions (groups) of the variance that are as homogeneous as possible in relation to the response variable. The algorithm identifies the best candidate split (parent node) that minimizes the mean impurity of the two derived child nodes. When no improvement can be made in a given partition a terminal node is defined, thus ending that branch. Splitting is terminated when all observations within each node have the identical distribution of independent variables, making splitting impossible. Cumulatively, nodes represent a set of rules that can be run down the tree to make a prediction.

The hierarchical nature of CARTs is an attractive quality of derived predictions. Local variation is represented within each branch, whereas the global trend is accounted for when the prediction is voted down the tree (Breiman et al. 1983). Although not empirically demonstrated in the literature, we postulate that this balance between global and local trend can account for non-stationarity and anisotropy. There are many additional advantages of CARTs, including the fact that CARTs are non-parametric and not subject to distributional assumptions; do not require transformations; can use categorical, ordinal, and continuous data simultaneously; are invariant to outliers; are capable of identifying and incorporating complex variable interaction and; are capable of handling high-dimensional data. With these advantages come three major drawbacks: (1) CARTs are subject to severe over-fit, (2) due to the sensitivity of CARTs to the complexity (α) parameter, the final tree may not be the optimal solution (Sutton 2005), and (3) CARTs can exhibit high variance and small changes in the data can result in different splits making interpretation somewhat unstable (Hastie et al. 2009). In addition to these two issues, colinearity between independent variables can cause difficulties in interpretation and bias in the Gini index (Sutton 2005). To address these limitations, ensemble learning approaches including, Bagging (Breiman 1996), Boosting (Freund and Schapire 1996; Friedman 2001; De'ath 2007), and Random Forest (Breiman 2001b) were developed.

8.3.2 Random Forest Algorithm

Random Forest (Breiman 2001b) is an algorithm that developed out of CART and bagging approaches. This algorithm is gaining prominence in remote sensing (Lawrence et al. 2006), forestry (Falkowski et al. 2009), ecology (Cutler et al. 2007; Evans and Cushman 2009; Murphy et al. 2010), and climate change (Prasad et al. 2006; Rehfeldt et al. 2006). By generating a set of weak-learners based on a bootstrap of the data, the algorithm converges on an optimal solution while avoiding issues related to CARTs and parametric statistics. Breiman (2001b) defines Random Forest as a collection of tree-structured weak learners comprised of identically distributed random vectors where each tree contributes to a prediction for x. Ensemble-based weak learning hinges on diversity and minimal correlation between learners. Diversity in Random Forest is obtained through a Bootstrap of training, randomly drawing selection of M (independent variables) at each node (defined as m), and retaining the variable that provides the most information content. To calculate variable importance, improvement in the error is calculated at each node for each randomly selected variable and a ratio is calculated across all nodes in the forest (Fig. 8.1).

The algorithm can be explained by:

1. Iteratively construct N Bootstraps (with replacement) of size n (36%) sampled from Z, where N is number of Bootstrap replicates (trees to grow) and Z is the population to draw a Bootstrap sample from.
2. Grow a random-forest tree T_b at each node randomly select m variables from M to permute through each node to find best split by using the Gini entropy index to assess information content and purity. Grow each tree to full extent with no pruning (e.g., no complexity parameter).
3. Using withheld data (OOB, out-of-bag) to validate each random tree T_b (for classification OOB Error; for regression pseudo R^2 and mean squared error).
4. Output ensemble of random-forest trees

$$\{T_b\}_1^B$$

5. To make a prediction for a new observation x_i:

   Regression:

   $$\hat{f}_{rf}^B(x) = \frac{1}{B}\sum_{b=1}^{B} T_b(x)$$

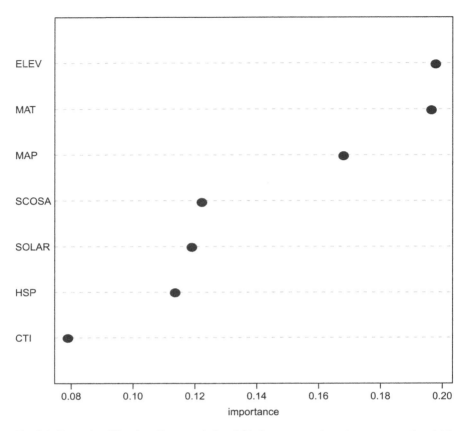

Fig. 8.1 Example of Random Forest scaled variable importance plot using permutated variable mean increase in error for all x variables in *A. lasiocarpa* model

Classification: Let $\hat{C}_b(x)$ be the class prediction of the *B*th random-forests tree then

$$\hat{C}_{rf}^{B}(x) = \text{majorityvote}\left\{\hat{C}_b(x)\right\}\frac{1}{B}$$

Even though the performance of an individual tree will improve given an increase of variables randomly permutated through a node (m), the correlation of trees is increased, reducing the overall performance of the model. Commonly, the optimal m is defined for classification problems as sqrt (M); and for regression $M/3$, where M is a pool of independent variables. It has been demonstrated that Random Forest is robust to noise even given a very large number of independent variables (Breiman 2001a; Hastie et al. 2009). Hastie et al. (2009) showed that with six relevant and 100 noise variables the probability of selecting a relevant variable at any given split is $p = 0.46$. Because the algorithm's power is not affected by degrees of freedom it is possible to specify considerable more independent variables (x) than observations of the dependent variable (y). Whereas Random Forest is not subject to over-fit (Breiman 2001a), caution should be made to not over correlate or inflate the variance of a Random Forest ensemble.

8.3.3 Model Selection

Parsimony is an underlying requirement in the frequentist modeling effort; however, it is rarely addressed in machine learning. The primary motivations for parsimony in frequentist approaches are maintaining well defined hypothesis and reducing the risk of over-fit; the impetus for parsimony in machine learning is motivated by model performance and interpretability (Murphy et al. 2010). Models with fewer variables are much easier to interpret and easier to apply *post-hoc* exploratory techniques to. We have observed another reason for seeking parsimony in a Random Forest models – model performance. In applying model selection we have seen a marked improvement in model fit and predictive performance. There are two explanations for this. First, when Random Forest is run with a large M, but the number of variables that actually provide signal to the data is relatively small, Random Forest is likely to perform poorly with a small m (Hastie et al. 2009). However, if you arbitrarily increase m you risk correlating the ensemble learners. By reducing M to a subset of variables with a signal you improve the overall performance of the model. Second, as spurious variables are removed, trees become much shallower (simpler). This in turn reduces the size of the plurality vote matrix by reducing votes that account for noise, resulting in a higher signal to noise ratio and overall reduction in error (Evans and Cushman 2009; Falkowski et al. 2009; Murphy et al. 2010). Colinearity and multi-colinearity problems can also influence model performance and interpretability (Murphy et al. 2010).

We draw a distinction between variable and model selection. For example, in studies aimed at gene expression, the final goal of utilizing Random Forest is to identify a subset of genes that best describe a particular trait (Díaz-Uriarte and Alvarez de Andrés 2006). In short, the results of the analysis are the final selected variables and not inference or prediction. In ecological models we are not only interested in describing a process but also in inference and prediction. The distinction is drawn around the fact that variable selection approaches are not seeking parsimony but rather reductionism. These approaches are overly aggressive and often result in too few variables to explain a process comprehensively. If a variable has very high explanatory power, Random Forest can exhibit a good fit given a single variable. It is important to seek a parsimonious set of variables, that when predicted to a landscape not only will provide a good fit but also adequately represent the complexities of spatial pattern.

Murphy et al. (2010) developed a model selection approach that uses the permuted variable importance measures and model optimization to select a parsimonious model. The procedure standardizes the importance values to a ratio and iteratively subsets variables within a given ratio, running a new model for each subset of variables. Each resulting model is compared with the original model, which is held fixed. Model selection is achieved by optimizing model performance based on a minimization of both "out-of-bag" error and largest "within-class" error for classification or maximizing variance explained and minimizing mean squared error in regression. There is also an optional penalty for the number of parameters that will select the model with the fewest number of parameters from closely competing models. There are other simple approaches, including a "leave one variable out" and test performance of sub-models, or simply sub-setting variables based on the importance values and running a single new model with the strongest variables. This may not be as empirically driven as a formalized model selection procedure, but will often result in a more interpretable model.

8.3.4 Imbalanced Data

One issue with Random Forest arises within classification problems when classes in categorical response variables are imbalanced (Chen et al. 2004). Imbalances in the response variable result in biased classification accuracy. This is due to the bootstrap over-representing the majority class, leading to under-prediction of the minority class. The resulting model fit is deceptive – exhibiting very small overall OOB error due to very small errors in the majority class as a result of extremely high cross-classification error from the minority-class. With highly skewed data there is a possibility that this same problem could arise in regression problems, but to date there is no published work that has tested this. However, Jiménez-Valverde and Lobo (2006) imply that unbalanced samples are not as serious a problem as historically thought. Despite this, the Bootstrap approach to generate weak-learners in Random Forest causes additional issues not seen in other modeling approaches. Due to minority samples not being drawn with the same frequency as the majority class, a prediction bias is given to the majority class, thus an adequate picture of model fit is not provided. Historically, there are three common ways to address imbalanced data: (1) assign a high cost to misclassification of the minority class (2) down-sample the majority class (Kubat et al. 1998), and (3) over-sample the minority class (Chawla et al. 2002). Chen et al. (2004) proposed the addition of class weights, making Random Forest cost sensitive.

To address this problem, Evans and Cushman (2009) developed a novel approach to balance the response variable that iteratively down-samples the majority class by randomly drawing $2*[n$ of minority$]$ from the majority class and running a new Random Forest model iteratively using different random subsets while holding the sample-size of the minority-class constant. To ensure that the distributions of the independent variables in each sub-sample matched distribution in the full data, the covariance of each sub-sampled model is tested against the covariance in full data using a matrix equality test (Morrison 2002). The final ensemble model is built by combining trees from all the resulting down-sampled Random Forest models. Because the underlying theory of Random Forest is ensemble learning, it is possible to combine trees from different models to make a prediction using the combined plurality votes-matrix of the full ensemble (Breiman 2001a).

8.3.5 Model Validation

Two common methods for evaluation models in machine learning classification problems are the Kappa statistic and the Area Under the Curve (AUC) of a Receiver Operator Characteristic (ROC) (Fawcett 2006). ROC is defined as the sensitivity plotted against [1 − specificity]. Sensitivity indicates the proportion of true positives [a/(a+c)] and specificity the proportion of false negatives (commission error) [b/(b+d)]. The balance between sensitivity and specificity is a indication of model performance at a class level. The AUC indicates the area under a ROC, ranging from 0 to 1 (so that 0.5 indicates no discrimination and 1.0 perfect classification). Caution should be used when interpreting the ROC/AUC because; (1) error components are equally weighted (Peterson et al. 2008), (2) models can over-value models of rare species (Manel et al. 2001), and (3) certain models do not predict across the spectrum of probabilities violating the assumption that the specificity spans the entire range of probabilities (0–1). Peterson et al. (2008) proposed modifications to ROC by calculating a partial ROC that limits the x-axis to the domain to each specific model. Manel et al. (2001) recommends the Kappa statistic as an alternative to ROC/AUC. The Kappa (Cohen 1960; Monserud and Leemans 1992) evaluates the agreement between classes (binary or multiple) adjusting random chance agreement. Because the Kappa does not account for the expected frequency of a class and does not make distinctions among various types and sources of disagreement the weighted Kappa was developed (Cohen 1968) The incorporation of a weight allows for near agreement and adjusts for expectation in the frequency of observations.

It has continually been stated that one compelling component of Random Forest is that there is no need for independent validation. The model error is assessed against the OOB data in each Bootstrap replicate, providing an error distribution. Although we agree with the robustness of this approach to test model sensitivity to sample distribution, we also advocate the addition of simple data-withhold cross validation techniques. An additional validation procedure we commonly apply is a randomization procedure where the independent data are randomized, and the model is run and the error tabulated. This is performed a large number of times (i.e., 1,000), providing an error distribution that the model under scrutiny can be compared to, thus providing a significance value (Evans and Cushman 2009; Murphy et al. 2010).

8.3.6 Visualization

There are many visualization techniques that can be employed to explore mechanistic relationships, variable interaction, and model performance. Conditional density estimates (Hall et al. 1999; Falkowski et al. 2009) (Fig. 8.2a), partial dependence plots (Friedman 2001) (Fig. 8.2b), bivariate kernel density estimates (Simonoff 1998) (Fig. 8.3), and multi-dimensional scaling (Cox and Cox 1994) plots to explore mechanistic relationships, both before and after analysis. With discrete data, conditional density plots can be used to explore the influence of a given independent variable on a set of responses. In our example we use elevation

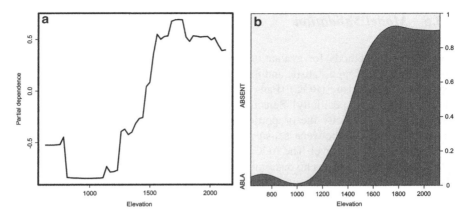

Fig. 8.2 Methods for exploring mechanistic relationships and variable interaction in categorical data. (**a**) Partial plot of elevation and presence of *A. lasiocarpa* and (**b**) conditional density plot of elevation of presence/absence for *A. lasiocarpa*

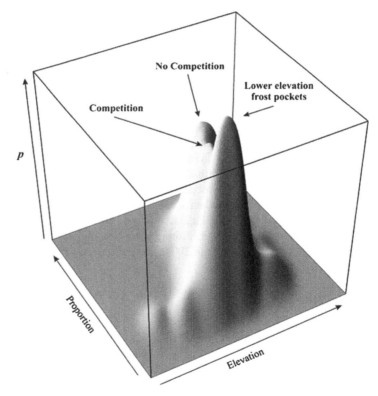

Fig. 8.3 Method for exploring the mechanistic relationships in continuous data. Non-parametric bivariate Kernel Density Estimate (KDE) plot of proportion of *A. lasiocarpa* and elevation, illustrating competition with *Picea engelmannii* (Engelmann spruce) across an elevational gradient

and examine its influence on presence/absence of *Abies lasiocarpa* (subalpine fir) (Fig. 8.2b). With continuous independent and response variables, nonparametric bivariate and multivariate kernel density estimates can be used (Fig. 8.3). This is an effective exploratory method for visualizing gradient relationships.

By partialing out the average effect of all other variables we can explore the influence of a given variable on the probability of occurrence of this species (Fig. 8.2a). This method can also be applied to multiple independent variables to explore variable interaction (Cutler et al. 2007). A proximity matrix is created within Random Forest by running in-bag and out-of-bag down each tree, if cases 1 and 2 both result in the same terminal node, then the proximity is increased by one.

The normalized proximity is calculated by dividing by the number of trees in the ensemble. The pairwise values in the proximity matrix can be treated as a dissimilarity or distance measure. This measure of dissimilarity can be used to visualize the separability of classes using multi-dimensional scaling plots. Crookston and Finley (2008) derived a method for using the scaled proximities to perform nearest neighbor multiple-imputation.

8.3.7 Spatial Structure

It should be noted that although non-parametric models do not assume independence and thus, are not affected by spatial-autocorrelation, they also do not explicitly incorporate spatial structure. There are a few proposed methods to account for spatial structure that range from naïve trend to direct incorporation of spatial structure. The most simple approach is to incorporate geographic coordinates or a variable that indicate the trend of geographic space (Legendre and Legendre 1998; Chefaoui and Lobo 2007). More complex approaches include incorporating a function into a model that allows for a spatial lag, effectively acting as a die-off function (Mouer and Riemann 1999), or adding a distance matrix of observed samples to act as a spatial weight (Allouche et al. 2008).

8.4 Data Suitability

8.4.1 Dependent Variable

Statistical issues relating to spatial aggregated data, such as photo-interpreted stands or landcover classifications based on remote sensing efforts, have been emerging (Chesson 1981; Cressie 1996; Dungan et al. 2002; Cushman et al. 2008), justifying a trend toward individual species gradient models. Attempting to use polygon data to build niche models overly smoothes or misrepresents the underlying variation of variables used to construct the niche hypervolume. To support a gradient modeling approach, spatially-referenced plot level data are necessary

(Evans and Cushman 2009). A sample design that captures spatial and statistical variability in both dependent and independent variables is critical to ensure that the model provides an adequate representation of the ecological niche and is capable of landscape level prediction. Unfortunately, few data-collection efforts are comprehensive or designed to capture fine-scale spatial variability over entire landscapes, making it necessary to plan extensive data collection efforts into a study.

8.4.2 Independent (Predictor) Variables

In mountainous terrains patterns of precipitation and topography are the primary drivers of species occurrence and of the formation of plant communities (Whittaker and Niering 1975; Costa et al. 2008). Precipitation and timing interact with a variety of processes (i.e., temperature, solar radiation) to determine temperature and moisture regimes. These, in combination with biotic factors, competition, and disturbance, largely determine vegetation composition and structure.

Independent variables that influence species occurrence can be grouped into direct and indirect predictors (Guisan and Zimmermann 2000). Direct predictors include soil characteristics, temperature, precipitation and solar radiation – which directly influence characteristics of the physical environment and its suitability for individual species. Indirect predictors include geomorphometric surrogates such as elevation, aspect, slope, and slope position and are effective surrogates for some of the driving physical variables that influence vegetation distribution and abundance. Direct predictors have several critical advantages over indirect measures. First, using indirect geomorphometric surrogates rather than limiting variables such as temperature add an extra inferential step in interpreting species–environment relations. Second, and most importantly; the use of direct measures allow for projection into future climate space.

8.5 Prediction of Current and Future Species Distributions

In this case study, we demonstrate the application of the Random Forest algorithm for predicting current and potential future distribution of plant species (Prasad et al. 2006; Rehfeldt et al. 2006). We focus on two plant species, *A. lasiocarpa* (subalpine fir) and *Pseudotsuga menziesii* (Douglas fir). We used 30-year normalized mean annual temperature (MAT) and precipitation (MAP) predictions from the spline climate model presented in Rehfeldt et al. (2006). We combine the two climate variables with variables describing slope position (HSP, Murphy et al. 2010), long-wave solar radiation (INSO, Fu and Rich 1999), a slope/aspect transformation (SCOSA, Stage 1976), and a wetness index (CTI, Moore et al. 1993) to develop a limiting-factor niche model for these two species. Finally, we apply these niche models across a topographically complex landscape to predict the occurrence of these species. We characterize the structure of the realized niche of these species and develop fine-scale species distribution maps based on predicted climate–species relationships. We focus on species-level predictions because multi-species, community level analyses fail to optimally predict any given species and may not be useful for extrapolation into novel climates where communities plausibly disassemble due to the trajectories of individual species responses.

8.5.1 Study Area

We use species occurrence and abundance data from 411 vegetation plots systematically spaced at 1.7-km intervals across the Bonners Ferry Ranger District, on the Panhandle National Forest in Northern Idaho, USA. These plots were established by the US Forest Service as part of a pilot project aimed to intensify the USFS Forest Inventory and Analysis grid. Our 3,883 km^2 study area encompasses portions of the Selkirk and Purcell mountain ranges, with elevations ranging from 630 to 2,600 m. *Pinus ponderosa* and *P. menziesii* occur at lower elevations and south-facing slopes and *Picea engelmanii* and *A. lasiocarpa* occupy higher elevation sites. The study area and vegetation sampling methods are described in detail by Evans and Cushman (2009).

8.6 Methods

We built our Random Forest model using the Random Forest (Liaw and Wiener 2002) package available in R (R Development Core Team 2009). This allowed us the flexibility to program customized model selection (Murphy et al. 2010), validation (Evans and Cushman 2009), and visualization routines. We focused directly on the key limiting environmental gradients of temperature, solar energy, and water availability, rather than variables that would provide indirect measures of process. We predicted the distribution of the two tree species at fine spatial scales (30 m^2) corresponding to the dominant scale at which species interact with limiting environmental resources. By constructing species-level, fine-grain, limiting-factor predictive models, we hope to contribute to improved accuracy of predicting the distributional shifts in species occurrence with the advance of changing climate.

To predict the distribution of species under potential future climates, we fixed the contemporary (2000) climate-niche model and predicted the model into new climate space (2040 and 2080). We perturbed our two climate variables using a weighted mean from 20 GCMs (General Circulation Models) (McGuffie and Henderson-Sellers 1997). For the 2040 climate we increased the temperature by 2°C and precipitation by 2%, for the 2080 climate we increased temperature by 3.39°C and precipitation by 4.3%. We are, in effect, shifting the climate niche space and projecting the resulting changes in distribution.

8.6.1 Uncertainty

There is very large variation in GCMs ranging from positive to negative change. This provides considerably uncertainty in future climate spaces. A common approach to address this variation is to derive a weighted mean across all 20 GCMs. Although justifiable, the weighted mean collapses variability among model scenarios, thus affecting model sensitivity. To evaluate model sensitivity in future climate spaces, we implemented a Monte Carlo simulation, generating a simulation envelope through randomization of climate variables. We used the range of 20 GCMs to parameterize the simulation for each year; 2040 min-temp=+1.26, 2040 max-temp=+2.9, 2040 min-precipitation=−8.9%, 2040 max-precipitation=+9%, 2080 min-temp=+1.91, 2080 max-temp=+5.93, 2080 min-precipitation=−9.9%, 2080 max-precipitation=+19.6%. The climate data were randomized ($n=9,999$) with the GCM ensemble range for each year and the Random Forest model representing the contemporary climate re-predicted into the randomized climate space. The simulation envelope was created using the predicted probability from each randomization.

8.7 Results

The results for the contemporary distribution model (Fig. 8.4a, b) provided very good model fits for both species. The OOB error for the *A. lasiocarpa* model was 10.11% with classification error equally balanced between presence and absence classes. Back-prediction to the data demonstrated an almost perfect fit with AUC of 0.99 and a 1% error rate. The *P. menziesii* model exhibited a 14.83% OOB error with a slightly higher (4%) error rate for the presence class. Back-prediction provided an AUC of 0.98 and a 3% error rate. One-thousand cross-validations with a 10% data-withhold provided <2% error rates for both models.

The two future-climate projections provided intuitive results very consistent with the ecological gradient of each species (Fig. 8.4). Both species demonstrate contraction of the predicted distribution (Fig. 8.4). The 2040 *A. lasiocarpa* projection (Fig. 8.4c) exhibited a 27.45% contraction and a 36.40% contraction for 2080 (Fig. 8.4e). Both time-steps show *A. lasiocarpa* receding into higher elevations. *A. lasiocarpa* is highly influenced by both temperature and moisture, as it prefers cold temperatures with high levels of precipitation. With increased temperature, *A. lasiocarpa* is constrained to higher elevations than its current range. The *P. menziesii* results are considerably more dramatic, showing a 66% decrease in 2040 (Fig. 8.4d) and an 80.92% decrease in 2080 (Fig. 8.4f). Across its range *P. menziesii* is fairly opportunistic, capable of occupying a wide range of environment conditions. In our study area *P. menziesii* occurs in a very hot portion of its gradient. With increased temperatures, *P. menziesii* exceeds it temperature tolerance in short order, showing considerable loss by 2040 (Fig. 8.4d).

The Monte Carlo simulation indicated that there is considerable uncertainty using the weighted-mean of the GCMs in the 2040 time period (Fig. 8.5a, c).

Fig. 8.4 (**a**) *A. lasiocarpa* contemporary (2000) climate presence, (**b**) *P. menziesii* contemporary (2000) climate presence, (**c**) Projected *A. lasiocarpa* 2040 climate presence, (**d**) Projected *P. menziesii* 2040 climate presence, (**e**) Projected *A. lasiocarpa* 2080 climate presence, (**f**) Projected *P. menziesii* 2080 climate presence

The 2040 predicted probability distribution is very poorly matched for the *P. menziesii* model (Fig. 8.5c) exhibiting an extremely mismatched distribution well outside the simulation envelope. The 2040 *A. lasiocarpa* model exhibited the correct distributional shape but falls outside the simulation envelope (Fig. 8.5a). In the 2080 *P. menziesii* projection, the simulation envelope shows considerable stochasticity with the projected probability distribution matching the shape but falling outside the envelope (Fig. 8.5d). In contrast, the *A. lasiocarpa* model for 2080 shows a projection well matched with the simulation envelope (Fig. 8.5b). Overall results demonstrate that there is a certain amount of uncertainty in using the weighted-mean GCM values in all but the *A. lasiocarpa* 2080 model (Fig. 8.5b).

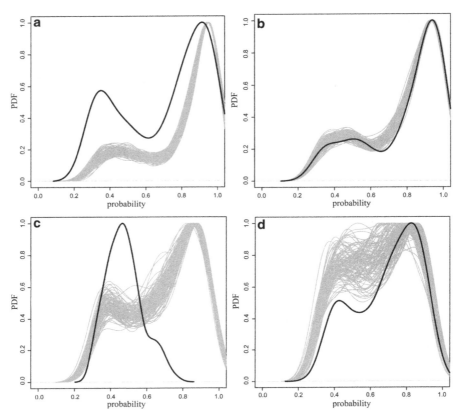

Fig. 8.5 Monte Carlo simulation of GCM's to quantify uncertainty. (**a**) *A. lasiocarpa* 2040 simulated probabilities, (**b**) *A. lasiocarpa* 2080 simulated probabilities, (**c**) *P. menziesii* 2040 simulated probabilities, (**d**) *P. menziesii* 2080 simulated probabilities. The black line represents the probability distribution using the weighted-mean GCM and the grey lines represent the simulation envelope probability distributions from the Monte Carlo

This uncertainty is acceptable in all but the *P. menziesii* 2040 model (Fig. 8.5c) that exhibits a very poorly matched distribution. This simple implementation of Monte Carlo simulation allowed us to examine the stability of our projections into future climate space. These results illustrate the necessity of examining the effect of selected values used to perturb climate.

To illustrate the utility of nonparametric bivariate kernel density estimates we examined an environmental gradient of *A. lasiocarpa* proportion and elevation (Fig. 8.3). It is clear that elevation highly influences the proportion of *A. lasiocarpa*; however, there is an obvious discontinuity in the mid-range of the elevational gradient. Exploration of this discontinuity uncovered a competition with *Picea engelmannii* (Engelmann spruce). Occurrence of mid-elevation frost pockets exclude *P. engelmannii* in the lower elevation limits of *A. lasiocarpa*. The conditional density and partial plots (Fig. 8.2) both support this inference.

8.8 Discussion

The details of landscape structure influence relationships between forest ecosystems, climate, and disturbance regimes in complex and interacting ways. Fine-scale environmental structure has a strong influence on species distribution, dominance, and succession (Whitaker 1967; Tilman 1982; ter Braak and Prentice 2004). The biophysical context of a location within a landscape also strongly influences growth rates and regeneration (Bunn et al. 2005). Furthermore, the probability of disturbances (Runkle 1985; Risser 1987), and the patterns of recovery (Finegan 1984; Glenn and Collins 1992) are strongly dependent on the pattern of environmental variation across the landscape. In addition, each species has an ecological response to variations in these environmental conditions, characterized by its ecological niche.

In this case study, we adopt a gradient perspective in lieu of hierarchical models of system organization (McGarigal et al. 2009). With a gradient perspective, a system of hierarchically organized aggregate subsystems is not assumed. Rather, emphasis is on directly measuring the response variables and the factors that drive their behavior, and modeling the relationships between them across space at the dominant scale of their operational environment.

By focusing directly on fine-scale, species-level responses to limiting ecological factors, we can describe interesting details of the niche structure of these species. For example, our analysis shows that, contrary to much of classic ecological theory, many species exhibit multimodal niche structure along limiting environmental gradients. In addition, our analysis enables quantification of the degree of niche overlap and environmental partitioning among species. By focusing on the species-level to identify limiting ecological variables, we feel these results allow for more reliable projection into future environments with altered climate regimes. Had we only focused on measures of the *abiotic* substraight rather than *biotic* limiting factors such as temperature, it would have been virtually impossible to perturb these variables to represent future climate space. It should be noted that our analysis was designed to illustrate the utility of nonparametric modeling methods. There are severe limitations to envelope approaches that are related to scale mismatches between fine-scale species responses and coarse-scale GCM climate models (Randin et al. 2009; Willis and Bhagwat 2009). These limitations are somewhat mitigated by the inclusion of geomorphometric variables that represent micro-topographic influences that interact with climate so as to better represent fine-scale spatial variation.

Resource management decisions should be made in the context of potential future climate change impacts. Dynamic, climate-adapted, spatial predictions of vegetation, based on the physical variables that ultimately drive species occurrence, are needed to understand and predict changes in future vegetation. The analysis presented here represents an important step toward developing tools that will allow us to accurately predict changes in species distribution with continuing climate change. Vegetation distribution is already changing (Rehfeldt et al. 2006; McKenney et al. 2007; Iverson et al. 2008) in response to changes in climate. However, precisely how an individual species will move across the landscape at local scales while interacting with the environment and disturbance is unknown. A key step in this direction is developing high-resolution models that can integrate highly accurate, fine-scale, species-level vegetation predictions with associated future climate projections. Whereas the spatial distributions of vegetation will certainly shift, many variables may interact in nonlinear ways to influence the potential suitability of the biophysical environment for an individual plant species. In the two species presented in this chapter, we observed that responses to climate change can be highly variable and species specific (Willis and Bhagwat 2009), contradicting the notion that vegetation communities respond to climate change in unison. Machine learning approaches that can account for the many complexities in these systems will allow us to explore the potential impacts of changing climates on future tree species distributions at scales that match the dominant biological governing processes, such as dispersal, initiation, and growth.

Acknowledgments Funding for this research was provided by the USDA Forest Service, Rocky Mountain Research Station and The Nature Conservancy. The authors would like to thank G. Rehfeldt, A. Hudak, N. Crookston, L. Iverson, and A. Cutler for valuable discussion on Random Forest and species distribution modeling and A. Prasad, J. Kiesecker and two anonymous reviewers for comments that strengthened this chapter. Additionally we would like to thank the editors for their patience and perseverance in seeing this book published.

References

Allouche O, Steinitz O, Rotem D, Rosenfeld A, Kadmon R (2008) Incorporating distance constraints into species distribution models. J Appl Ecol 45:599–609.
Austin M (2007) Species distribution models and ecological theory: a critical assessment and some possible new approaches. Ecol Modell 200:1–19.
Breiman L, Friedman JH, Olshen RA, Stone CJ (1983) Classification and regression trees. Wadsworth, London.
Breiman L (1996) Bagging predictors. Mach Learn 24:123–140.
Breiman L (2001a) Statistical modeling: the two cultures. Stat Sci 16:199–231.
Breiman L (2001b) Random forests. Mach Learn 45:5–32.
Bunn AG, Graumlich LJ, Urban DL (2005) Trends in twentieth-century tree growth at high elevations in the Sierra Nevada and White Mountains, USA. Holocene 15:481–488.
Chawla NV, Bowyer KW, Hall LO, Kegelmeyer WP (2002) SMOTE: synthetic minority oversampling technique. J Artif Intell Res 16:321–357.
Chefaoui RM, Lobo JM (2007) Assessing the conservation status of an Iberian moth using pseudo-absences. J Wildl Manage 71:2507–2516.
Chen C, Liaw A, Breiman L (2004) Using random forest to learn imbalanced data. Technical Report 666. Statistics Department, University of California, Berkeley.
Chesson PL (1981) Models for spatially distributed populations: the effect of within-patch variability. Theor Popul Biol 19:288–325.
Cohen J (1960) A coefficient of agreement for nominal scales. Educ Psychol Meas 20:37–46.
Cohen J (1968) Weighted kappa: nominal scale agreement with provision for scaled disagreement or partial credit. Psychol Bull 70:213–20.
Cook TD, Campbell DT (1979) Quasi-experimentation: design and analysis issues for field settings. Houghton Mifflin, Boston.
Costa GC, Wolfe C, Shepard DB, Caldwell JP, Vitt LJ (2008) Detecting the influence of climate variables on species distribution: a test using GIS niche-based models along a steep longitudinal environmental gradient. J Biogeogr 35:637–646.

Cox TF, Cox MAA (1994) Multidimensional scaling. Chapman and Hall, Boca Raton.
Cressie N (1996) Change of support and the modifiable areal unit problem. Geogr Syst 3:159–180.
Cressie N, Calder CA, Clarke JS, Ver Hoef JM, Wikle CK (2009) Accounting for uncertainty in ecological analysis: the strengths and limitations of hierarchical statistical modeling. Ecol Appl 19:553–570.
Crookston NL, Finley AO (2008) yaImpute: an R package for kNN imputation. J Stat Softw 23:1–16.
Curtis JT, McIntosh RP (1951) An upland forest continuum in the prairie-forest border region of Wisconsin. Ecology 32:476–496.
Cushman SA, McKelvey K, Flather C, McGarigal K (2008) Do forest community types provide a sufficient basis to evaluate biological diversity? Front Ecol Environ 6:13–17.
Cutler DR, Edwards TC Jr, Beard KH, Cutler A, Hess KT, Gibson J, Lawler J (2007) Random forests for classification in ecology. Ecology 88:2783–2792.
De'ath G (2007) Boosted trees for ecological modeling and prediction. Ecology 88:243–251.
De'ath G, Fabricius KE (2000) Classification and regression trees: a powerful yet simple technique for ecological data analysis. Ecology 81:3178–3192.
Díaz-Uriarte R, Alvarez de Andrés SA (2006) Gene selection and classification of microarray data using random forest. BMC Bioinformatics 7:3.
Dungan JL, Perry JN, Dale MRT, Legendre P, Citron-Pousty S, Fortin MJ, Jakomulska A, Miriti M, Rosenberg MS (2002) A balanced view of scale in spatial statistical analysis. Ecography 25:626–240.
Evans JS, Cushman SA (2009) Gradient modeling of conifer species using random forests. Landsc Ecol 24:673–683.
Falkowski MJ, Evans JS, Martinuzzi S, Gessler PE, Hudak AT (2009) Characterizing forest succession with lidar data: an evaluation for the inland Northwest, USA. Remote Sens Environ 113:946–956.
Fawcett T (2006). An introduction to ROC analysis. Pattern Recognit Lett 27:861–874.
Finegan B (1984) Forest succession. Nature 312:109–114.
Freund Y, Schapire RE (1996) Experiments with a new boosting algorithm. In: Saitta L (ed) Machine learning: proceedings of the thirteenth international conference. Morgan Kaufmann, San Francisco.
Friedman JH (2001) Greedy function approximation: a gradient boosting machine. Ann Stat 29:1189–1232.
Fu P, Rich PM (1999) Design and implementation of the Solar Analyst: an ArcView extension for modeling solar radiation at landscape scales. In: Proceedings of the 19th annual ESRI User Conference, San Diego.
Gleason HA (1926) The individualistic concept of the plant association. Bull Torrey Bot Club 53:7–26.
Glenn RH, Collins SL (1992) Effects of scale and disturbance on rates of immigration and extinction of species in prairies. Oikos 63:273–280.
Guisan A, Zimmermann NE (2000) Predictive habitat distribution models in ecology. Ecol Modell 135:147–186.
Hall P, Wolff RCL, Yao Q (1999) Methods for estimating a conditional distribution function. J Am Stat Assoc 94:154–163.
Hastie T, Tibshirani R, Friedman J (2009) The elements of statistical learning: data mining, inference, and prediction. 2nd edition. Springer, New York.
Hutchinson GE (1957) Concluding remarks. Cold Spring Harb Symp Quant Biol 22:415–427.
Iverson LR, Prasad AM, Matthews SN, Peters M (2008) Estimating potential habitat for 134 eastern US tree species under six climate scenarios. For Ecol Manage 254:390–406.
Jiménez-Valverde A, Lobo JM (2006) The ghost of unbalanced species distribution data in geographic model predictions. Divers Distrib 12:521–524.
Kubat M, Holte RC, Matwin S (1998) Machine learning for the detection of oil spills in satellite radar images. Mach Learn 30:195–215.
Lawrence RL, Wood SD, Sheley RL (2006) Mapping invasive plants using hyperspectral imagery and Breiman Cutler classifications (randomForest). Remote Sens Environ 100:356–362.

Legendre P, Legendre L (1998) Numerical ecology. Elsevier, Amsterdam.
Lele SR, Dennis B (2009) Bayesian methods for hierarchical models: are ecologists making a Faustian bargain? Ecol Appl 19:581–584.
Liaw A, Wiener M (2002) Classification and regression by Random Forest. R News 2:18–22.
Manel S, William HC, Ormerod SJ (2001) Evaluating presence-absence models in ecology: the need to account for prevalence. J Appl Ecol 38:921–931.
McGarigal K, Cushman SA (2005) The gradient concept of landscape structure. In: Wiens J, Moss M (eds) Issues and perspectives in landscape ecology. Cambridge University Press, Cambridge.
McGarigal K, Tagil S, Cushman SA (2009) Surface metrics: an alternative to patch metrics for the quantification of landscape structure. Landsc Ecol 24:433–450.
McGuffie K, Henderson-Sellers A (1997) A climate modelling primer. John Wiley & Sons, Chichester.
McKenney DW, Pedlar JH, Lawrence K, Campbell K, Hutchinson MF (2007) Potential impacts of climate change on the distribution of North American trees. BioScience 57:939–948.
Millar CI, Stephenson NL, Stephens SL (2007) Climate change and forests of the future: managing in the face of uncertainty. Ecol Appl 17:2145–2151.
Monserud RA, Leemans R (1992) Comparing global vegetation maps with the Kappa statistic. Ecol Modell 62:275–293.
Moore ID, Gessler P, Nielsen GA, Peterson GA (1993) Terrain attributes: estimation and scale effects. In Jakeman AJ, Beck MB, McAleer M (eds) Modelling change in environmental systems. John Wiley & Sons, Chichester.
Morrison D (2002). Multivariate statistical methods. 4th edition. McGraw-Hill series in probability & statistics. McGraw-Hill, New York.
Mouer MH, Riemann R (1999) Preserving spatial and attribute correlation in the interpolation of forest inventory data. In: Lowell K, Jaton A (eds) Spatial accuracy assessment: land information uncertainty in natural resources. Ann Arbor Press, Chelsea.
Murphy MA, Evans JS, Storfer AS (2010) Quantifying *Bufo boreas* connectivity in Yellowstone National Park with landscape genetics. Ecology 91:252–261.
Olden JD, Lawler JJ, Poff NL (2008) Machine learning methods without tears: a primer for ecologists. Q Rev Biol 83:171–193.
Park YS, Chon TS (2007) Biologically inspired machine learning implemented to ecological informatics. Ecol Modell 203:1–7.
Peters RH (1991) A critique for ecology. Cambridge University Press, Cambridge.
Peterson AT, Papes M, Soberón J (2008) Rethinking receiver operating characteristic analysis applications in ecological modelling. Ecol Modell 213:63–72.
Prasad AM, Iverson LR, Liaw A (2006) Newer classification and regression tree techniques: bagging and random forests for ecological prediction. Ecosystems 9:181–199.
R Development Core Team (2009). R: A language and environment for statistical computing. R Foundation for Statistical Computing, Vienna, Austria. ISBN 3-900051-07-0, URL http://www.R-project.org.
Randin CF, Engler R, Normand S, Zappa M, Zimmermann N, Pearman PB, Vittoz P, Thuller W, Guisan A (2009) Climate change and plant distribution: local models predict high-elevation persistence. Glob Chang Biol 15:1557–1569.
Rehfeldt GE, Crookston NL, Warwell MV, Evans JS (2006) Empirical analyses of plant-climate relationships for the western United States. Int J Plant Sci 167:1123–1150.
Risser PG (1987) Landscape ecology: state of the art. In: Turner MG (ed) Landscape heterogeneity and disturbance. Springer-Verlag, New York.
Robinson WS (1950) Ecological correlations and the behavior of individuals. Am Sociol Rev 15:351–357.
Rogan J, Franklin J, Stow D, Miller J, Woodcock C, Roberts D (2008) Mapping land-cover modification over large areas: a comparison of machine learning algorithms. Remote Sens Environ 112:2272–2283.
Runkle JR (1985) Disturbance regimes in temperature forests. In: Pickett STA, White PS (eds) The ecology of natural disturbance and patch dynamics. Academic Press, New York.

Simonoff JS (1998) Smoothing methods in statistics. Springer-Verlag, New York.
Stage A (1976) An expression for the effect of aspect, slope and habitat type on tree growth. For Sci 22:457–460.
Sutton CD (2005) Classification and regression trees, bagging, and boosting. In: Rao CR, Wegman EJ, Solka JL (eds) Handbook of statistics: data mining and data visualization, Volume 24. Elsevier, Amsterdam.
ter Braak CJF, Prentice IC (2004) A theory of gradient analysis. Adv Ecol Res 34:235–282.
Tilman D (1982) Resource competition and community structure. Princeton University Press, Princeton.
Whitaker RH (1967) Gradient analysis of vegetation. Biol Rev 42:207–264.
Whittaker RH, Niering WA (1975) Vegetation of the Santa Catalina mountains, Arizona. V. biomass, production and diversity along the elevation gradient. Ecology 56:771–790.
Wiens JA (1989) Spatial scaling in ecology. Funct Ecol 3:385–397.
Willis KJ, Bhagwat SA (2009) Biodiversity and climate change. Science 326:806–807.

Chapter 9
Genetic Patterns as a Function of Landscape Process: Applications of Neutral Genetic Markers for Predictive Modeling in Landscape Ecology

Melanie A. Murphy and Jeffrey S. Evans

9.1 Introduction

Integrating landscape ecology and population genetics (Manel et al. 2003), *landscape genetics*[1] aims to link observed patterns of genetic variation to underlying landscape process(es) (Storfer et al. 2007). Landscape genetics is a useful, emerging approach with the potential to develop new understanding of ecological theory and improve management decisions (Balkenhol et al. 2009). Current applications that incorporate both genetic data and landscape variables develop hypotheses based on two fundamentally different approaches: (1) assessing patterns of adaptive traits (selection) (Holderegger et al. 2006) and (2) using *gene flow* and *drift* (*neutral* variation) to ask questions about ecological processes (Holderegger and Wagner 2008). While genetic data that measure adaptive traits have valuable applications in landscape ecology (for review, see Holderegger et al. 2006), the intent of this chapter is to demonstrate the promise of using neutral genetic *markers* for predictive modeling in landscape ecology, to identify the potential pitfalls of these approaches, and to provide references for additional resources that may be valuable for researchers and managers.

Many methods that make use of neutral genetic data can be applied to answer a variety of questions in landscape ecology. Conceptually, these methods build on a history of investigating the effects of spatial process on genetic variation (geographical genetics; Epperson 2003; Scribner et al. 2005), and relate landscape heterogeneity to observed patterns of genetic variation (Storfer et al. 2007). The type of genetic data employed and the modeling approach implemented are dependent on both

---

[1] Italicized terms are defined in the glossary (found at the end of the chapter).

M.A. Murphy (✉)
Renewable Resources, Ag. 2010, University of Wyoming Dept. 3354,
1000 E University Ave., Laramie, WY 82071, USA
e-mail: Melanie.murphy@uwyo.edu

research questions and management goals. Therefore, we discuss three general objectives that can be addressed by incorporating genetic and landscape data:

1. Quantifying landscape processes that influence patterns of species occurrence.
2. Evaluating species status and landscape quality by assessing genetic diversity.
3. Estimating landscape influences on functional connectivity using gene flow metrics.

Whereas these objectives are not an exhaustive list of all of the ways that genetic data can be exploited to address ecological questions, they do encompass a large array of potential applications. We selected the above objectives because they demonstrate three different types of information derived from genetic data: (1) occurrence using a binary point measure, (2) point measures that summarize genetic variability, and (3) pairwise measures of genetic differentiation. For each objective, we discuss appropriate genetic data, present potential models and model validation, provide examples that quantify landscape influences on a genetically measured response variable, and suggest future applications. Finally, we illustrate predictive landscape genetics using a case study of Boreal toad (*Bufo boreas*) from Yellowstone National Park.

9.2 Management, Communication, Analytical, and Data Availability Challenges

9.2.1 *Management Challenges*

Management decisions based on species presence, conservation status, and functional connectivity often need to be made quickly in data-poor systems. Through the development of empirical relationships between neutral genetic data and environmental factors, landscape genetic models can provide information regarding species occurrence, genetic diversity, and gene flow; thus providing considerable insight for research studies and management decisions. Although molecular methods do not provide the depth of information available through demographic-based approaches such as mark-recapture, these data summarize an average condition over time and can be collected in a single field effort. Genetic data are also intrinsically valuable as genetic diversity is vital for population sustainability. Additionally, unique groups qualifying for legal protection can be identified (Young and Clark 2000). Combining molecular and demographic data results in the most complete picture of species ecology and conservation needs (Riley et al. 2006; Howeth et al. 2008).

9.2.2 *Communication Challenges*

While landscape genetics is a very promising approach for answering ecological questions and addressing management concerns, cross-disciplinary communication challenges need to be addressed for application of this approach to be successful. Landscape genetics is a highly technical and integrated approach (Fig. 9.1) that

Fig. 9.1 Landscape genetics pulls on a broad range of fields. Highly technical skills from these fields are required for robust landscape genetic analysis including: genetic laboratory methods, population genetic analytical tools, in-depth understanding of spatial statistics and GIS/Remote Sensing, use of landscape metrics, and adaptation of ecological models. Researchers and managers interested in applying landscape genetic approaches should foster strong collaborative working groups covering these skills and academic programs that build a comprehensive skill set in graduate education.

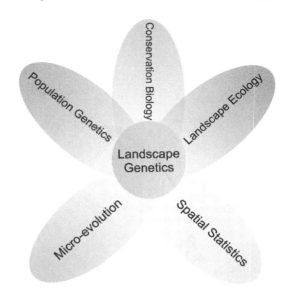

requires a broad range of skills – including analytical expertise in both population genetics and spatial analysis (Storfer et al. 2007). Landscape ecologists and population geneticists need to work together in interdisciplinary teams to construct robust analyses with promising management applications (Fig. 9.2). Within these teams, communication can be a major hurdle as the terminology from population genetics can be unfamiliar to landscape ecologists and familiar terms (e.g., "*migration*") are often used in a slightly different manner. In an effort to increase interdisciplinary communication, we include a glossary of common terms at the end of the chapter; these terms are *italicized* in the text at first use.

9.2.3 *Analytical Challenges*

Genetic data are unique because they assay differences in genetic composition collected over time due to DNA mutations. Unlike many variables measured in the field which have direct interpretation as a matrix of traits (i.e., tree height, species composition), raw genetic data often have little direct ecological interpretation (Storfer et al. 2007; Murphy et al. 2008). To draw inferences, these data are digested into identification of a group, measures of genetic diversity, or relative differentiation between groups or individuals. The time-scale measured by the genetic data is dependent on marker choice (Table 9.1); it may extend over evolutionary time periods appropriate for biogeographical questions or may measure fine-scale differences suitable for landscape-level questions and relevant to contemporary management concerns (Avise 2004; Storfer et al. 2007). The lack of direct interpretation, the potential for current and historic processes generating *genetic patterns* to be convolved, and the pairwise nature of commonly applied genetic metrics leads to numerous analytical challenges (Murphy et al. 2008; Balkenhol et al. 2009).

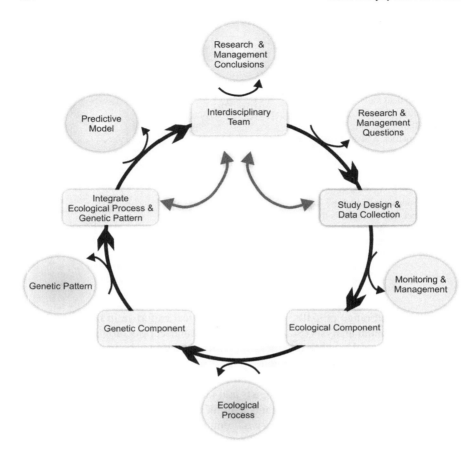

Fig. 9.2 Conceptual diagram of steps for effective interdisciplinary work in landscape genetics. (a) Interdisciplinary team: build collaborations that collectively have the skills needed for landscape genetics (Fig. 9.1). (b) Field data collection: based on identified study questions, design and implement a study. At this point, the interdisciplinary team can reconvene and further refine the research and management questions. (c) Ecological process: calculate or compile independent variables of ecological processes in the landscape with a hypothesized relationship with genetic patterns. (d) Genetic pattern: collect and analyze genetic data. (e) Integration: quantify the relationship between ecological process and observed genetic pattern as an interdisciplinary team. Use the relationship between pattern and process to develop a predictive model. (f) Conclusions: interdisciplinary team works together to draw research and management conclusions

9.2.4 Challenges Acquiring Data

To meet the goals of a given landscape genetics study, new genetic data will usually need to be generated (Figs. 9.1 and 9.2; However, see GenBank for *sequence* data, http://www.ncbi.nlm.nih.gov/). Although production of genetic data requires a

9 Genetic Patterns as a Function of Landscape Process

Table 9.1 General overview of genetic markers applied in landscape genetics studies

	Sequence (mtDNA or cDNA)	Sequence (nDNA)	Allozymes[a]	AFLP	RFLP[b]	μsat	SNP
Required tissue	Low	Low	High	Low	High[b]	Low	Low
Inheritance	Single parent	Bi-parental	Bi-parental	All DNA[c]	Varies	Bi-parental	Varies
Nature	Single copy	Dominant	Dominant	Dominant	Codominant	Codominant	Codominant
Neutrality	Varies, treated as neutral	Varies	Likely under selection	Unknown, generally treated as neutral	Varies	Neutral	Varies
Development	Low	Low-moderate	Low	Low	Moderate	High	High
Cost	Moderate	Moderate	Low	Moderate	Low	High	Moderate
Spatial/temporal scale	Broad	Broad	Broad	Fine	Mid	Fine	Fine
Loci	1	2 to 10+	10 to 20+	50 to 500+	1 to 20[d]	5 to 20+	10 to 100s

There may be special cases for species or systems, but Table 9.1 summarizes typical markers, their applications and related guidelines. Markers differ in their mode of inheritance, scale, difficulty of use, and cost (Avise 2004). Therefore, in applying genetic methods it is important to match the marker to the underlying research or management question (DeYoung and Honeycutt 2005). *Required tissue* is the amount of tissue needed to use a particular marker. Most modern methods rely on PCR, which allows for analysis of small amounts of DNA (a few cells). *Inheritance* is the mode of inheritance in species with sexual reproduction (i.e., single parent or bi-parental). *Nature* is the overall type of marker where dominant markers are those where a heterozygote (A_1A_2) and homozygote (A_1A_1) for an allele are indistinguishable. In co-dominant markers, all genotypes at a locus are distinguishable (A_1A_1, A_1A_2, and A_2A_2 for example at a bi-allelic locus). *Markers:* mitochondrial DNA sequence (mtDNA), chloroplast DNA sequence (e.g., plants, cDNA), nuclear DNA sequence (nDNA), allozymes (also referred to as protein electrophoreses), Amplified Fragment Length Polymorphisms (AFLPs), Restriction Fragment Length Polymorphism (RFLPs), microsatellites (μsat, referring to nDNA for our purposes), and Single-Nucleotide Polymorphisms (SNP). *Development* represents the cost and difficulty of marker development in the target species. *Cost* is the relative cost of use and analysis of a developed marker. *Resolution* is the ability to distinguish genetic difference, in general for the type of marker. *Loci* is the number of loci typically used or needed for analysis

[a] Also referred to a protein electrophoresis because analysis is on protein variation, not DNA
[b] Unless the method is applied to *PCR* product
[c] Amplification is applied to all DNA and origin of resulting amplifications (bands) is generally unknown
[d] Single locus could be used for species identification, more loci needed for genetic relationships

specific set of expertise and equipment in a laboratory accustomed to population genetic work, it is feasible for researchers and managers whose expertise is not population genetics to incorporate genetic data in their work. The necessary skills and laboratory equipment may be accessed in conjunction with collaborators or data can be generated through a commercial laboratory.

Costs associated with generating genetic data have historically been a challenge to the wide-scale application of molecular markers in ecology and conservation. However, the expense of genetic work has declined dramatically in recent years and will continue to do so with future technological advancement. Total cost is highly variable, depending on (1) type of marker (Table 9.1), (2) availability of *primers* (Table 9.1), (3) type of labor (academic vs. service laboratory), (4) number of samples processed, (5) number of *loci*, and (6) required turn-around time. Researchers and managers should be aware that robust ecological analysis of genetic data commonly requires large datasets in both sample size (100s to 1,000s of individuals or 10 to 100s of groups) and number of markers (10–15 *microsatellite* loci; Murphy et al. 2008). Excluding labor costs, a "rule-of-thumb" for standard fragment analysis is approximately US $1/sample/locus plus approximately US $2 for DNA extraction. For example, 1,000 samples at ten microsatellite loci would cost approximately US $13,000 plus labor and an initial optimization expense (~ US $2,000 – US $4,000 if markers do not need to be developed). We should note that when researchers invest in thorough optimization procedures (e.g., multiplex *polymerase chain reactions* (PCRs), efficient use of a 5-dye system,), the cost per sample can be substantially reduced (Murphy, unpublished data).

It is essential for researchers and managers to evaluate data quality, whether it is in their own facility or through a commercial lab. At a minimum, data should be accompanied by: (1) a protocol for monitoring contamination (i.e., negative controls for DNA extractions and any amplifications); (2) a random subset of data (5–10%) generated in duplicate to screen for human and genotyping errors; (3) known sample(s) included in each PCR and sequencer run to verify size consistency (in the case of fragment analyses, see Table 9.1), and (4) calculation of genotyping error rate for the dataset (see Murphy et al. 2010). As many data concerns are specific to the molecular markers and type of DNA, we discuss specific data availability and suitability for each landscape genetic question.

9.3 Species Identification

9.3.1 Motivations for Application of Genetic Data for Species Identification

Data on the presence and spatial locations of a species are important for addressing questions of distribution (Araújo and Guisan 2006), response to disturbance (Reed and Merenlender 2008), and spread (Wagner et al. 2006; Archie et al. 2009). Accurate occurrence data are required to answer these questions, but acquiring such data may be a limitation for some critical management applications. Species, or other genetically identifiable groups of interest, may be morphologically indistinguishable due to physical similarities (Bickford et al. 2007) or genetic introgression (Miller et al. 2003). For species that are morphologically distinguishable, collecting occurrence data may still be difficult due to cryptic behavior and/or small population sizes (Eggert et al. 2003). In both these cases, genetic data can be used to accurately identify species.

9.3.2 Suitable Data and Availability

Species identification can be achieved using most molecular markers (Table 9.1), however *mitochondrial DNA* (mtDNA) fragment analysis (Murphy et al. 2000), identification of restriction enzyme sites (Adams and Waits 2007), or single nucleotide polymorphisms (*SNPs*) (Belfiore et al. 2003) are cost-efficient methods for processing large numbers of samples taken across the landscape (Table 9.1). In addition to these species identification approaches, quantitative PCR estimates the number of copies of target DNA and can assess the prevalence of a disease within a host (Table 9.2; e.g., Goldberg et al. 2010). Development of the above tests for a focal species is generally fairly fast and inexpensive (Table 9.1). However closely related species, hybrids, or populations within a species may require a more in-depth approach. Methods that assign individual multi-locus *genotypes* to a group ("*assignment tests*"; Table 9.2) are effective when focal group(s) are not identifiable using simple tests (Miller et al. 2003).

Elusive species can be identified through genetic analyses of non-invasive samples (Waits and Paetkau 2005; Broquet et al. 2007) such as hair (Riddle et al. 2003), feces (Murphy et al. 2000), feathers (Rudnick et al. 2007), skin (Valsecchi et al. 1998), and water (Ficetola et al. 2008). However, DNA from non-invasive samples tends to be of low quality and quantity: low quality/quantity DNA may be susceptible to genotyping error and sensitive to low-level contamination (Taberlet et al. 1999). To produce reliable results, special measures need to be taken including use of dedicated laboratory space (separated from concentrated DNA and PCR products) and implementation of strict data validation protocols (Waits and Paetkau 2005). Excellent reviews of non-invasive sampling issues (Broquet et al. 2007; Ruell and Crooks 2007) and data standards (Waits and Paetkau 2005) are available for those interested in applying non-invasive techniques.

9.3.3 Past, Current, and Future Applications

Species identification using molecular markers to address ecological questions was initially implemented to describe current spatial distributions (Sneath and Sokal 1973), mostly for rare and elusive species. In an early application of molecular-based species identification, mtDNA from fecal samples was used to monitor brown bears

Table 9.2 Introduction to select genetic metrics commonly applied in landscape genetic analyses by objective

Objective	Measure	Metric	Type	Marker	Advantages	Disadvantages	References
ID	Species ID	Sequence concordance	Point	mtDNA	Assess long-term biogeographic patterns	Higher cost for $>n$	Avise (2004)
		Fragment length	Point	RFLP mtDNA	Fast, lower cost	Low resolution	Murphy et al. (2000)
		Assignment test	Point	μsat AFLP	High resolution, good for closely related groups	Higher cost, >loci	Miller et al. (2003)
	Pathogen presence		Point	Fragment – PCR	Inexpensive	Presence only	Annis et al. (2004)
			Quantify	Fragment – qtPCR	Quantify level of infection	Higher cost than above	Goldberg et al. (2010)
Diversity		H_e	Point	Any	Standard measure, simple calculation	Not sensitive to recent change	Wright (1951)
		Alleles/locus	Point	μsat AFLP Allozymes	Sensitive to recent change	Reaches an asymptote in hyper-allelic systems.	Leberg (2002), Van Loon et al. (2007)
		N_e	Point	μsat	Detect source sink populations	Assumes mutational mechanism	DeYoung and Honeycutt (2005), Wang (2005)
	Bottleneck	Wilcoxon sign-rank test	Point	μsat	Detect recent decline	Can lack power	Cornuet and Luikart (1996)
Functional connectivity	Genetic distance	F_{ST} (G_{ST})	Pairwise, group	mtDNA μsat AFLP	Standard metric, theoretically based, robust	Low resolution (power) at fine-scales	Wright (1951)
		R_{ST}	Pairwise, group	μsat	F_{ST} analogue specific to μsats	Potential bias w/o loci and samples	Balloux and Lugon-Moulin (2002), Slatkin (1995)

9 Genetic Patterns as a Function of Landscape Process

Metric	Type	Marker	Advantages	Disadvantages	References
D statistics	Pairwise	mtDNA μsat AFLP	Shorter time-scale, higher resolution than F_{ST}	Pairwise metric presents statistical challenges	Dieringer and Schlötterer (2003)
D_N	Pairwise, group, or individual	mtDNA μsat AFLP	Theoretically based	Based on equilibrium assumptions	Nei (1972)
Genetic chord (D_c)	Pairwise, group or individual	mtDNA μsat AFLP	Minimal assumptions	Pairwise	Cavelli-Sforza and Edwards (1967)
D_{ps}	Pairwise, group or individual	mtDNA μsat AFLP	No equilibrium assumptions	Pairwise	Bowcock et al. (1994)
Relatedness	Pairwise, individual	μsat	Individual based	Pairwise metric presents statistical challenges	Anderson and Weir (2007), Belkhir et al. (2002)
Rousset's a	Pairwise, individual	μsat	Individual based	Pairwise gradient of membership	Rousset (2008)
Assignment tests	Point/surface	μsat AFLP	Individual based	Requires clustering data	Murphy et al. (2008)
Assignment tests	Pairwise, population	μsat AFLP	Identify real-time dispersers	Groups must be identifiable	Manel et al. (2005)
Probability of parentage	Pairwise	μsat AFLP[1]	Identify real-time gene flow	Requires intensive sampling	Jones and Ardren (2003)

Proportion membership

ID dispersers

Parentage analysis

See main text for abbreviations. *Objective* is the landscape genetic objective addressed. *Measure* is what is being assessed related to the overall objective. *Metric* is a statistic of a particular measure. *Type* is if the metric is a point or pairwise measure of genetic variation. *Marker* lists the markers that are most commonly used in calculation of a given metric (Table 9.1). *Advantages* are the pros of using a particular metric. *Disadvantages* are the cons of using a particular metric. *References* lists select reference(s), including further review of particular approaches. Genetic distance metrics vary in performance and should be carefully considered (Goldstein et al. 1995; Paetkau et al. 1997). In addition, there are many methods and freely available programs that have been developed for spatial and landscape analyses of genetic data (Guillot et al. 2009)

[1] Codominant marker - heterozygosity has to be estimated and parentage estimated with some probability.

(*Ursus arctos*) that were reintroduced into the French Pyrenees (Taberlet et al. 1997). Expanding on this approach, restriction enzyme tests from fecal DNA have been applied to identify multiple species and their distributions within a study area (Riddle et al. 2003). Molecular-based species identification can also be used to describe the distribution of a difficult to detect invasive species. In one particularly creative approach, species presence of an invasive amphibian was reliably determined by sequencing a short mtDNA fragment from water samples (Ficetola et al. 2008).

Molecular-based species identification data are a measure of presence and can be used for a variety of ecological applications analogous to field-based presence data, potentially minimizing the likelihood of pseudo-absences. For example, arctic and red foxes use similar ecological resources but have overlapping distributions, an unexpected observation in the context of niche exclusion theory (Dalén et al. 2004). One potential explanation is temporal avoidance during times of resource scarcity. Using mtDNA fragment analyses from fecal samples, Dalén et al. (2004) found that while arctic and red foxes are sympatric during the winter, they are allopatric during the breeding season. The authors demonstrated that genetically-based presence data can be used to elucidate species ecology and illustrate effects of competition on a niche of a species. Shifts in species fine-scale distribution may also occur due to human habitat use. Based on restriction fragment length polymorphism (RFLP) analyses from fecal samples, "low impact" recreation was shown to alter the distribution of carnivores within a protected area (Reed and Merenlender 2008).

Detecting disease presence is important for resource management and understanding ecosystem dynamics (Ostfeld et al. 2008) as disease presence may be related to ecosystem condition. For example, using a fragment amplification test (amplification=presence, no amplification=absence), tick-borne disease occurrence was associated with moisture (Foley et al. 2005). In addition to environmental factors, anthropogenic impacts may increase disease prevalence (Goldberg et al. 2010). Also based on a fragment amplification test, Goldberg et al. (2010) hypothesized that amphibian disease (*Batrachochytrium dendrobatidis*) is associated with human access along roads that provides a conduit for disease transmission into novel habitats.

In general, predictive models using molecular identification are currently underutilized in ecological investigations. However, predictive models have been applied in ecological applications of spatial epidemiology (Archie et al. 2009). The temporal component does not require additional genetic data or analysis beyond disease identification, but samples are collected over time so that the rate of spread can be estimated (Real et al. 2005; Blanchong et al. 2008). A statistical relationship is then built between disease occurrence over time and landscape condition in order to predict spread. These predictions are used to make management decisions such as vaccination of wild populations to control outbreaks (Real and Biek 2007). In addition, simulations using the influence of landscape barriers can be implemented to predict spread. In the case of rabies, researchers were able to determine that rivers impose natural barriers to raccoon dispersal that can potentially contain outbreaks in a localized area (Rees et al. 2008). Expanding on these ideas, existing models of invasive species spread could be parameterized using genetic data (Ferrari and Lookingbill 2009).

9.4 Genetic Diversity

9.4.1 Motivations for Application of Genetic Data to Assess Genetic Diversity

Based on ecological theory, high-quality habitats should support more individuals than low-quality habitats (Hanski and Gaggiotti 2004). Under this paradigm, poor quality habitats are expected to have a smaller *effective population size* (N_e), lower genetic diversity and potentially be in decline in comparison to stable populations in high quality habitats (Bellinger et al. 2003). To estimate effective population size using demographic data, information on mating structure, sex ratio, and average age of reproduction are required. In order to estimate population trajectory based on these demographic data, long-term monitoring datasets which include a measure of abundance are essential (Green 2003). However, effective population size (Wang 2005) and signatures of population *bottlenecks* can be identified with a single field sample based on *allele* frequency distributions (Cornuet and Luikart 1996).

9.4.2 Suitable Data and Availability

Although most markers can be used to assess genetic diversity (Table 9.1), multi-locus genotypes from markers such as microsatellites, *AFLPs*, or SNPs are most appropriate for fine-scale landscape studies (Storfer et at. 2010). These markers have higher statistical power and measure a shorter time-scale than less variable marker types (Avise 2004; see glossary for explanation of maker types). Microsatellites are hyper-variable, neutral co-dominant markers that meet assumptions of neutral population genetics models. However, they have to be identified within a specific species and may not be available for rare species without developing new primers (Table 9.1). In contrast, AFLPs are random amplifications across the genome where presence/absence of bands is recorded; primer sets are not species-specific (Table 9.1). However, AFLPs are dominant markers meaning that *homozygotes* and *heterozygotes* are indistinguishable, limiting the available analytical approaches (Avise 2004). In addition, AFLP primers may require thorough optimization to identify repeatable polymorphic bands (Avise 2004). SNP loci are emerging as an alternative marker type to microsatellites and AFLPs (Morin et al. 2004). They are co-dominant markers based on single *base-pair* changes anywhere in the genome (Table 9.1). Due to the low number of alleles per locus (usually 2), more loci are required compared to microsatellites and AFLPs (Table 9.1). However, loci are easily to identify using high-throughput sequencing (Rokas and Abbot 2009).

Genetic diversity can be assessed by measuring heterozygosity, allelic richness, effective population size, and the presence of population bottlenecks (Table 9.2). Heterozygosity (H_e) is the proportion of individuals with two alleles at a locus averaged across all loci and is a measure of overall genetic diversity. Allelic richness, the number of alleles per locus averaged across all loci, is more sensitive to loss of genetic diversity than heterozygosity when using multi-allelic markers as isolated populations lose alleles more quickly than overall heterozygosity (Luikart and Cornuet 1998). Effective population size can be estimated from the number of alleles present in sampled individuals (DeYoung and Honeycutt 2005). Population bottlenecks can be detected with a single post-bottleneck sample based on a shift in the expected allele frequency distribution (Luikart and Cornuet 1998).

9.5 Past, Current, and Future Applications

Genetic diversity has been quantified in relation to landscape variables as a measurement of habitat quality. Initial studies of landscape effects on genetic diversity compared presumptive low quality and high quality habitats. For example, black grouse were found to have lower genetic diversity (heterozygosity and allelic diversity) in fragmented versus unfragmented habitats (Caizergues et al. 2003). Although identifying this pattern is worthwhile, management decisions are easier to justify if the relationship is quantified.

Several examples quantify diversity. Based on previously published *allozyme* data, genetic diversity was shown to be positively correlated with land protection status (Ji and Leberg 2002). This result has major management implications because it suggests that protecting habitat does in fact preserve genetic diversity. Latitude and elevation isolate populations and therefore influence genetic diversity. Using linear regression between elevation and heterozygosity, Columbia spotted frog microsatellite diversity was found to be negatively correlated with elevation (Funk et al. 2005). Local extinction is likely common in this habitat, with recolonization coming only from low elevations. A similar pattern has been seen with plants across latitude – intra-population genetic diversity is negatively correlated with latitude based on microsatellite data (Hu et al. 2008).

To date, genetic diversity is underutilized in predictive landscape genetic modeling. Effective population size could be used as an index of habitat quality. Recent population bottlenecks may be the result of habitat loss, disease (Rees et al. 2009), or population sinks (Murphy et at. 2010a). Effective population size or presence of a bottleneck signature could be modeled as a function of landscape condition (Murphy et at. 2010a; Rees et al. 2009) using point-pattern statistics (Fortin and Dale 2005).

9.6 Functional Connectivity

9.6.1 Motivations for Application of Genetic Data for Quantifying Functional Connectivity

Functional connectivity postulates that the landscape facilitates or provides resistance to individual movement and is commonly assessed using estimates of dispersal

(Taylor et al. 2006). Animal dispersal models are traditionally based on mark-recapture data, radio telemetry, or expert opinion (Crooks and Sanjayan 2006). While useful, these methods have limitations. For some taxonomic groups, more than 30 years may be required for stable estimates of dispersal derived from mark-recapture data (Green 2003). Radio telemetry provides extremely detailed information on individual movements that are invaluable for resource selection models (see Chap. 3 in this volume), but radio telemetry is expensive, logistically complicated to apply to the large sample sizes needed for comprehensive estimates of dispersal, and researchers have to correct for telemetry bias (Frair et al. 2004). Although expert systems may be tremendously valuable in informing management decisions (see Chap. 12 in this volume), empirically based alternatives are advantageous especially when there are legal considerations. Finally, quantification of movement that results in successful breeding is fundamental for estimating functional connectivity and meeting management goals. Movement, as quantified above, may not result in successful breeding. Genetically based data are an alternative for parameterizing functional connectivity models that can be estimated quickly, can be applied to large sample sizes, are empirically based, and quantify dispersal that results in breeding (Storfer et al. 2007).

9.6.2 Suitable Data and Availability

Multi-locus genotypic data (e.g., microsatellite, AFLP, SNP) are most appropriate for functional connectivity questions (Table 9.1). Genetic data can be used to assess functional connectivity in two major ways: (1) calculating genetic distance among individuals/groups or (2) using molecular markers to directly identify migrants or offspring of migrants. Although both of these approaches are pairwise estimates of functional connectivity, genetic distance is an indirect estimate of migration over time while migrant identification is a real-time estimate of functional connectivity (Table 9.2).

Genetic distances are an index of genetic differentiation with values ranging from 0 to 1; 0 indicates no differentiation (*admixture*) and 1 indicates no similarity (completely unrelated). The absolute values are dependent on the overall genetic variation in the species, markers used, and evolutionary history which makes the values directly comparable only for the same species and loci (Avise 2004). Genetic distances can be measured between groups or individuals, do not require sampling of all individuals at a given location, and are normalized over time. One caveat is that sites with no current connectivity may appear to have a high degree of similarity due to past gene flow (Gauffre et al. 2008).

Table 9.2 lists and briefly explains a few of the many available genetic distance statistics, and focuses on those utilized most frequently in landscape genetics studies. Due to the long history and theoretical basis of the metric, F_{ST} is commonly used. Based on heterozygosity (Wright 1951), F_{ST} measures longer time scales than metrics based on allele frequencies and can be insensitive to recent landscape change (Murphy et al. 2008; Murphy et al. 2010b). Multiple metrics based on allele frequency distributions or allele sharing have more recently been developed using a range of mathematical foundations (Table 9.2). Genetic chord distance (Cavelli-Sforza and Edwards 1967) and proportion of shared alleles (Bowcock et al. 1994) are two genetic distance statistics that are appealing for landscape genetic applications as they make no assumptions of mutational mechanisms or equilibrium conditions (Table 9.2). Mutational mechanisms of microsatellites are complex and difficult to model (Colson and Goldstein 1999); assumptions of genetic equilibrium are likely violated in rapidly changing landscapes or declining populations (Maruyama and Fuerst 1985).

Functional connectivity can also be measured by counting the number of dispersers between sites (Manel et al. 2005) or offspring of dispersers (Jones and Ardren 2003). To assess number of dispersers between sites, individual multi-locus genotypes are clustered into groups using assignment tests; the statistical basis for this clustering varies depending on clustering algorithm (for an excellent overview of the pros and cons of particular methods, see Manel et al. 2005). Assignment tests identify real-time movement, are less likely to be influenced by past connectivity than genetic distance statistics, do not require sampling of all of the individuals in the study area, and groups do not have to be identified *a priori*. Once individuals are assigned to groups (when possible), a matrix is created containing the number of individuals in each group that were assigned to each other group as an estimate of connectivity. However, the observed movement does not necessarily result in gene flow and sufficient genetic structure must be present to differentiate origin and destination sites. Depending on the scale and amount of genetic variability, this may require a large number of markers (e.g., >15 microsatellite loci) or may be impossible in the presence of high levels of gene flow (Manel et al. 2005).

Parentage analysis (Jones and Ardren 2003) can be used to detect real-time gene flow, an approach often applied to estimate functional connectivity of plants. In this type of analysis, offspring are assigned to potential parents. If parents are from different sites, the offspring are direct evidence of current gene flow (Jones and Ardren 2003). While this is an extremely powerful method (e.g., Epperson 2003), many analytical methods require sampling the majority of potential parents for meaningful results that may be difficult for non-sessile organisms.

All of these measures of functional connectivity are pairwise; no relationship is expected between genetic data and landscape characteristics at the sample location. The independent variables can be any measure of the landscape matrix between sample locations that has a hypothesized relationship with functional connectivity. To avoid spurious correlations, researchers and managers should constrain analyses to include only variables with strong *a priori* ecological justifications (e.g., Murphy et al. 2010a, b).

9.6.3 Past, Present and Future Applications for Predicting Functional Connectivity

Functional connectivity across a landscape assessed using genetic markers was initially investigated by overlaying genetically identified groups on landscape features to identify potential dispersal barriers. For example, an overlay of wolverine microsatellite data on a terrain map suggested that population structure followed mountain ranges with valleys functioning as barriers (Cegelski et al. 2003). Using similar methodology, more genetic structure was observed in fragmented habitats than in intact habitats for black grouse – suggesting habitat fragmentation decreased functional connectivity (Caizergues et al. 2003). Overlay analyses are useful and serve to develop informed hypotheses of functional connectivity. The major drawback of these approaches is that the inferred relationship is not quantified and may be spatially coincident rather than the result of the hypothesized landscape process (Storfer et al. 2007). For example, sharp differentiation of brown bears observed in northern versus southern Scandinavia is due to recent recolonization after species extirpation and not due to a landscape barrier (Waits et al. 2000).

One analytical approach to test hypotheses of functional connectivity is graph theory. The null hypothesis is that straight-line distance between sites explains the most variation in genetic distance. An alternative hypothesis is created where the graph is constrained to only include a subset of connections (edges), such as eliminating edges that cross a putative dispersal barrier. The graph that explains the most variation in genetic distance is selected. For example, using allozymes and microsatellite loci, a land snail was found to be more likely to disperse via hedgerows than via straight-line distance (Arnaud 2003). The advantage of this approach is that it clearly tests competing hypotheses of functional connectivity; however, occupied areas and intervening habitat must be clearly delineated.

Well designed least-cost path analyses can also test hypotheses of functional connectivity between groups (Spear et al. 2005) or individuals (Cushman et al. 2006). Least-cost paths use weights associated with landscape variables of interest to predict areas of high and low "cost" to dispersal. Optimal paths between groups or individuals are then calculated, representing an "effective distance" (Michels et al. 2001). Once optimized, least-cost paths can be applied to design corridors between core habitats (Epps et al. 2007).

Due to the distance dependency in all alternative models, many of the tested models may be statistically significant. "Causal modeling" has been implemented as a way to circumvent this issue (Legendre 1993). Based on the dependency among models, researchers can develop a matrix of supported/unsupported models for each alternative hypothesis. For example, Cushman et al. (2006) tested multiple hypotheses of black bear connectivity using microsatellite data. Although many models were supported, landscape resistance was selected due to statistical support for the set of predicted models (i.e., isolation-by-distance, barrier, and landscape resistance models).

Although least-cost paths have clear expectations and testable hypotheses, researchers and managers are often interested in using genetic data to estimate limits to functional connectivity. However, an *a priori* understanding of landscape resistance is needed to estimate least-cost paths. Competing hypotheses of dispersal cost are usually based on either expert knowledge (McRae and Beier 2007) or iterative model fitting (Cushman et al. 2006; Epps et al. 2007). Alternatively, costs can be parameterized using observed habitat use as a reasonable estimate of dispersal habitat suitability (Wang and Summers 2010). While the strength of this approach is that costs are empirically based (Funk and Murphy 2010), dispersal habitat may have divergent requirements from habitat suitable for occupancy. Animals may actually disperse more quickly through unsuitable habitats than suitable habitats (Semlitsch 2008).

Another limitation of least-cost paths is that generally only the optimal path is used to calculated ecological distance, which may bias results (Rayfield et al. 2010). Circuit theory addresses this, as all possible paths through the landscape contribute simultaneously to functional connectivity (McRae 2006). However, the circuit theory algorithm is very processor intensive, currently limiting the size of the study area and data resolution that can be analyzed. In addition, circuit theory assumes that there is a potential for gene flow in each cell which renders it inappropriate for patch dependent species at fine spatial scales (McRae 2006).

Migrant detection can be used as an alternative to genetic distances within the same framework or to add supplemental information. For example, it can be used to identify population of origin and add directionality to functional connectivity (Janssens et al. 2008). This directionality can have ecological implications. For example, using a Bayesian approach to estimate number of recent migrants of common toads, high elevation sites were found to exchange fewer migrants than low-elevation sites (Martínez-Solano and González 2008).

The pairwise nature of genetically based functional connectivity data limits the analytical methods currently available (Balkenhol et al. 2009; Murphy et al. 2008). The data are often analyzed using Mantel, partial-Mantel tests, and multiple matrix regression (Mantel 1967; Legendre and Legendre 1998). Questions have been raised regarding the statistical power of the Mantel test, the implementation of permutation-based significance testing for Mantel and partial Mantel tests, and overall utility of matrix regressions for solving multivariate models (Castellano and Balletto 2002; Rousset 2002; Murphy et al. 2008). In addition, ecological data used to assess functional connectivity are often spatially autocorrelated with non-linear relationships between independent and dependent variables, thus violating assumptions of parametric models and rendering many standard analytical frameworks invalid (Wagner and Fortin 2005). Moreover, these data and corresponding measures of genetic distance may have a low statistical signal to noise ratio resulting in models with little predictive power.

Several analytical methods have been proposed to address these challenges. Wombling has been shown to accurately identify breaks in genetic structuring using multi-locus genotypic data (Cercueil et al. 2007), appropriate when putative landscape variables are abrupt features that can be associated with genetic discontinuities. If landscape variables represent a gradient, interpolated ancestry values from an assignment test have been shown to accurately identify underlying landscape process generating observed genetic patterns (Murphy et al. 2008). Spatial regression has potential, if methods for assigning spatial location to the pairwise data can be improved (Spear and Storfer 2008). In the case study below, we present another potential solution: using non-parametric classification and regression trees (Random Forests (RF)) to predict high dispersal areas Murphy et al. 2010b.

9.7 Case Study: Predictive Landscape Genetics of the Boreal Toad (*Bufo boreas*) in Yellowstone National Park

9.7.1 Overview of the Questions and Study Motivation

Demonstrating many of the concepts covered in this chapter, we present a case study where we incorporate genetic data to predict functional connectivity of boreal toads (*Bufo boreas boreas*) across Yellowstone National Park (Murphy et al. 2010b). Because boreal toads are in decline throughout portions of their range (Muths et al. 2003), managers in Yellowstone National Park wanted to know where hotspots of functional connectivity were located so potential impacts could be taken into account when making management decisions. However, breeding sites are sparsely dispersed across the landscape (Murphy 2008), recapture rates are low – making demographic studies difficult, and little is known about the non-breeding habitat requirements of the species (Muths 2003).

9.7.2 Suitable Data and Availability

We used 15 previously developed microsatellite loci (Simandle et al. 2006) to genotype 805 individuals across Yellowstone National Park (Murphy et al. 2010b). More than 90% of known boreal toad breeding sites in the study area were sampled, with the addition of new sites identified from surveying a random stratification of wetlands (strata: elevation, solar insolation, and precipitation). Negative controls were included in DNA extractions and PCR amplifications to monitor for contamination, a random subset of samples (5–10%) were reamplified, known samples were included in each PCR to verify consistency in fragment size, and genotyping error was less than 0.05% (Murphy et al. 2010b). We used the multilocus genotypic data to calculate a genetic distance measure – proportion of shared alleles (D_{ps}, Table 9.2).

Hypotheses of functional connectivity were built around three ecological processes identified in a previously published study of boreal toad functional connectivity in Yellowstone National Park (Murphy et al. 2010b): habitat permeability, topographic morphology, and temperature-moisture regime. Spatial data for independent variables were available through National Landcover Data (NLCD), a digital elevation model (DEM), a spline-based climate model for the Western United States (Rehfeldt 2006) and thermal feature data collected through the USGS for Yellowstone National Park (Murphy et al. 2010b). Independent variables were calculated along a saturated network connecting each site to all other sites (Murphy et al. 2010b). At 30 m resolution, all data were appropriate in scale for both the sample design and dispersal of boreal toads (Muths 2003).

9.7.3 Models Applied and Validation Techniques

We first estimated functional connectivity for boreal toads in Yellowstone National Park (Murphy et al. 2010) using the RF algorithm (see Chap. 8). RF is a statistically powerful ensemble-learner algorithm that is insensitive to autocorrelation, is non-parametric, has the ability to incorporate complex interactions among independent variables (Cutler et al. 2007), and can address the challenges presented by pairwise functional connectivity data. To increase ecological interpretability, we developed model selection and significance testing for RF (see Chap. 8; Murphy et al. 2010b).

In the selected model, we found functional connectivity among boreal toad breeding sites was explained by the following variables in order of importance: growing season precipitation (mean, positive), distance (negative), topographic roughness (mean, negative), slope*sin (aspect) (mean, negative), habitat burned in the 1988

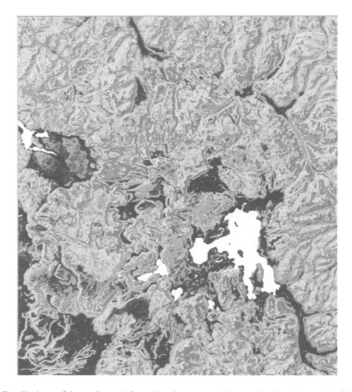

Fig. 9.3 Prediction of boreal toad functional connectivity to the landscape in Yellowstone National Park draped over topography (based on RF model from Murphy et al. (2010)). Color ramp represents resistance to dispersal with blue areas indicating high functional connectivity and red areas indicating low functional connectivity. White areas indicate major lakes and were omitted from the analyses

fires (count of cells, positive), impervious surfaces (mean percent, negative), canopy cover (mean percent, complex), elevation relief ratio (mean, negative), degree days greater than 5°C (mean, negative), and frost free period (mean, positive) (Murphy et al. 2010b). The selected model was significant from random and explained 56.28% of the variation in genetic distance (Murphy et al. 2010b).

Although our functional connectivity hypothesis tests were extremely informative, managers wanted to identify areas of high and low connectivity across the landscape to inform the decision making process. To address this need, we developed a method to predict a model of functional connectivity to the landscape. To account for the pairwise nature of the data, we applied a mean 50 m moving-window (the median connection distance between sites) to rasters of all variables retained in the selected RF model (see above). To incorporate the observed *isolation by distance* effect, we calculated distance to nearest occupied site/distance to nearest surveyed site for each 30 m cell in the study area (Murphy 2008). Error was assessed via comparison of predicted connectivity with observed genetic distance between breeding sites. The RF model was predicted to the derived rasters (Evans and Cushman 2009), providing a continuous surface of functional connectivity across the entire landscape.

We found that areas of high functional connectivity exhibit less topographic complexity and higher rainfall than areas of low functional connectivity (Fig. 9.3). Predicted lanscape resistance can be considered when managers weigh decisions about where to plan new development or focus restoration efforts. Flat areas are beneficial for toad functional connectivity but also are likely candidates when considering future road construction. If managers are concerned about maintaining functional connectivity of boreal toads, hotspots of connectivity can be avoided when considering development alternatives (Fig. 9.3). In the western areas of the park, functional connectivity appears to be elevated, even in the presence of high topographic complexity (Fig. 9.3). This is likely due to the distribution of burned areas from the 1988 Yellowstone fires which are associated with increased gene flow (Murphy et al. 2010b). Restoration of the natural fire cycle in Yellowstone National Park is likely beneficial for boreal toad functional connectivity.

Predicting functional connectivity to the landscape elucidates current landscape condition. Yet landscapes are rapidly changing due to climatic and development forces. Once a model of functional connectivity is fit, it is possible to explore potential effects of landscape change by employing an envelope approach that fixes the model, modifies the independent variables to represent alternative scenarios and then forecasts functional connectivity under hypothesized climate change or development scenarios (see Chap. 8).

In the case of boreal toads, we could predict functional connectivity under different models of climate change. Overall, we would expect functional connectivity to decline as Yellowstone becomes more hot and dry. We could also look at the effect of specific proposed development alternatives, such as road placement, on functional connectivity. One consideration in envelope approaches is that they assume a linear projection from the contemporary model, ignoring non-linear changes in the variable interactions and emergent influences not observed in the contemporary model. Because the statistical relationships are not refit, they can become decoupled and produce erroneous results. Error is also of concern as it is very difficult to track and it potentially assumes linear, additive or exponential forms when sub-models (e.g., climate change, connectivity) are combined. However, there is high potential for conservation and management application of these types of predictive models of functional connectivity in the future.

9.8 Conclusions

Landscape genetics is a powerful approach with the potential to provide data for predictive models in ecology. These data will allow researchers and managers to model species presence, genetic diversity, and functional connectivity. However, statistical approaches for landscape genetics are still being developed and require communication across disciplines (Figs. 9.1 and 9.2); robust "canned" methods that quantify effect of landscape data on observed genetic patterns are currently unavailable (Balkenhol et al. 2009). Collaboration among quantitative landscape ecologists, population geneticists, and managers is vital to avoid pitfalls and result in successful landscape genetics models (Fig. 9.2).

Acknowledgments We would like to thank: S. Cushman, R. Dezzani, D. Patla, C. Peterson, A. Storfer, and the Palouse Landscape Genetics reading group for feedback on data collection and analyses presented in this chapter. K. Sochi, Y. Wiersma, A. Drew, and three anonymous reviewers provided valuable comments on earlier versions of the chapter. Research in our case study was permitted by the Washington State University IACUC committee (ASAF #3378) and the Yellowstone Research Office (YELL-05452). Funding for the case study was provided by the EPA – STAR fellowship (FP-916695), NSF – DDIG (DEB-0608458), James King fellowship, Theodore Roosevelt Memorial Fund, Cannon National Parks, Sigma-Xi grants-in-aid (2004), Society for Wetland Science, and Graduate Women in Science. MA Murphy was supported by a Colorado State University Post-doctoral Fellowship (WC Funk).

Glossary

Landscape genetics is both interdisciplinary and highly technical. Terminology used in landscape genetics may be unfamiliar to landscape ecologists and managers. This glossary includes some of the more common terms used in population, conservation, and landscape genetics to improve cross-disciplinary communication. In addition to our glossary, there are many excellent books dedicated to population and/or conservation genetics which are recommended for in-depth explanation of terminology, additional concepts, historical development, methods, and techniques (Allendorf and Luikart 2007; Avise 2004; Clark and Russell 2005; Frankham et al. 2002; Hartl and Clark 1997). Words *italicized* within a definition are also defined in the glossary.

Term	Definition	References
Admixture	Individuals are "mixed" genetically, usually among 2 or more groups	Avise (2004)
AFLP	Amplified Fragment Length Polymorphisms – Non-specific *primers* are used in PCR to amplify random bands throughout the genome	Avise (2004)
Allele	Copy of a *gene* or *marker* at a *locus*	Frankham et al. (2002)
Allozyme	Proteins are run out on a gel using electrophoresis to quantify variation. Protein size, structure, and size affect how quickly they travel through the gel. Also called protein electrophoresis	Avise (2004)
Assignment test	A class of statistical methods used to assign individual *genotypes* to groups of most likely/probable membership	Manel et al. (2005)
Base-pair	A given nucleotide (and compliment) in the DNA	Clark and Russell (2005)
(Genetic) Bottleneck	A sudden decrease in *effective population size*, resulting in skewed *allele* frequency distributions	Allendorf and Luikart (2007)
Chloroplast DNA (cDNA)	Circular DNA found in the chloroplast with uni-parental inheritance. Generally does not undergo *recombination*	Allendorf and Luikart (2007)
Coding	DNA in a region that is thought to produce RNA and DNA. A coding region of DNA can also be thought of as a *gene*	Avise (2004)
(Genetic) drift	Random changes in allele frequencies over time due to sampling effects from one *generation* to the next	Allendorf and Luikart (2007)
Effective population size	The idealized population size (e.g., equal sex ratio, random mating, non-overlapping generations,) that would result in the observed level of *genetic drift*	Frankham et al. (2002)
Gene	A section of DNA that *codes* for a protein	Clark and Russell (2005)
Gene flow	Exchange of genetic information among populations/demes through the exchange of genetic *migrants*	Allendorf and Luikart (2007)
Generation	Average age of reproduction	Allendorf and Luikart (2007)
Genotype	An organism's genetic composition (at a *locus*, across *loci*, or as determined through expressed trait)	Clark and Russell (2005)

(continued)

(continued)

Term	Definition	References
Geographical (spatial) genetics	Use of genetic methods to investigate the effects of spatial process on genetic variation. Under our definition, this is distinguishable from *landscape genetics* because even though spatial process (e.g., spatial autocorrelation) is the focus, landscape process (e.g., landscape heterogeneity) is not explicitly included	Epperson (2003)
Genetic pattern	Distribution of a response variable across the landscape as measured or identified by genetic *markers*	Storfer et al. (2007)
Heterozygote	An individual with two different *alleles* at a *locus*	Allendorf and Luikart (2007)
Homozygote	An individual with two copies of the same *allele* at a *locus*	Frankham et al. (2002)
Isolation by distance	Individuals close together are more genetically similar than those far apart.	Nei (1972)
Landscape genetics	{R}esearch that explicitly quantifies the effects of landscape composition, configuration and/{or} matrix quality on *gene flow* and/{or} spatial genetic variation"	Storfer et al. (2007)
Locus (loci)	Location on a chromosome of a *gene* or *marker*	Allendorf and Luikart (2007)
Marker	Any trait (*coding* or non-*coding*) used to assay genetic variation	Avise (2004)
Microsatellite (μsat)	Tandem repeat in nucleotides of 1–6 bases, repeated generally 5–100 times. Non-*coding* and thought to be *neutral* in most cases	Allendorf and Luikart (2007)
Migration	Permanent movement of *genes* (or non-coding alleles) among groups	Hartl and Clark (1997)
Mitochondrial DNA (mtDNA)	Circular DNA found in the mitochondria with uni-parental inheritance. Does not undergo *recombination*	Allendorf and Luikart (2007)
Neutral	Genetic trait not under (significant) selection	Avise (2004)
Nuclear DNA (nDNA)	Bi-parentally inherited (in species with sexual reproduction) DNA found in the nucleus of cells	Clark and Russell (2005)
Phylogeography	Geographic analysis of DNA on evolutionary timescales. Distinguishable from *landscape genetics* by being similar in temporal/geographic scale to biogeography	Avise (2000), Storfer et al. (2007)

(continued)

(continued)

Term	Definition	References
Polymerase chain reaction (PCR)	Method of creating many copies of a targeted section of DNA from few (or one) copy of original DNA template. *PCR* is conducted over many cycles (~15–50) with an exponential relationship between number of copies of the target DNA and number of *PCR* cycles	Clark and Russell (2005)
(DNA) Primer	A short segment of DNA that binds to the DNA template in *PCR* to initiate synthesis of a new copy of a targeted section of DNA	Clark and Russell (2005)
Recombination	"Crossing over" of DNA during meiosis. Recombination is the process by which traits on the same chromosome are shuffled in offspring compared to the parental DNA	Hartl and Clark (1997)
(DNA) Sequence	Complete identification, in order, of all base-pairs for the target region	Clark and Russell (2005)
SNP	Single Nucleotide Polymorphism – specific *base-pair* variants	Morin et al. (2004)
RFLP	Restriction Fragment Length Polymorphism – restriction enzymes cut at any area with a particular sequence. The resulting banding pattern is compared among samples	Avise (2004)

References

Adams JR, Waits LP (2007) An efficient method for screening faecal DNA genotypes and detecting new individuals and hybrids in the red wolf (*Canis rufus*) experimental population area. Conserv Gen 8:123–131.

Allendorf FW, Luikart G (2007) Conservation and the genetics of populations. Blackwell Publishing, Oxford, UK.

Anderson AD, Weir BS (2007) A maximum-likelihood method for the estimation of pairwise relatedness in structured populations. Genetics 176:421–440.

Annis SL, Dastoor FP, Ziel H, Daszak P, Longcore J (2004) A DNA-based assay identifies Batrachochyytrium dendrobatidis in amphibians. J Wlf Diseases 40:420–428.

Araújo MB, Guisan A (2006) Five (or so) challenges for species distribution modelling. J Biogeogr 33:1677–1688.

Archie EA, Luikart G, Ezenwa VO (2009) Infecting epidemiology with genetics: a new frontier in disease ecology. Trends Ecol Evol 24:21–30.

Arnaud JF (2003) Metapopulation genetic structure and migration pathways in the land snail *Heliz aspersa*: influence of landscape heterogeneity. Landsc Ecol 18:333–346.

Avise JC (2000) Phylogeography: the history and formation of species. Harvard University Press, Cambridge, MA.

Avise JC (2004) Molecular markers, natural history, and evolution. Sinauer, Sunderland, MA.

Balkenhol N, Gugerli F, Cushman SA, Waits LP, Coulon A, Arntzen JW, Holderegger R, Wagner HH and Participants of the Landscape Genetics Research Agenda Workshop (2009) Identifying future research needs in landscape genetics: where to from here? Landsc Ecol 24:455–463.

Balloux F, Lugon-Moulin N (2002) The estimation of population differentiation with microsatellite markers. Mol Ecol 11:155–165.

Belfiore NM, Hoffman FG, Baker RJ, Dewoody JA (2003) The use of nuclear and mitochondrial single nucleotide polymorphisms to identify cryptic species. Mol Ecol 12:2011–2017.

Belkhir K, Castric V, Bonhomme F (2002) IDENTIX, a software to test for relatedness in a population using permutation methods. Mol Ecol Notes 2:611–614.

Bellinger MR, Johnson JA, Toepfer J and Dann P (2003) Loss of genetic variation in greater prairie chickens following a population bottleneck in Wisconsin, U.S.A. Conserv Biol 17:717–724.

Bickford D, Lohman DJ, Sodhi NS, Ng PKL, Meier R, Winker K, Ingram KK, Das I (2007) Cryptic species as a window on diversity and conservation. Trends Ecol Evol 22:148–155.

Blanchong JA, Samuel MD, Scribner KT, Weckworth BV, Langenberg JA, Filcek KB (2008) Landscape genetics and the spatial distribution of chronic wasting disease. Biol Lett 4:130–133.

Broquet T, Ménard N, Petit E (2007) Noninvasive population genetics: a review of sample source, diet, fragment length and microsatellite motif effects on amplification success and genotyping error rates. Conserv Gen 8:249–260.

Bowcock AM, Ruiz-Linares A, Tomfohrde J, Minch E, Kidd JR, Cavalli-Sforza LL (1994) High resolution of human evolutionary trees with polymorphic microsatellites. Nature 368:455–457.

Caizergues A, Rätti O, Helle P, Rotelli L, Ellison L, Rasplus JY (2003) Population genetic structure of male black grouse (*Tetrao tetrix* L.) in fragmented vs. continuous landscapes. Mol Ecol 12:2297–2305.

Castellano S, Balletto E (2002) Is the partial Mantel test inadequate? Evolution 56:1871–1873.

Cavelli-Sforza LL, Edwards AWF (1967) Phylogenetic analysis: models and estimation procedures. Am J Hum Gen 19:233–257.

Cegelski CC, Waits LP, Anderson NJ (2003) Assessing population structure and gene flow in Montana wolverines (*Gulo gulo*) using assignment-based approaches. Mol Ecol 12:2907–2918.

Cercueil A, François O, Manel S (2007) The genetical bandwidth mapping: a spatial and graphical representation of population genetic structure based on the Wombling method. Theor Popul Biol 71:332–341.

Clark DP, Russell LD (2005) Molecular biology made simple and fun. Cache River Press, St. Louis, MO.

Colson I, Goldstein DB (1999) Evidence for complex mutations at microsatellite loci in *Drosophila*. Genetics 152:617–627.

Cornuet JM, Luikart G (1996) Description and power analysis of two tests for detecting recent population bottlenecks from allele frequency data. Genetics 144:2001–2014.

Crooks KR, Sanjayan M (eds) 2006. Connectivity conservation. Cambridge University Press, Cambridge, UK.

Cushman SA, McKelvey KS, Hayden J, Schwartz MK (2006) Gene flow in complex landscapes: testing multiple hypotheses with causal modeling. Am Nat 168:486–499.

Cutler DR, Edwards Jr TC, Beard KH, Cutler A, Hess KT, Gibson J, Lawler JJ (2007) Random forests for classification in ecology. Ecology 88:2783–2792.

Dalén L, Elmhagen B, Angerbjörn A (2004) DNA analysis on fox faeces and competition induced niche shifts. Mol Ecol 13:2389–2392.

DeYoung RW, Honeycutt RL (2005) The molecular toolbox: genetic techniques in wildlife ecology and management. J Wild Manage 69:1362–1384.

Dieringer D, Schlötterer C (2003) Microsatellite analyser (MSA): a platform independent analysis tool for large microsatellite data sets. Mol Ecol Notes 3:167–169.

Eggert LS, Eggert JA, Woodruff DS (2003) Estimating population sizes for elusive animals: the forest elephants of Kakum National Park, Ghana. Mol Ecol 12:1389–1402.

Epperson BK (2003) Geographical genetics. Princeton University Press, Princeton and Oxford.

Epps CW, Wehausen JD, Bleich VC, Torres SG, Brashares JS (2007) Optimizing dispersal and corridor models using landscape genetics. J Appl Ecol 44:714–724.

Evans JS, Cushman SA (2009) Gradient modeling of conifer species using random forests. Landsc Ecol 24:673–683.

Ferrari JR, Lookingbill TR (2009) Initial conditions and their effect on invasion velocity across heterogeneous landscapes. Biol Invasions 11:1247–1258.

Ficetola GF, Miaud C, Pompanon F, Taberlet P (2008) Species detection using environmental DNA from water samples. Biol Lett 4:423–425.

Foley JE, Queen EV, Sacks B, Foley P (2005) GIS-facilitated spatial epidemiology of tick-borne diseases in coyotes (*Canis latrans*) in northern and coastal California. Comp Immunol Microbiol Infect Dis 28:197–212.

Fortin MJ, Dale M (2005) Spatial analysis: a guide for ecologists. Cambridge University Press, Cambridge, UK.

Frair JL, Nielsen SE, Merrill EH, Lele SR, Boyce MS, Munro RHM, Stenhouse GB, Beyer HL (2004) Removing GPS collar bias in habitat selection studies. J Appl Ecol 41:201–212.

Frankham R, Ballou JD, Briscoe DA (2002) Introduction to conservation genetics. Cambridge University Press, New York.

Funk WC, Murphy MA (2010) Testing evolutionary hypotheses for phenotypic divergence using landscape genetics. Mol Ecol 19:427–430.

Funk WC, Blouin MS, Corn PS, Maxell BA, Pilliod DS, Amish S, Allendorf FW (2005) Population structure of Columbia spotted frogs (*Rana luteiventris*) is strongly affected by the landscape. Mol Ecol 14:483–496.

Gauffre B, Estoup A, Bretagnolle V, Cosson JF (2008) Spatial genetic structure of a small rodent in a heterogeneous landscape. Mol Ecol 17:4619–4629.

Goldberg CS, Hawley TJ, Waits LP (2010) Local and regional patterns of amphibian chytrid prevalences on the Osa Peninsula, Costa Rica. Herpetol Rev 40:309–311.

Goldstein DB, Linares AR, Cavalli-Sforza LL, Feldman MW (1995) An evaluation of genetic distances for use with microsatellite loci. Genetics 139:463–471.

Green DM (2003) The ecology of extinction: population fluctuation and decline in amphibians. Biol Conserv 111:331–343.

Guillot G, Leblois R, Coulon A, Frantz AC (2009) Statistical methods in spatial genetics. Mol Ecol 18:4734–4756.

Hanski I, Gaggiotti OE (eds) (2004) Ecology, genetics, and evolution of metapopulations. Elsevier, Burlington, VT.

Hartl DL, Clark AG (1997) Principles of population genetics. Sinauer Associates, Sunderland, MA.

Holderegger R, Wagner HH (2008) Landscape genetics. BioScience 58:199–207.

Holderegger R, Kamm U, Gugerli F (2006) Adaptive vs. neutral genetic diversity: implications for landscape genetics. Landsc Ecol 21:797–807.

Howeth JG, McGaugh SE, Hendrickson DA (2008) Contrasting demographic and genetic estimates of dispersal in the endangered Coahuilan box turtle: a contemporary approach to conservation. Mol Ecol 17:4209–4221.

Hu LJ, Uchiyama K, Shen HL, Saito Y, Tsuda Y, Ide Y (2008) Nuclear DNA microsatellites reveal genetic variation but a lack of phylogeographical structure in an endangered species, *Fraxinus mandshurica*, across north-east China. Ann Bot 102:195–205.

Janssens X, Fontaine MC, Michaux JR, Libois R, de Kermabon J, Defourny P, Baret PV (2008) Genetic pattern of the recent recovery of European otters in southern France. Ecography 31:176–186.

Ji W, Leberg P (2002) A GIS-based approach for assessing the regional conservation status of genetic diversity: An example from the southern Appalachians. Environ Manage 29:531–544.

Jones AG, Ardren WR (2003) Methods of parentage analysis in natural populations. Mol Ecol 12:2511–2523.

Leberg PL (2002) Estimating allelic richness: effects of sample size and bottlenecks. Mol Ecol 11:2445–2449.

Legendre P (1993) Spatial autocorrelation: trouble or new paradigm? Ecology 74:1659–1673.

Legendre P, Legendre L (1998) Numerical ecology. Elsevier Science, Amsterdam.

Luikart G, Cornuet JM (1998) Empirical evaluation of a test for identifying recently bottlenecked populations from allele frequency data. Conserv Biol 12:228–237.

Manel S, Gaggiotti OE, Waples RS (2005) Assignment methods: matching biological questions with appropriate techniques. Trends Ecol Evol 20:136–142.

Manel S, Schwartz MK, Luikart G, Taberlet P (2003) Landscape genetics: combining landscape ecology and population genetics. Trends Ecol Evol 18:189–197.

Mantel N (1967) The detection of disease clustering and a generalized regression approach. Cancer Res 27:209–220.

Martínez-Solano I, González EG (2008) Patterns of gene flow and source-sink dynamics in high altitude populations of the common toad *Bufo bufo* (Anura: Bufonidae). Biol J Linn Soc 95:824–839.

Maruyama T, Fuerst PA (1985) Population bottlenecks and nonequalibrium models in population genetics. II. Number of alleles in a small population that was formed by a recent bottleneck. Genetics 111:675–689.

McRae BH (2006) Isolation by resistance. Evolution 60:1551–1561.

McRae BH, Beier P (2007) Circuit theory predicts gene flow in plant and animal populations. Proc Nat Acad Sci 104:19885–19890.

Michels E, Cottenie K, Neys L, DeGalas K, Coppin P, DeMeester L (2001) Geographical and genetic distances among zooplankton populations in a set of interconnected ponds: a plea for using GIS modeling of the effective geographical distance. Mol Ecol 10:1929–1938.

Miller CR, Adams JR, Waits LP (2003) Pedigree-based assignment tests for reversing coyote (*Canis latrans*) introgression into the wild red wolf (*Canis rufus*) population. Mol Ecol 12:3287–3301.

Morin PA, Luikart G, Wayne R (2004) SNPs in ecology, evolution, and conservation. Trends Ecol Evol 19:208–216.

Murphy MA (2008) New approaches in landscape genetics and niche modeling for understanding limits to anuran distributions, PhD Thesis, School of Biological Sciences, Washington State University: Pullman, WA. pp. 252.

Murphy MA, Waits LP, Kendall KC (2000) Quantitative evaluation of fecal drying methods for brown bear DNA analysis. Wildl Soc Bull 28:951–957.

Murphy MA, Evans JS, Cushman S, Storfer A (2008) Representing genetic variation as continuous surfaces: an approach for identifying spatial dependency in landscape genetic studies. Ecography 31:685–697.

Murphy M, Dezzani R, Pilliod D, Storfer A (2010a) Landscape genetics of high mountain frog Metapopulations. Mol Ecol 19, 3634–3649.

Murphy MA, Evans JS, Storfer A (2010b) Quantifying *Bufo boreas* connectivity in Yellowstone National Park with landscape genetics. Ecology 91:252–261.

Muths E (2003) Home range and movements of boreal toads in undisturbed habitat. Copeia 1:160–165.

Muths E, Corn PS, Pessier AP, Green DE (2003) Evidence for disease-related amphibian decline in Colorado. Biol Conserv 110:357–365.

Nei M (1972) Genetic distance between populations. Am Nat 106:283–292.

Ostfeld RS, Glass GE, Kessing F (2008) Spatial epidemiology: an emerging (or re-emerging) discipline. Trends Ecol Evol 20:328–336.

Paetkau D, Waits LP, Clarkson PL, Craighead L, Strobeck C (1997) An empirical evaluation of genetic distance statistics using microsatellite data from bear (Ursidae) populations. Genetics 147:1943–1957.

Rayfield B, Fall A, Fortin MJ, Fall A (2010) The sensitivity of least-cost habitat graphs to relative cost surface values. Landsc Ecol 25:519–532.

Real LA, Biek R (2007) Spatial dynamics and genetics of infectious diseases on heterogeneous landscapes. J R Soc Interface 4:935–948.

Real LA, Henderson JC, Biek R, Snaman J, Jack TL, Childs JE, Stahl E, Waller L, Tinline R, Nadin-Davis S (2005) Unifying the spatial population dynamics and molecular evolution of epidemic rabies virus. Proc Nat Acad Sci 102:12107–12111.

Reed SE, Merenlender AM (2008) Quiet, noncomsumptive recreation reduced protected area effectiveness. Conserv Lett 1:146–154.

Rees EE, Pond BA, Cullingham CI, Tinline R, Ball D, Kyle CJ, White BN (2008) Assessing a landscape barrier using genetic simulation modelling: implications for raccoon rabies management. Prev Vet Med 86:107–123.

Rees EE, Pond BA, Cullingham CI, Tinline RR, Ball D, Kyle CJ, White BN (2009) Landscape modelling spatial bottlenecks: implications for raccoon rabies disease spread. Biol Lett 5:387–390.

Rehfeldt GE (2006) A spline model of climate for the Western United States, RMRS-GTR-165. U.S. Department of Agriculture, Forest Service, Rocky Mountain Research Station, Fort Collins, CO.

Riddle AE, Pilgrim KL, Mills LS, McKelvey KS, Ruggiero LF (2003) Identification of mustelids using mitochondrial DNA and non-invasive sampling. Conserv Genet 4:241–243.

Riley SPD, Pollinger JP, Sauvajot RM, York EC, Bromley C, Fuller TK, Wayne RK (2006) A southern California freeway is a physical and social barrier to gene flow in carnivores. Mol Ecol 15:1733–1741.

Rokas A, Abbot P (2009) Harnassing genomics for evolutionary insights. Trends Ecol Evol 24:192–200.

Rousset F (2002) Partial Mantel tests: reply to Castellano and Balletto. Evolution 56:1874–1875.

Rousset F (2008) Genepop '007: a complete re-implementation for Genepop software for Windows and Linux. Mol Ecol Resour 8:103–106.

Rudnick JA, Katzner TE, Bragin EA, DeWoody JA (2007) Species identification of birds through genetic analysis of naturally shed feathers. Mol Ecol Notes 7:757–762.

Ruell EW, Crooks KR (2007) Evaluation of noninvasive genetic sampling methods for felid and canid populations. J Wildl Manage 71:1690–1694.

Scribner KT, Blanchong JA, Bruggeman DJ, Epperson BK, Lee CY, Pan YW, Shorey RI, Prince HH, Winterstein SR, Luukkonen DR (2005) Geographical genetics: conceptual foundations and empirical applications of spatial genetic data in wildlife management. J Wildl Manage 69:1434–1453.

Semlitsch RD (2008) Differentiating migration and dispersal process for pond-breeding amphibians. J Wildl Manage 72:260–267.

Simandle ET, Peacock MM, Zirelli L, Tracy CR (2006) Sixteen microsatellite loci for the *Bufo boreas* group. Mol Ecol Notes 6:116–119.

Slatkin M (1995) A measure of population subdivision based on microsatellite allele frequencies. Genetics 139:457–462.

Sneath PHA, Sokal RR (1973) Identification and discrimination. In: Sneath PHA, Sokal RR (eds) Numerical taxonomy: the principles and practice of numerical classification. W. H. Freeman and Company, San Francisco, CA.

Spear SF, Storfer A (2008) Landscape genetic structure of coastal tailed frogs (*Ascaphus truei*) in protected vs. managed forests. Mol Ecol 17:4642–4656.

Spear SF, Peterson CR, Matocq M, Storfer A (2005) Landscape genetics of the blotched tiger salamander (*Ambystoma tigrinum melanostictum*). Mol Ecol 14:2553–2564.

Storfer A, Murphy MA, Evans JS, Goldberg CS, Robinson S, Spear SF, Dezzani R, Demmelle E, Vierling L, Waits LP (2007) Putting the 'landscape' in landscape genetics. Heredity 98:128–142.

Storfer A, Murphy MA, Holderegger R, Spear SF, Waits L (2010). Landscape genetics, where are we now? Mol Ecol 19, 3496–3514.

Taberlet P, Luikart G, Waits LP (1999) Noninvasive genetic sampling: look before you leap. Trends Ecol Evol 14:323–327.

Taberlet P, Camarra JJ, Griffin S, Uhrès E, Hanotte O, Waits LP, Dubois-Paganon C, Burke T, Bouvet J (1997) Noninvasive genetic tracking of the endangered Pyrenean brown bear population. Mol Ecol 6:869–876.

Taylor PD, Fahrig LF, With KA (2006) Landscape connectivity: a return to the basics. In: Crooks KR, Sanjayan MA (eds) Connectivity conservation. Cambridge University Press, Cambridge, UK.

Valsecchi E, Glockner-Ferrari D, Ferrari M, Amos W (1998) Molecular analysis of sloughed skin sampling in whale population genetics. Mol Ecol 7:1419–1422.

Van Loon EE, Clearly DFR, Fauvelot C (2007) ARES: software to compare allelic richness between uneven samples. Mol Ecol Notes 7:579–582.

Wagner HH, Werth S, Kalwij JM, Bolli JC, Scheidegger C (2006) Modelling forest recolonization by an epiphytic lichen using a landscape genetic approach. Landsc Ecol 21:849–865.

Wagner HH, Fortin MJ (2005) Spatial analysis of landscapes: concepts and statistics. Ecology 86:1975–1987.

Waits LP, Paetkau D (2005) Noninvasive genetic sampling tools for wildlife biologists: a review of applications and recommendations for accurate data collection. J Wildl Manage 69:1419–1433.

Waits L, Taberlet P, Swenson JE, Sandegren F, Fransén R (2000) Nuclear DNA microsatellite analysis of genetic diversity and gene flow in the Scandinavian brown bear (*Ursus arctos*). Mol Ecol 9:421–431.

Wang IJ, Summers K (2010) Genetic structure is correlated with phenotypic divergence rather than geographic isolation in the highly polymorphic strawberry poison-dart frog. Mol Ecol 19:447–458.

Wang J (2005) Estimation of effective population sizes from data on genetic markers. Phil Trans R Soc B 360:1395–1409.

Wright S (1951) The genetical structure of populations. Annal Eugen 15:323–354.

Young AG, Clark GM (eds) (2000) Genetics, demography, and viability of fragmented populations. Cambridge University Press, Cambridge, UK.

Chapter 10
Simplicity, Model Fit, Complexity and Uncertainty in Spatial Prediction Models Applied Over Time: We Are Quite Sure, Aren't We?

Falk Huettmann and Thomas Gottschalk

10.1 Introduction

There is a strong need to assess impacts on wildlife and their habitats before they occur, and to act proactively to avoid "costs" (e.g., loss of species, wilderness, ecological services, human lives or money; see Nielsen et al. 2008 for an applied example). Although the concept of being proactive has been known for decades, the global climate change discussion has brought these concepts to the forefront. Proactive action also has use in impact studies of stochastic catastrophes such as floods or hurricanes. Simulations and predictions across time can help to mitigate or even resolve current problems (Fig. 10.1). Such techniques are widely used to assess risks and they have evolved into industrial standards elsewhere, such as in operations research (Fuller et al. 2008), within the pharmaceutical, insurance, and car industries, and in public health (Herrick et al. in press). Weather forecasts have already shown the value and power of such models, driving many day-to-day decisions. In ecology, a good forecast system, based on state-of-the-art modeling techniques, and applied in an adaptive management framework could help to achieve sustainability worldwide (Braun 2005; Huettmann 2007a; Gottschalk et al. 2010). Historical models are further helpful for understanding the past, the present, the future, and for getting a sense of variability of the system overall (see Wickert 2007 for an example predicting historical bird distributions). On such large scales in time and space, even tiny (decimal) model improvements, and beyond the traditional 95% significance level, can have huge impact. However, there is an ongoing discussion about which model and assessment approach is best suited to predict species distribution (Manel et al. 1999; Elith et al. 2006; Guisan et al. 2007). The traditional frequentist view is still applied, but is of limited value (Breiman 2001a).

F. Huettmann (✉)
EWHALE lab, Institute of Arctic Biology, Biology and Wildlife Department, University of Alaska, Fairbanks, AK 99775, USA
e-mail: fhuettmann@alaska.edu

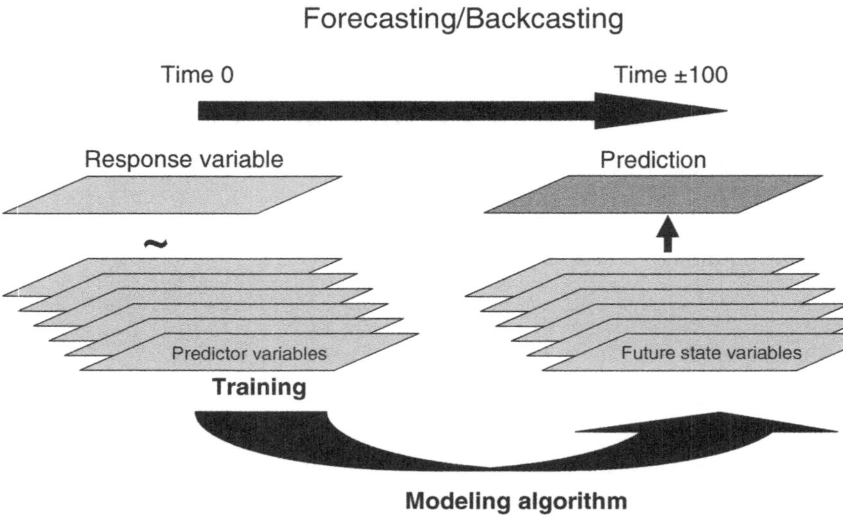

Forecasting/Backcasting

Fig. 10.1 Steps involved in how to model across time (Fore- and Backcasting)

It has long been believed and promoted that simple, parsimonious models would perform "best" to generalize accurately over time and space (Burnham and Anderson 2002; Braun 2005). Alternatively, dynamic or process-based models and models involving biotic interactions, feedbacks, and adaptations can become extremely complex when attempting to mimic nature (Austin 2002) and are still a challenge (Araújo and Guisan 2006). Traditionally, simple linear functions, often linked with maximum likelihood theory, have been used at the core of many of these models (e.g., Manly et al. 2002). Model averaging has been promoted as a solution to generalize from competing models (Burnham and Anderson 2002). However, examinations of the predictive performance and real-world applications of a variety of such modeling techniques show that (1) best-possible predictions become a science objective of global interest and application, (2) parsimony is not the main theme anymore, (3) digital data quality and its documentation is at the core of good modeling, (4) precision does not always guarantee predictive accuracy, and (5) that, depending on the underlying data structure itself, different modeling algorithms and philosophies perform heterogeneously, but a coherent group of algorithms (especially algorithms named as BRT (Boosted Regression Trees), MAXENT (Maximum Entropy), GDM (Generalised Dissimilarity Modelling), BRUTO (Adaptive Backfitting), RF (RandomForest) and MARS (Multivariate Adaptive Regression Splines) consistently achieved better (e.g., Elith et al. 2006)).

This presents a paradigm shift beyond modeling, and affects how we draw inferences and perform research for conservation and wildlife management applications. It further affects many aspects of the modern sciences (e.g., statistics, ecology, informatics, wildlife management, landscape ecology, conservation biology), and how to study and obtain valid inference (Breiman 2001a; Oppel et al. 2009; Cushman and Huettmann 2010). One conclusion from this, so far, is that complex, blackbox, ensemble models (e.g., Araújo and New 2007; Desktop GARP www.nhm.ku.edu/desktopgarp/UsersManual.html, Open Modeler http://openmodeller.sourceforge.net/) should always be considered if high predictive performance is required. After all, who wants poorly performing models when one has to use a model to convince the public on sustainability management issues?

Another conclusion is that an alternative (digital) data set available for accuracy assessment of predictions must always be part of the actual research design. It is tragic that this specific need to collect alternative data for model validation has not yet become clear to the research design community as a requirement in field data collection (Braun 2005; Magness et al. 2010). Performance metrics should be robust, coherent, and have an international standard. Surprisingly, a shift of performance metrics has many implications, and can shift scientific findings. With the shift away from traditional performance metrics (such as R^2s and p-values at the 95% threshold) toward metrics based on a confusion matrix (e.g., Area Under the Receiver Operating Characteristic curve (AUC), Kappa; Fielding and Bell 1997; Pearce and Ferrier 2000a), uncertainty can arise (Myers 1997; Boyce et al. 2002). For one, the community of managers and scientists, and the general public at large, is still widely entrenched in the 95% probability concept and hardly trusts anything else. This comes from a notion that been introduced since the 1930s (Fisher 1930) and is still maintained in relevant science journals, their boards, institutions and co-workers (O'Connor 2000; Oppel et al. 2009 for a statistical critique). A second paradigm to overcome is parsimony (Burnham and Anderson 2002), which suggests that the preferred solution should always be the most simple ("Occams Razor"), as promoted in the Akaike Information Criterion (AIC). Third, aside from the fact that many predictive models still lack alternative datasets for any relevant (quantitative) accuracy assessment, models that predict forward and backward in time suffer inherently from this provided notion of "trust," which is an accepted metric of their predictive accuracy (Lawler et al. 2006; Wickert 2007). Without a quantitative certainty measure, modeling can drift into an unproven subjective exercise, at which point it cannot be used for a serious and objective science-based management (Araújo and Guisan 2006; Lozier et al. 2009). In such cases, modeling wanders into the realm of "subjective story telling with numbers," to be easily dismissed by the public at large and by decision-makers (for example in court decisions, where the acceptance of modeling is critical). Here, we try to give an overview of this accuracy testing problem in modeling, how it currently is addressed, what can and should be done about it, and what the future will hold to improve the situation and establish modeling as a "sound science." (For terminology used in this chapter, please see Table 10.1).

Table 10.1 Terminology related to modeling

Term	Explanation	Application	References
Prediction	Predict a model outcome to a given location or time	Distribution maps	Guisan and Thuiller (2005)
Extrapolation	Fill adjacent space for the same time window	Spatial extrapolations, predicted distribution and population estimate maps	Yen et al. (2004) and Graeber (2005)
Forecast	Apply a model forward in time	Future climate change and landcover change applications	Lawler et al. (2009)
Back-/hind-casting	Same as forecasting, but applied backwards in time	Historical distributions, palaeo-modeling	Martínez-Meyer et al. (2004), Martínez-Meyer and Peterson (2006) and Wickert (2007)
Simulation	Projecting the potential effects of alternative management activities on wildlife	Landcover change Climate change Management change	Larson et al. (2004) and Ziv (1998)

10.2 How Do Temporal Dynamics Create Challenges in Predictive Landscape Ecology? How Are They Addressed?

Most niches are rather robust in time (Wiens and Graham 2005; Martínez-Meyer and Peterson 2006). However, over long time frames, the ecological niche can change (Fig 10.2). This is specifically true when niche competition and its framework is altered. These changes are part of the global evolutionary process, which work by awarding fitness in future generations (Hickerson et al. 2009; Cushman 2010). There are "natural" changes, but also "anthropogenic" changes in the ecological framework; they tend to act at different time-scales, and the human role of this component has dramatically increased over the last 50 years (Braun 2005; Foley et al. 2005; Chapin et al. 2010). A response to these changes is addressed through evolutionary processes. It is widely believed now that humans have increased the time intervals of evolutionary steps (Chapin et al. 2010). Predictive modelers deal with change in the environment, quantified as the ecological niche, which is used to predict species and events. In most cases of predictive modelling the general concept of spatial extrapolation has been used (Table 10.1). This is a "spread" of the described ecological niche in space (Fig. 10.3) using known spatial attributes. It assumes temporal stability (Manly et al. 2002). In the short term, such predictions (either forward or backward) might work well if the ecological niche is not rapidly evolving. Alternative approaches, not yet well recognized but rather powerful, come through the use of many "proxies," which also allowing for statistical interactions. Proxies involve the use of "placeholder" variables that compensate for the lack of the original,

10 Simplicity, Model Fit, Complexity and Uncertainty in Spatial Prediction Models 193

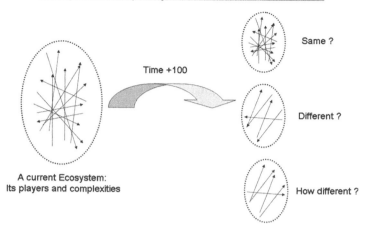

Fig. 10.2 Ecosystem complexity and model forecasting

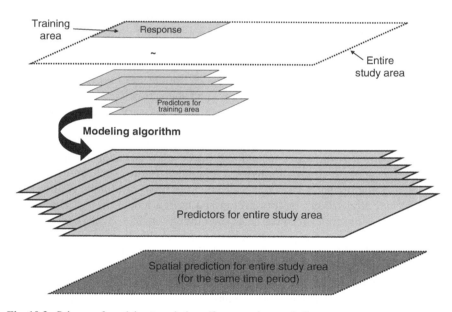

Fig. 10.3 Scheme of spatial extrapolations (for same time period)

best explanatory, predictor (e.g., if spatially explicit information on climate conditions are not available, altitude values from a digital elevation model are used as a "proxy" for regional difference in climate; Koerner and Paulsen 2004). Such approaches are often the only approach when the "true" predictor is unknown or not available. The reason why this overall approach can still achieve robust predictions is that many proxies are virtually able to mimic, and subsequently, to compensate, for the original predictor that is believed to explain a response. This is referred to as "swapping predictiors" and will become a huge statistical and management topic in the future. Testing these assumptions can lead to surprising results, and thus, should always be done.

The growing availability of digital data has made it possible to use many (10–6,000) more predictor variables than the "traditional five predictors" (Manly et al. 2002) that have been used in species distribution models (Magness 2009). The positive effects on the prediction accuracy can be amazing, and are not well acknowledged (rather the opposite attitude still prevails widely; Burnham and Anderson 2002 for instance still refer to "data mining" in a derogatory sense). When many predictors are used, their interactions must be addressed. These interactions can be multi-dimensional, but are often ignored, misunderstood or simply mislabeled as "bad." In contrast, explicit incorporation of ecological interactions can improved prediction accuracy. In addition, many interactions remain untested, and are often wrongly assigned to an actual predictor (see Oppel et al. 2009). The notion that interactions would make models generally worse should be re-considered (see also Breiman 2001a). The current attitude originates in traditional, "linear curve-fitting" modeling culture ("frequentists") where parametric assumptions such as independence are required, and all features of a model have to be fully understood first (an approach that unfortunately is still widely promoted, Murray and Conner 2009; but see Wittingham et al. 2006). However, in most ecological applications, such notions are limited in value, and not realistic. Examples of such problems, and their effective solutions, can be seen in Ritter (2007; for achieving accurate solutions very fast) and Craig and Huettmann (2008; for good accuracies from noisy data).

If models are applied beyond spatial extrapolation, then the traditional approach to forecasting has two options: (a) develop a model on the "current regime" and apply it to conditions of a simulated future, such as with future landscapes (Huettmann et al. 2005; Nielsen et al. 2008) or climate (Fuller et al. 2008; Lawler et al. 2009), or (b) develop and train a model on past conditions, and then move the trend forward (e.g., Lemoine et al. 2007). It is self-evident that such concepts can be applied backward in time (hind-casting), as well (see Table 10.1, and publications in palaeo-modeling, e.g., Nogués-Bravo et al. 2008; Hickerson et al. 2009).

These modeling techniques raise four questions: (1) are the described niches accurate, (2) are modeled processes constant over time, (3) do these models extrapolate accurately within these processes (both mathematically and biologically), and (4) how can the predictive accuracy performance be assessed in reliable and quantitative terms, and globally, across projects and applications? The obvious lack of a traditional, international, cross-species benchmark prediction accuracy in such studies is dealt with in the following sections. The obvious dilemma is that such models are crucial though for our decision-making, but that there is no truth really to assess them with!

10.3 Very Long-Range Forecasting Is Difficult (Virtually Impossible) to Validate in the Field Directly; So What Strategies Are Used to Test and/or Chose Among Predictive Models?

Because one cannot "know truth," future models cannot directly be assessed for their accuracy using traditional indices, such as with R^2 as a measure of fit or based on metrics from the confusion matrix (sensitivity, specificity, Cohen's Kappa statistic, TSS, ROC (Allouche et al. 2006) for categorical predictions), and mean residual error (Pineiro et al. 2008; for continuous predictions). Some of these traditional measures might work for predictions close-by in time, but are assumed not to work well for predictions far removed in time and additionally for large areas.

In order to provide trust, several strategies have been applied (Table 10.2). But because the traditional approach of confronting the model with alternative data can usually not be applied, a variety of methods, sometimes in concert, are frequently used to compensate.

It is worth pointing out that another modeling fact has rarely been considered in wildlife and landscape studies. That is, with niche conservatism a tight mathematical link should remain between the species and the identified predictor (e.g., climate). This specific mathematical link, often referred to in other disciplines as "coupling," will more or less remain constant over time. Thus, such models that predict far over time, either backward or forward, shall have a very high accuracy as long as the underlying predictors (often also modeled) are accurate. With that, it is possible that a highly accurate model can be derived even far into the future (or far back in history), although the underlying predictor models are not always well assessed (Fig. 10.4). Climate models present a nice example for such situations. When IPCC models get adjusted and updated, the actual modeled wildlife-climate niche remains unaffected but the final model changes. Many models and predictions rely on underlying models to begin with (see also Lawler et al., Chap. 13 this volume). Most DEMs and their derived layers (i.e., slope, aspect and sun incidence) consist of (interpolation) models and algorithms, potentially adding inherent and accumulating model errors.

With so many options and no clear guidelines, what works best? In ecology, it appears that a wide paradigm shift is approaching, and the traditional concept of "expertism" is virtually outdated (see Chap. 11 this volume). This can be shown when comparing experts with models. Strictly traditional quantitative approaches have been demonstrated to always be successful with regard to model fits, parsimony, and back-casting (Table 10.1). Concepts built on biological mechanisms are losing ground as well, due to a lack of generalizability and scale problems that affect the nature of the relationship to be inferred upon and generalized. Mechanistic models often are favored for explanatory and teaching reasons (Euskirchen et al. 2006, 2009). Meta-analysis approaches, however, appear to be very powerful and provide for a more honest and transparent overview in heterogeneous situations. They are widely used in the social sciences to provide

Table 10.2 Approaches to verify the validity of spatial models applied across time

Name	Description	Pro	Con	References
"Exploratory model"	No statement and effort is made regarding accuracy and trustworthiness of a model	Fast, no effort	Does not adhere to scientific principles of repeatability, transparency or testability	NA
Model fit	Metrics are provided that show how well the data fit the model and assumptions	Provides estimates that training data fit the underlying assumptions and statistics	Inductive. Does not relate to prediction accuracy	Manly et al. (2002)
Model predicted onto itself	Metrics are provided that show how well the algorithm can mimic its own data	Provides trust that the model can predict its current state	Does not provide any information on model accuracy for predictions outside of the training data	Manly et al. (2002)
Parsimony	The model is made as simplistic as possible to explain most variance and avoid overfitting	Model is forced to be "simplistic," and usually linear and easy to understand	"Simplistic" and linear models are a man-made, artificial and inappropriate description of a complex reality and ecology	Burnham and Anderson (2002)
Match with alternative data	The model is confronted with alternative testing data	A fair and powerful test when good alternative data exist	Such alternative data are usually lacking. Alternative data can be different in space, time or in regards to methods used	Pearce and Ferrier (2000a, b)
Backcasting and related	A model gets trained on the current situation, and then applied to a known, historic situation (these steps can get exchanged). The idea is that if the model would be robust over two known time steps, its predictions should be valid in the longer range also	Concept appears robust for the tested time frame	New factors can enter the system, making the previous version obsolete. (Humans affecting the atmosphere and climate over the last 50 years are good examples for such a situation)	IPCC http://www.ipcc.ch/
Expert knowledge	An expert for the biological system will look at the model, and often at its components and underlying algorithm, and judge whether is matches his/her experience how a system would behave	General agreement within the scientific community	Many experts differ in their opinion and often, no agreement can be found. For unknown ecological phenomena, no experts might be available	IPCC http://www.ipcc.ch/

Multiple model agreement	Identical or similar starting points get used, and it is assessed whether models converge, or at least show the same trends	Consensus	This approach suffers when all models are wrong, because they are based on similar data or assumptions	Marmion et al. (2009)
Metaanalysis	Many studies, often with different underlying data, qualities and approaches, are compared, and a general trend gets estimated	Provides a true overview, picture and summary	Non-quantitative, not very strong for situations with a small effect size	Worm and Myers (2003)
Mechanisms	Lab studies describing biological mechanisms and processes are used to "make sense" of the model results	Conceptually, models and biology should match and provide for best traditional inference and assessment	Provides just the aspatial results from the lab. Potential mismatch due to differences in scale, and underlying mechanisms tested	Austin (2002)

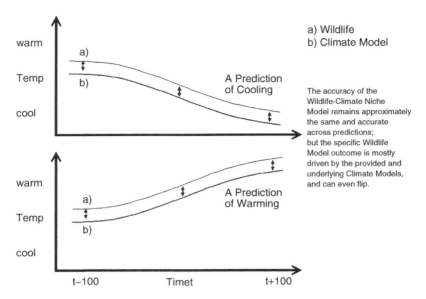

Fig. 10.4 Coupling of an accurate Wildlife-Climate Niche Model with a Climate Model, but with flipping prediction outcomes and showing the very high dependence on good and provided base models. The Wildlife response (**a**) is directly coupled with the Climate model (**b**) and reflecting all its variations. Reliable Climate Model are needed as reliable drivers for valid Wildlife-Climate Niche Model outcomes

"evidence," and are gaining recognition in ecology and science-based management such as fisheries research (Worm and Myers 2003) (e.g., Millennium Ecosystem Assessment; www.millenniumassessment.org/en/index.aspx). With no other available options, this approach may currently be among the best when conclusions need to be drawn from a variety of data sets, methods, and result formats. Second, continued data collection (study and modeling) should be the prime goal for providing "truth" (defined as the best available knowledge which is continually improving over time).

10.4 How Important Is Temporal Variability and Timing in Predictive Modeling, or: How Important to Predict Where and What, but Without When?

If a model system consists of a closed set of predictors, and these do not change their behavior and interactions over time (so if the model system behavior is constant), it should be relatively easy to model forward and backward across time, and also to other areas. However, ecological systems are rarely static, and new factor combinations can become significant over time. The recent discussion about the behavior of the atmosphere, and the role of natural versus anthropogenic factors demonstrate this point. The atmosphere of 1 million years ago was very different from that of today and thus cannot be truly compared, nor modeled, without relevant correction factors and new variables. Many similar examples can be presented. Just consider, for instance, evolutionary processes, migration timing, changes in food and shelter, anadromous spawning due to lunar and temperature changes, effects caused by colder years versus warmer years, boom and bust cycles, or sources and sinks related to density dependence.

Many of the modern model applications deal with prediction of distributions, based on presence/absence data or population densities (Yen et al. 2004; Booms et al. 2009). However, it is usually a better practice, and more relevant for adaptive management, to model population productivity instead of just densities, or simply presence and absence (as an index of abundance); one should always keep in mind that density can be a misleading metric (Van Horne 1983). The use of indices in general, and indices of abundance specifically, should be replaced with better concepts (Burnham and Anderson 2002; Anderson et al. 2003). The habitat of species that need certain stages of succession or periodicity of disturbance cannot directly be mapped from a single landcover map. The existence of detailed information about habitat is strongly related to the resolution of the underlying landcover map, as it affects the number of land-use types (Chen et al. 2004; Gottschalk et al. 2005) and landscape indices (Wickham and Rhitters 1995).

Statistical "noise" that gets reported in the results of spatial model studies can often be attributed to mismatches caused by time (see Betts et al. 2009 for autocorrelation). Due to "pooling data across time," it is not unusual to see training data collected during biological cycles (e.g., samples pooled across regime shifts and then applied to out-of-cycle events). According to theory, this will result in models with a lower prediction accuracy, and consequently, with a poor ability to explicitly predict spatially across time. Yen et al. (2004; Marbled Murrelet) and Huettmann (2007b; Short-tailed Albatross) present such an example for modeling an ecosystem that is known to be affected by El Niño and regime shifts. However, such applications can still achieve good model accuracy, if the temporal link is not determined "by year" but via long-term averaged data.

On the other hand, niche conservatism should always be considered. This is when a niche is very robust and unchanging (Wiens and Graham 2005). This can been seen with endemic species. Niche conservatism is best suited to species with a relatively slow natural evolution that does not change drastically over less than 50 generations.

Spatial models are frequently set for "the current time" (Guisan and Thuiller 2005); this helps to make best possible use of available, and often limited, data. It then sets a platform for modeling, and further improvements in time and space. In such applications (and when based on ocurrence [i.e., presence-absence, or abundance]) time is usually assumed to be stable, or data gets pooled across periods. To our knowledge, no study exists to date that has assessed quantitatively any of the relevant impacts of pooling on model prediction accuracy through time. Determining "best practices" is an important avenue for future work, some of the identification of best practices can be done by using simulations with virtual species (Hirzel et al. 2001).

10.5 What Information, Model Algorithms, Statistics and Ecological Theory Are Available and Being Used to Address Temporal and Predictive Concerns?

10.5.1 Information Used to Assess Predictive Models

There are two general types of information that can be used to assess a model outcome: databases and knowledge. Information from databases either makes use of the (large) data body and splits it into two sets (jackknifing) – training or testing – or it draws random samples (bootstrap approaches). These are standard techniques and are well described (Efron and Tibshirani 1998). In terms of using knowledge to assess model performance, Traditional Knowledge (TK) is a common technique used in the social sciences that often focuses on local people as being representative for "the site." Another technique is to carry out interviews in person, via mail, or the internet (Magness 2009). All of these techniques meant to provide "truth" have pros and cons. We emphasize again that the truth, and the data creating it, is a critical feature to assess models, and for using them in the real world.

10.5.2 Statistical Models and Algorithms Used to Forecast

There have been three major approaches to a successful forecast (see Chapter). These are Linear "Curve-fitting" models, Markov Chains, and Machine Learning (Hegel et al. 2010 for overview). Initially, linear models have been popular because they seem to provide an easy to understand, multiplicative approach to assessing the future. But they do not break out of the trained pattern because they behave in a "linear" fashion, which makes their prediction results almost circular (because nothing but the expected outcome can get predicted). Markov Chain models are frequently used in Landscape Ecology (e.g., Onyeahialam et al. 2005), and can handle spatial complexities. But eventually these behave as linear models too. They also do not deal so well with interactions, non-linearity, and prediction accuracy. Finally, machine-learning methods, such as "bagging," have become popular in recent years to forecast over time (Prasad et al. 2006; Lawler et al. 2009; for methods see Breiman 2001b). This is because the algorithm is fast, publicly available, well tested, and does virtually not overfit. This is due to the nature of the algorithm. In RandomForest, many initial "trees" are produced and optimized ("bagging") from a random draw of columns (samples) and rows (predictors) never using the complete data set. Then, an "average tree" is finally presented based on "voting." RandomForest is also part of ensemble models (Araújo and New 2007).

10.5.3 Statistics Used to Assess Predictive Models

Although initially somewhat ignored, more studies now focus on predictive model assessment metrics and statistics. Model assessment is key for a science-based adaptive management strategy that makes best possible use of (predictive) models in space and time. Pearce and Ferrier (2000b), as well as Fielding and Bell (1997) present measures and techniques based on the confusion matrix and GIS concepts (see Fig 10.5 for GIS concepts). Manel et al. (2001) show how prevalence affects such indices. For several cases we would like to emphasize where behavior and robustness of these metrics is not well known: (1) ROC and Kappa with regard to autocorrelation, spatial sampling strategies, and spatial sample coverage, and (2) the use of the Boyce Index (Hirzel et al. 2006) for "presence only" models with an unknown fraction of truly confirmed absence (Engler et al. 2004). To our knowledge, there is currently no good and widely accepted measure that expresses quantitative certainty for either of the model assessment indices (but see B. Schroeder for ROC with confidence intervals and thresholds http://brandenburg.geoecology. uni-potsdam.de/users/schroeder/download.html). Pseudo-confidence intervals (ones that appear mathematically correct but which are biologically meaningless) can probably be computed. A robust bootstrap approach should be the method of choice (Efron and Tibshirani 1998).

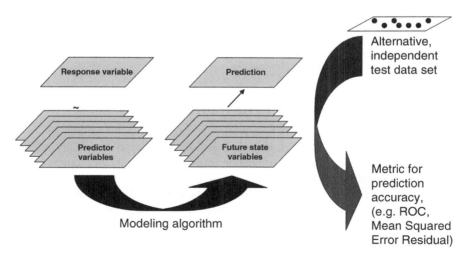

Fig. 10.5 Steps involved for accuracy assessment

10.5.4 Use of Underlying Ecological Theory

The use of ecological theory is often promoted in order to make models more realistic, to help interpret models better, to make models perform better, to assess their performance, and to help models attain wider use overall (Austin 2002). Ecological theory is rather rich (e.g., Rickleffs and Miller 1999) and it should provide a framework, biologically and quantitatively, for modeling in space and time. Although ecological modeling established itself early, it entered the mainstream relatively recently, particularly with regard to spatial applications (e.g., Austin 2002). Neural networks are a good representation of the concept of trying to mimic nature and its architecture and processes, that is, to model nature.

However, many cases can be found where the application of ecological theory and its variables does not always increase predictive model performance, and thus, ecology was ignored. The major scheme taken from ecological theory and applied in predictive modeling is the one dealing with the ecological niche: ENFA (Ecological Niche Factor Analysis; Hirzel et al. 2002). This idea is based on the rather powerful principle that geographic data can be transferred from the geographic space into a (quantitative) ecological space (Peterson 2001). This is achieved through the use of geo-referenced data linked with ecological attributes. Several algorithms and settings exist for this data linking, and these details are factors contributing to how well the ecological niche of a species can be characterized, and how well the model predictions perform.

Another large assumption made by many modeling studies regarding ecological theory is the one on "ecosystem stability." It assumes that ecosystems and niches are stable and do not change in space, time, and scale (Wiens and Graham 2005). These (wrong) assumptions might be the key reason for model assessment mismatches. Many model–prediction studies deal with short-term periods, and possible changes can therefore often be ignored. However, the census period when species data were sampled frequently does not fit with the actual time frame that the environmental variables were sampled in (Gottschalk et al. 2005). This occurs for instance when an easily available but old landcover map is used.

10.6 Open Questions Yet to Be Addressed

Other than perhaps metaanalysis, there are no fully satisfactory and really reliable and convincing methods to assess model performance for spatial predictions across time. Although the effects of temporal autocorrelation on models still needs further investigation (Araújo and Guisan 2006), and the continued collection of design-based data is a good means by which to fill data and modeling gaps and move toward better assessments. There also is a general lack of emphasis for more research on indices that can describe certainty (i.e., with 95% confidence). Further, one needs to know *"How long do I have to sample, and how many samples are necessary, until I know I am right or wrong for obtaining reliable predictions?"* This applies for instance in cases where biological cycles do occur but are not known well enough to be extrapolated to a wider study area. This question on sensitivity goes hand in hand with the need to know the spatial extent of the effect (Betts et al. 2009), as well as the effect size and the underlying variability for being "certain" (Thomas 1997). Although these questions are fairly well addressed for linear statistics, we still widely lack answers when using machine learning and ensembles.

If uncertainty is lacking, the amount of precaution needed must be further assessed. The provision of maximum transparency (i.e., via open-access models) (Huettmann 2009; Ohse et al. 2009) and repeatability is a must. The uncertainty leads to the underlying and wider question: can we truly generalize beyond our spatial sample across time, and outside of our obtained training data and ecological niche? Existing modeling applications provide first answers on that (*"yes we generally can predict beyond our sample"*); and so do the wide popularities of niche modeling as such (which should come used for a good reason). "Time" will further help to provide a detailed answer when looking back on predictions made earlier.

Finally, looking back on previous modeling achievements, a question should then be answered how inference improves with precision and accuracy? And does this get us closer to known mechanisms, and can it contribute to improving the state of the environment? We see so many modeling applications by now, but do we have less endangered species?

10.7 Toward A Modeling Culture: On the Use of Digital Data and Models in Court Cases, for Adaptive Management, in Risk Assessment, Cumulative Impact Studies, and Elsewhere

Models exist to be used, assessed, and improved. They present a quantified hypothesis, to be confronted with data. Setting a (digital) modeling culture is crucial (Huettmann 2007b), and it requires skilled individuals, institutions, and infrastructure. Gains are large, and include cost savings, reduction of error, improved sustainability (Cushman and Huettmann 2010), and maintenance of ecological services. Selected criteria and examples where models have been used for relevant cases are provided in Table 10.3. Additional examples are presented in Chapin et al. (2010). It is crucial to maintain such a (digital) culture, enhance it further, and also address global governance issues during this process (Huettmann 2007a). A basic change in infrastructure, goals and set up of ministries, and legal policies and procedures is required if adaptive management and sustainability is to be truly achieved. Further, communication is needed to increase acceptance and to draw attention on possible misinterpretations and common flaws of model results. Freely available internet-based model approaches (e.g., http://openmodeller.sourceforge.net/) can greatly facilitate the (global) acceptance of modeling and its results and applications.

Table 10.3 Selected models used in policy

Name of application	Name and type of model	Application details	Pros	Cons
Northern spotted owl conservation management	Resource selection probability function (RSPF), Manly et al. (2002)	Decision-making Re. cutting patterns	Quantitative, spatially explicit	Lack of ecological mechanism
Forest planning in New Brunswick, Canada	Spatial optimization, Woodstock/Stanley (http://www.remsoft.com/) Huettmann et al. (2005)	Decision-making Re. cutting patterns and budget planning	Explicit in space and time	Incomplete forest inventory data
ESA listing of polar bears in the U.S.	Demography, Population Viability Analysis (PVA), Hunter et al. (2007)	Listing decision and population sizes	Statistically accurate (frequentist)	Not truly spatial

10.8 Conclusion

This investigation deals with an important research question: how to improve predictive accuracy of ecological phenomena into the future? The study, and provision of answers, is urgent for global sustainability. Poor models might lead to incorrect management decisions. This fear is inherent to any decision-making and a precautionary principle and maximal transparency must be used. A public modeling culture should form a foundation for decision-making and sustainable management. Studies should be reasonably quantitative, take advantage of existing computing and data power, and involve time and space based on the ecological niche and evolution. Anthropogenic changes in ecosystem functioning over time are not well addressed in classic modeling and they therefore require further investigation. Further work is needed to address temporal uncertainty, how to communicate model findings, and what to base actions on. The topics raised in this chapter open up new research questions, and consolidate predictions as a solid research discipline to focus on. We need a new infrastructure and approaches, driven by creativity and synergy, and pushed forward through all available (spatial) data and tools via open access and global data sharing.

Existing solutions using metaanalysis to summarize a complex situation currently provide for the best available science. Proactive concepts offer a wise approach. Erring on the safe side is better than acting when it is too late. Modeling provides for a rather appropriate science for the provision of guidance for the future and as a base for adaptive management for a sustainable globe. Changes in policy, infrastructure and funding are necessary and should reflect these new priorities crucial for the global village.

Acknowledgments We appreciate the initial discussion (and idea) for this MS with T. Edwards. Additional help came from our colleagues A. Drew, Y. Wiersma, students in the EWHALE lab and co-workers world-wide. The important work by the R. O'Connor (quantitative ornithologist;

deceased) and B. Ripley (S-Plus and R) and L. Breiman (machine learning modeler; deceased) cannot be emphasized highly enough. We greatly value the software support and kind cooperation with Salford Systems, Dan Steinberg and his team. Comments from Y. Wiersma and two anonymous reviewers helped to significantly improve the manuscript. This is EWHALE lab publication #95.

References

Allouche O, Tsoar A, Kadmon R (2006) Assessing the accuracy of species distribution models: prevalence, kappa and the true skill statistic (TSS). J Appl Ecol 3:1223–1232.

Anderson DR, Cooch EG, Gutierrez RJ, Krebs CJ, Lindberg MS, Pollock KH, Ribic CA, Shenck TM (2003) Rigorous science: suggestions on how to raise the bar. Wildl Soc Bull 31: 296–305.

Araújo MB, Guisan A (2006) Five (or so) challenges for species distribution modelling. J Biogeogr 33:1677–1688.

Araújo MB, New M (2007) Ensemble forecasting of species distributions. Trends Ecol Evol 22:42–47.

Austin MP (2002) Spatial prediction of species distribution: an interface between ecological theory and statistical modelling. Ecol Model 157:101–118.

Betts MG, Ganio L, Huso M, Som N, Huettmann F, Bowman J, Wintle BA (2009) Comment on "Methods to account for spatial autocorrelation in the analysis of species distributional data: a review". Ecography 32:374–378.

Booms T, Huettmann F, Schempf P (2009) Gyrfalcon nest distribution in Alaska based on a predictive GIS model. Polar Biol 33:1601–1612.

Boyce MS, Vernier PR, Nielsen SE, Schmiegelow FKA (2002) Evaluating resource selection functions. Ecol Model 157:281–300.

Braun CE (2005) Techniques for wildlife investigations and management. The Wildlife Society (TWS), Bethesda, MD.

Breiman L (2001a) Statistical modelling: the two cultures. Stat Sci 16:199–231.

Breiman L (2001b) Random forests. Mach Learn 45:5–32.

Burnham KP, Anderson DR (2002) Model selection and multimodel inference: a practical information-theoretic approach. 2nd edition. Springer, New York.

Chapin FS, Kofinas GP, Folke C (eds) (2010) Principles of ecosystem stewardship: resilience-based natural resource management in a changing world. Springer, New York.

Chen D, Stow DA, Gong P (2004) Examining the effect of spatial resolution and texture window size on classification accuracy: an urban environment case. Int J Rem Sens 25:2177–2192.

Craig E, Huettmann F (2008) Using "blackbox" algorithms such as TreeNet and Random Forests for data-mining and for finding meaningful patterns, relationships and outliers in complex ecological data: an overview, an example using golden eagle satellite data and an outlook for a promising future. In: Wang H (ed) Intelligent data analysis: developing new methodologies through pattern discovery and recovery. IGI Global, Hershey, PA.

Cushman S (2010) Space and time in ecology: noise or fundamental driver? In: Cushman S, Huettmann, F (eds) Spatial complexity, informatics and wildlife conservation. Springer, Tokyo.

Cushman S, Huettmann F (2010) Spatial complexity, informatics and wildlife conservation. Springer, Tokyo.

Efron B, Tibshirani RJ (1998) An Introduction to the bootstrap. Chapman & Hall, New York.

Elith J, Graham CH, Anderson RP, Dudík M, Ferrier S, Guisan A, Hijmans RJ, Huettmann F, Leathwick JR, Lehmann A, Li J, Lohmann LG, Loiselle BA, Manion G, Moritz C, Nakamura M, Nakazawa Y, Overton JMcC, Peterson AT, Philips SJ, Richardson K, Scachetti-Pereira R, Schapire RE, Soberón J, Williams S, Wisz MS, Zimmerman NE (2006) Novel methods improve prediction of species' distributions from occurrence data. Ecography 29:129–151.

Engler R, Guisan A, Rechsteiner L (2004) An improved approach for predicting the distribution of rare and endangered species from occurrence and pseudo-absence data. J Appl Ecol 41:263–274.

Euskirchen ES, McGuire AD, Kicklighter DW, Zhuang Q, Clein JS, Dargaville RJ, Dyek G, Kimball JS, McDonald KC, Mellioz JM , Romanovsky VE, Smith NV (2006) Importance of recent shifts in soil thermal dynamics on growing season length, productivity, and carbon Sequestration in terrestrial high-latitude ecosystems. Global Change Biol 12:1–20.

Euskirchen ES, McGuire AD, Chapin III FS, Yi S, Thompson CC (2009) Changes in vegetation in northern Alaska under scenarios of climate change, 2003–2100: implications for climate feedbacks. Ecol Appl 19:1022–1043.

Fielding AH, Bell JF (1997) A review of methods for the assessment of prediction errors in conservation presence/absence models. Environ Conserv 24:38–49.

Fisher RA (1930) Inverse probability. Proc. Cambridge PhD.

Foley JA, DeFries R, Asner GP, Barford C, Bonan G, Carpenter SR, Chapin FS, Coe MT, Daily GC, Gibbs HK, Helkowski JH, Holloway T, Howard EA, Kucharik CJ, Monfreda C, Patz JA, Prentice IC, Ramankutty N, Snyder PK (2005) Global consequences of land use. Science 306:570–572.

Fuller T, Morton DP, Sarkar S (2008) Incorporating uncertainty about species' potential distributions under climate change into the selection of conservation areas with a case study from the Arctic Coastal Plain of Alaska. Biol Conserv 41:1547–1559.

Gottschalk TK, Huettmann F, Ehlers M (2005) Thirty years of analysing and modelling avian habitat relationships using satellite imagery data: a review. Int J Rem Sens 26:2631–2656.

Gottschalk TK, Dittrich R, Diekötter T, Sheridan P, Wolters V, Ekschmitt K (2010) Modelling land-use sustainability using farmland birds as indicators. Ecol Indic 10:15–23.

Graeber R (2005) Towards a biodiversity assessment of the Pacific Rim: predictive large-scale GIS-modelling of Brown Bear distribution (Canada, Alaska, Russian Far East And Japan) in estuaries using compiled coastal data. MSc thesis, University of Hannover, Germany.

Guisan A, Thuiller W (2005) Predicting species distribution: offering more than simple habitat models. Ecol Lett 8:993–1009.

Guisan A, Zimmermann NE, Elith J, Graham CH, Phillips S, Peterson AT (2007) What matters for predicting the occurrences of trees: techniques, data or species' characteristics? Ecol Monogr 77:615–630.

Hegel T, Cushman SA, Evans J, Huettmann F (2010) Current state of the art for statistical modelling of species distributions. In: Cushman S, Huettmann F (eds) Spatial complexity, informatics and wildlife conservation. Springer, Tokyo.

Herrick K, Huettmann F, Runstadler J, Chernetsov N, Antonov A, Valchuk O, Gerasimov Y, Matsyna E, Matsyna A, Markovets M, Druzyaka A, Saito K (in press) Predictive RISK modeling of avian influenza in the Pacific Rim and beyond. In: Kremers H (ed) RISK modeling proceedings, CODATA Berlin Germany.

Hickerson MJ, Carstens BC, Cavender-Bares J, Crandall KA, Graham CH, Johnson JB, Rissler L, Victriano PF, Yoder AD (2009) Phylogeography's Past, present, and future: 10 years after Avise, 2000. Mol Phylogenetics Evol 54:291–301.

Hirzel A, Helfer V, Metral F (2001) Assessing habitat-suitability models with a virtual species. Ecol Model 145:111–121.

Hirzel AH, Hauser J, Chessel D, Perrin N (2002) Ecological-niche factor analysis: how to compute habitat-suitability maps without absence data? Ecology 83:2027–2036.

Hirzel AH, Le Lay G, Helfer V, Randin C, Guisan A (2006) Evaluating the ability of habitat suitability models to predict species presences. Ecol Model 199:142–152.

Huettmann F (2007a) Modern adaptive management: adding digital opportunities towards a sustainable world with new values. Forum Public Policy 3:337–342.

Huettmann F (2007b) Constraints, suggested solutions and an outlook towards a new digital culture for the oceans and beyond: experiences from five predictive GIS models that contribute to global management, conservation and study of marine wildlife and habitat. In: Vanden Berghe E, Appeltans W, Costello MJ, Pissierssens P (eds) Proceedings of ocean biodiversity informatics: an international conference on marine biodiversity data management. Hamburg,

Germany, 29 November–1 December, 2004. IOC Workshop Report, 202, VLIZ Special Publication 37.

Huettmann F (2009) The global need for, and appreciation of, high-quality metadata in biodiversity work. In: Spehn E, Koerner C (eds) Data mining for global trends in mountain biodiversity. CRC Press, Taylor & Francis, Boca Roton, FL.

Huettmann F, Franklin SE, Stenhouse GB (2005) Predictive spatial modeling of landscape change in the Foothills Model Forest. Forest Chron 81:1–13.

Hunter CM, Caswell H, Runge MC, Amstrup SC, Regehr EV, Stirling I (2007) Polar bears in the Southern Beaufort Sea II: demography and population growth in relation to sea ice conditions. Administrative report, USGS Alaska Science Center, Anchorage, AK.

Koerner C, Paulsen D (2004) A world-wide study of high altitude treeline temperatures. J Biogeogr 31:713–732.

Larson MA, Thompson FR, Millspaugh JJ, Dijak WD, Shifley SR (2004) Linking population viability, habitat suitability, and landscape simulation models for conservation planning. Ecol Model 180:103–118.

Lawler JJ, White D, Neilson RP, Blaustein AR (2006) Predicting climate-induced range shifts: model differences and model reliability. Glob Chang Biol 12:1568–1584.

Lawler JJ, Shafer SL, White D, Kareiva P, Maurer EP, Blaustein AR, Bartlein PJ (2009) Projected climate-induced faunal change in the western hemisphere. Ecology 90:588–597.

Lemoine N, Bauer HG, Peintinger M, Boehning-Gaese K (2007) Effects of climate and land-use change on species abundance in a central European bird community. Conserv Biol 21:495–503.

Lozier JD, Aniello P, Hickerson MJ (2009) Predicting the distribution of Sasquatch in western North America: anything goes with ecological niche modeling. J Biogeogr 36:1623–1627

Magness D (2009) An assessment of the U.S. Wildlife Refuge System management in times of climate change. PhD thesis, University of Alaska, Fairbanks, AK.

Magness D, Morton JM, Huettmann F (2010) How spatial information contributes to the management and conservation of animals and habitats. In: Cushman S, Huettmann F (eds) Spatial complexity, informatics and wildlife conservation. Springer Tokyo, Japan.

Manel S, Dias JM, Ormerod SJ (1999) Comparing discriminant analysis, neural networks and logistic regression for predicting species' distributions: a case study with a Himalayan river bird. Ecol Model 120:337–347.

Manel S, Williams HC, Ormerod SJ (2001) Evaluating presence-absence models in ecology: the need to account for prevalence. J Appl Ecol 38:921–931.

Manly FJ, McDonald LL, Thomas DL, McDonald LT, Erickson WP (2002) Resource selection by animals. Kluwer Academic Publishers, Dordrecht NL.

Marmion M, Parviainen M, Luoto M, Heikkinen RK, Thuiller W (2009) Evaluation of consensus methods in predictive species distribution modeling. Divers Distrib 15:59–69.

Martínez-Meyer, E, Peterson AT, Hargrove WW (2004) Ecological niches as stable distributional constraints on mammal species, with implications for Pleistocene extinctions and climate change projections for biodiversity. Glob Ecol Biogeogr 13:305–314.

Martínez-Meyer E, Peterson AT (2006) Conservatism of ecological niche characteristics in North American plant species over the Pleistocene-to-recent transition. J Biogeogr 33:1779–1789.

Murray K, Conner MM (2009) Methods to quantify variable importance. Implications for the analysis of noisy ecological data. Ecology 90:348–355.

Myers JC (1997) Geostatistical error management: quantifying uncertainty for environmental mapping. Van Nostrand Reinhold, New York.

Nielsen SE, Stenhouse GB, Beyer HL, Huettmann F, Boyce MS (2008) Can natural disturbance-based forestry rescue a declining population of grizzly bears? Biol Conserv 141:2193–2207.

Nogués-Bravo D, Rodríguez J, Hortal J, Batra P, Araújo M (2008) Climate change, humans, and the extinction of the woolly mammoth. PLoS Biol 6:e79.

O'Connor, R (2000) Why ecology lags behind biology. The Scientist 14:35.

Ohse B, Huettmann F, Ickert-Bond S., Juday G (2009) Modeling the distribution of white spruce (*Picea glauca*) for Alaska with high accuracy: an open access role-model for predicting tree species in last remaining wilderness areas. Polar Biol 32:1717–1724.

Onyeahialam A, Huettmann F, Bertazzon S (2005) Modeling sage grouse: Progressive computational methods for linking a complex set of local biodiversity and habitat data towards global conservation statements and decision support systems. Lecture Notes in Computer Science (LNCS) 3482, International Conference on Computational Science and its Applications (ICCSA) Proceedings Part III:152–161.

Oppel S, Strobel C, Huettmann F (2009) Alternative methods to quantify variable importance in ecology. Technical Report Number 65, Department of Statistics, University of Munich, Germany.

Pearce J, Ferrier S (2000a) Evaluating the predictive performance of habitat models developed using logistic regression. Ecol Model 133:225–245.

Pearce J, Ferrier S (2000b) An evaluation of alternative algorithms for fitting species distribution models using logistic regression. Ecol Model 128:127–147.

Peterson AT (2001) Predicting species' geographic distributions based on ecological niche modeling. Condor 103:599–605.

Pineiro G, Perelman S, Guerschman JP, Parelo JM (2008) How to evaluate models: observed vs. predicted or predicted vs. observed ? Ecol Model 216:316–322.

Prasad AM, Iverson LR, Liaw A (2006) Newer classification and regression tree techniques: bagging and random forests for ecological prediction. Ecosystems 9:181–199.

Rickleffs RE, Miller GL (1999) Ecology, 4th edition, Freeman & Co Publishers, New York.

Ritter J (2007) Species distribution models for Denali National Park and Preserve, Alaska. M.S. thesis, University of Alaska-Fairbanks, AK.

Thomas L (1997) Retrospective power analysis. Conserv Biol 11:276–280.

Van Horne B (1983) Density as a misleading indicator of habitat quality. J Wildl Manage 47:893–901.

Wickert C (2007) Breeding White Storks (*Ciconia ciconia*) in former East Prussia: comparing predicted relative occurrences across scales and time using a stochastic gradient boosting method (TreeNet), GIS and public data. M.Sc. Thesis, University of Potsdam, Germany.

Wickham JD, Rhitters KH (1995) Sensitivity of landscape metrics to pixel size. Int J Rem Sens 16:3585–3594.

Wiens JJ, Graham CH (2005) Niche conservatism: integrating evolution, ecology, and conservation biology. Ann Rev Ecol Evol System 36:519–539.

Wittingham MJ, Stephens PA, Bradbury RB, Freckleton RP (2006) Why do we still use stepwise modeling in ecology and behaviour? J Anim Ecol 75:1182–1189.

Worm B, Myers R (2003) Meta-analysis of cod-shrimp interactions reveals top-down control in oceanic food webs. Ecology 84:162–173.

Yen P, Huettmann F, Cooke F (2004) Modelling abundance and distribution of Marbled Murrelets (*Brachyramphus marmoratus*) using GIS, marine data and advanced multivariate statistics. Ecol Model 171:395–413.

Ziv Y (1998) The effect of habitat heterogeneity on species diversity patterns: a community-level approach using an object-oriented landscape simulation model (SHALOM). Ecol Model 111:135–170.

Chapter 11
Variation, Use, and Misuse of Statistical Models: A Review of the Effects on the Interpretation of Research Results

Yolanda F. Wiersma

11.1 Introduction

11.1.1 Predictive Habitat Modeling: No Unifying Route

The field of predictive habitat modeling evolved somewhat separately within the sub-disciplines of theoretical ecology, wildlife management, and landscape ecology. This chapter suggests that this is due to slightly different worldviews, cultures, and research applications within each subfield (Table 11.1). Within the theoretical ecology literature, models of all kinds (e.g., movement, foraging, competition, demographic) have been widespread for many years. The evolution from descriptive models of habitat quality (e.g., Whittaker and McCuen 1976), to mathematical formulations of niche (e.g., Austin 1985), to spatially-explicit predictive habitat models (e.g., Saarenmaa et al. 1988) was a gradual one. The driving force in this literature appears to be underlying theoretical formulations of a host of ecological processes and interactions (e.g., population dynamics, movement, predation, competition). In the wildlife biology literature, predictive habitat models are generally termed "resource selection functions" and their origin can be traced to Manly's work on natural selection (Manly 1985). Underpinning models in the wildlife management literature is a focus on species life-history and resource use. Rigorous methods for carrying out resource selection functions were documented in the wildlife literature in 1993 (Manly et al. 1993). Landscape ecology, a relatively young discipline compared with wildlife biology, began with spatially explicit models that placed emphasis on pattern description (e.g., Ribe et al. 1998) and issues of scale (e.g., Naugle et al. 1999). Here, the relationship between landscape pattern and process (one of the fundamental tenets of the sub-discipline) underpins much of the modeling work. Applications for predictive habitat models vary in the literatures as well. In wildlife management literature, the main application for models appears to be to improve

Y.F. Wiersma (✉)
Department of Biology, Memorial University St. John's, NL Canada
e-mail: ywiersma@mun.ca

Table 11.1 An overview of the underlying frameworks, models, and applications for predictive models within three distinct sub-fields

	Theoretical ecology	Wildlife biology	Landscape ecology
Underlying theoretical framework	Mathematical models	Single-species life history/ resource use	Links between landscape pattern and process
Data model	Equal emphasis on species and environmental data	More emphasis on species data	More emphasis on environmental/ habitat data
Ecological model	Tightly linked with existing models for foraging, dispersal, population dynamics, etc.	Tightly linked to life history/foraging models	Loosely linked to ecological models
Application/ main purpose of predictive models	Better theoretical models	Improved species management	Improved landscape management

species management, whereas landscape ecology literature focuses on applying models to better manage landscapes (often for a specific species or group of species). This summary is by no means comprehensive. There exists within the literature variation within different sub-disciplines (e.g., vegetation vs. wildlife vs. disease modeling) and within regions (European vs. North American vs. Australian). These various "schools" (North American vs. European vs. Australian) have contributed different approaches to statistical models (Manly et al. 1993; Boyce et al. 2002; Guisan and Zimmermann 2000; Pearce and Ferrier 2000). This chapter is not intended to provide an historical analysis of regional differences in the field of predictive modeling; rather, the focus is on the application of a variety of statistical models to the problem of predicting species occurrences across broad spatial scales. The statistical models posited by each of these schools have been applied across the globe, and this chapter will focus on their application, and not on the regional or historical context within which they were initially developed and/or applied.

This assessment of the use of specific statistical models for predictive habitat modeling is carried out through a close examination of their use in three distinct sub-fields: theoretical ecology, landscape ecology, and wildlife biology. These fields have been chosen for closer examination because all three have applied predictive habitat models extensively; however, the theoretical frameworks underpinning models and the statistical applications of models has varied between them. The definition of these as sub-fields is somewhat arbitrary, as there is obviously overlap between all three. However, the emphasis on the roles that statistical models play is considerably different between the sub-fields, and this diversity serves to make a comparative assessment interesting and worthwhile. An examination of how statistical models have been applied to similar types of predictive habitat models across these three subfields may assist in inferring the strengths and weaknesses of the different statistical approaches.

To assess the historical development of predictive habitat models within these three sub-fields, I carried out a literature search of one of the flagship journals for the relevant societies of each subfield - *Ecological Modelling* (a publication of the Ecological Society of America (ESA)), the *Journal of Wildlife Management* (a publication of The Wildlife Society (TWS)), and *Landscape Ecology* (a publication of the International Association for Landscape Ecology). Although ESA and TWS are American-based societies, the journals publish research from around the world. All journals have reasonably strong impact factors (2008 values; 2.176 for *Ecological Modelling*, 1.323 for the *Journal of Wildlife Management* and 2.453 for *Landscape Ecology*). The choice of these three journals is somewhat arbitrary, since there are many other excellent journals that regularly publish articles on predictive habitat modeling. However, I chose these three journals because they vary in what the emphasis of journal content is (e.g., wildlife management applications, versus studies conducted at landscape scales), although there is obviously overlap in the type of papers published in each (e.g., a landscape-scale study of a wildlife management issue might easily be published in either *Landscape Ecology* or the *Journal of Wildlife Management*). Moreover, these journals were among the top journals in terms of number of hits across all keyword searches. I used the Web of Science database to search all issues from 1998–2007 for the following keywords: "predictive habitat model*", "habitat model*", "resource selection function", "predictive model*", "distribution model*", "habitat suitability index", "HSI". In addition, I examined the Table of Contents for all issues to identify relevant titles. In all cases, I chose articles for further analysis that presented an original research paper outlining the development and application of a predictive habitat model. I did not include papers that simply described species-habitat associations, even where these made use of sophisticated statistical tools. The emphasis was on *predictive* models, for species of any kind. I also excluded papers that looked at the effects of habitat on ecological processes (population dynamics, dispersal, predation, etc.) other than species distribution. I complied a database with information on the type of statistical and data models used, the method for model selection (if applicable), and the species modeled. In a separate analysis, I compiled a separate database of papers that were more reflective. This included reviews, synthesis papers, critiques, editorials, and commentaries. Finally, I tallied up the number of papers in each journal that were aimed at presenting a new model with simulated data.

The field of predictive habitat modeling has passed through several stages. Development of predictive habitat models was facilitated by advances in computing power, particularly the increased use and power of Geographic Information Systems (GIS) and statistical software packages. This is evident in an examination of the publication of papers on predictive habitat modeling (Fig. 11.1). A keyword search in "Web of Science" for "(predictive) habitat model*" and "resource selection function" returned no papers prior to 1980 and very few papers throughout the 1980s. A rapid increase was seen starting in the mid-1990s, continuing into the 21st century (Fig. 11.1). Across the three journals, predictive models first appear in the early to mid-1990s (Fig. 11.2). By the late 1990s and early part of the 21st century, these journals were publishing at least one paper/year that presented original research on predictive habitat modeling. This number expanded by the middle of the first

---

*Indicates the wildcard term to allow for papers containing terms like models, modeling, and to account for variation between american and british spelling.

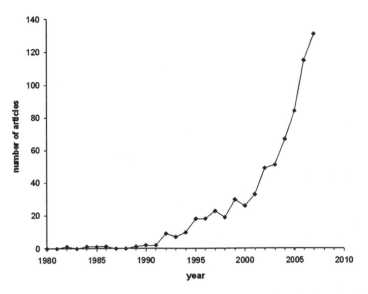

Fig. 11.1 Publication rate of papers corresponding to keywords: predictive habitat model*, habitat model*, and resource selection function across all journals in the *Web of Science* database for all years (* indicates a wildcard)

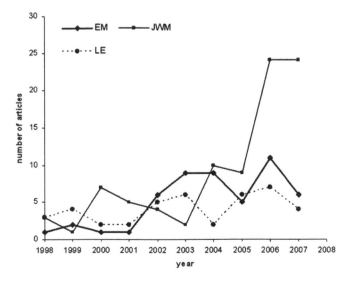

Fig. 11.2 Publication rate of original research papers describing predictive habitat models (that made use of real data) in three journals (EM: *Ecological Modelling*, JWM: *Journal of Wildlife Management*, LE: *Landscape Ecology*) between 1998 and 2007

decade of the twenty-first century, and recent years suggest a further increasing trend. There are spikes in the data in 2006 (Fig. 11.2), which indicate special issues devoted to predictive habitat models in both *Ecological Modeling* (vol. 199(2)) and *Journal of Wildlife Management* (vol. 70(2)). An examination of the publication of "reflective" papers that provided critique, commentary, or review (Fig. 11.3) suggests that modeling has progressed to a point where "meta-analyses" and assessments of techniques are possible, and perhaps necessary, to move forward. Most articles suggesting new models or techniques were published in *Ecological Modelling* (Fig. 11.4). This is not surprising, given that *Ecological Modelling* is the most theoretical of the three journals examined, whereas *Landscape Ecology*, and especially *Journal of Wildlife Management*, have a much more applied focus.

Despite differences in how predictive modeling originated within each of these sub-disciplines, it is evident that they followed similar uptake rates in the literature (Fig. 11.2). It is interesting to note the range of variation in statistical models employed to develop predictive habitat models within each of these sub-fields (Fig. 11.5). In all cases, regression-based models represent the largest share of statistical model types. Papers in *Ecological Modelling* employ the widest diversity of statistical models, whereas a greater number of papers in *Landscape Ecology* use tree-based models (e.g., Classification and Regression Tress (CART)) than in the *Journal of Wildlife Management*. The following section will discuss the implications of choosing one statistical model over another. This chapter is not intended to provide a detailed overview of specific statistical models as numerous articles exist

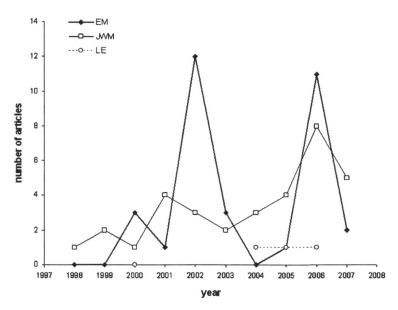

Fig. 11.3 Publication rate of "reflective/critique" papers on the topic of predictive habitat modelling in three journals (EM: *Ecological Modelling*, JWM: *Journal of Wildlife Management*, LE: *Landscape Ecology*) between 1998 and 2007

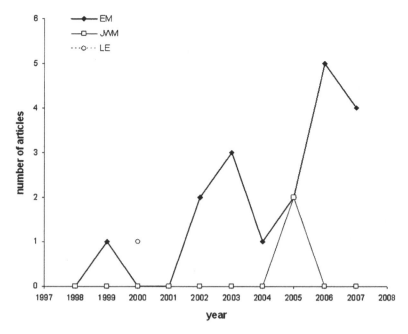

Fig. 11.4 Publication rate of articles proposing a new method or technique for predictive habitat modeling in three journals (EM: *Ecological Modelling*, JWM: *Journal of Wildlife Management*, LE: *Landscape Ecology*) between 1998 and 2007

on this topic (Guisan and Zimmermann 2000; Manly et al. 2003; MacKenzie et al. 2006; Elith et al. 2008). Rather, the intent of this chapter is to take a reflective look at how statistical models are applied.

11.2 Management Challenge, Ecological Theory, and Statistical Framework

Managers who rely on predictive habitat models to make decisions either for wildlife or landscape management (see Table 11.1) may not understand the underlying statistical models. Hooten's chapter in this volume provides a road-map of the different types of statistical models available for predictive habitat modelling (Chap. 3). An important consideration of all models, as outlined in the chapter by Laurent et al. (Chap. 5), is that all models carry with them different assumptions, both stated and unstated. These assumptions may be exclusionary (i.e., parameters are not considered in the predictive model) or inclusionary (i.e., assumptions for including given parameters). Whereas not all assumptions need to be stated, developers and users of predictive habitat models need to understand the stated and unstated assumptions inherent in their models in order for management to be effective.

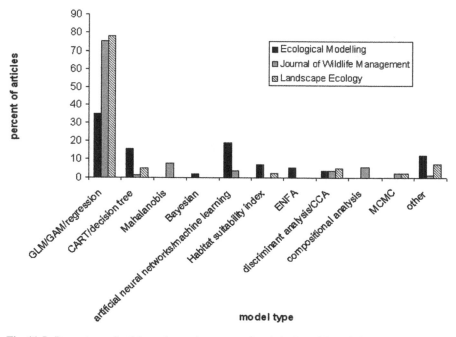

Fig. 11.5 Percentage of articles using various types of statistical models and algorithms in predictive modeling papers in three journals from 1998 to 2007. Sample size in each journal: *Ecological Modelling* (37), *Journal of Wildlife Management* (84), *Landscape Ecology* (41)

Laurent et al. (Chap. 5) present several common assumptions inherent in predictive habitat models. This chapter builds on these earlier chapters that outlined the current suite of models and considerations for applying models to the problem of predicting habitat. This chapter will compare and contrast the outcomes for predictive habitat models when different statistical models are applied,

Generally, predictive models use presence and/or absence data collected by telemetry, or via direct or indirect observation. The data are then used to build statistical models based on a suite of environmental, habitat, and resource values. Chapter 4 in this volume by Zuckerberg and Huettmann discusses the data considerations that should be taken into account when using either presence-only or presence-absence data. They also point out that much of the widely available data for predictive habitat modeling varies in its quality, and modelers should be cognizant of the limitations of the data sets they use. Different models may emphasize different attributes (abiotic vs. biotic, resources that are consumed vs. resources for other life history characteristics). Moreover, how to manage either species or landscapes is based on the assumption that distributions from predictive habitat models identify optimal habitat. Most models do not include information on species abundance, or discriminate differences in habitat quality that may affect variation in population dynamics (e.g., source vs. sink habitats).

11.2.1 Statistical Models

11.2.1.1 Distance-Based Models

Habitat models are commonly confounded by the fact that "unoccupied" habitat does not necessarily indicate "unsuitable" habitat. In the case where observed species absences were not recorded, such as with the radio collar data, a random selection of locations on the landscape may or may not have had the species present. Early envelope-based models such as the BIOCLIM model make use of presence-only data (Busby 1991). Distance-based models (such as maximum entropy, or MAXENT models) also require only data from "occupied" habitat areas, and thus avoid these ambiguities (Clark et al. 1993; Rotenberry et al. 2002; Phillips et al. 2006). Other statistical models that are based on distance-dissimilarity include the DOMAIN model (Carpenter et al. 1993), the geometric mean algorithm (Hirzel and Arlettaz 2003), Genetic Algorithm for Rule-Set Prediction (GARP; Stockwell and Peters 1999), and the more recent LIVES model (Li and Hilbert 2008). A comparison of the performance of various presence-only models using data for 42 species showed that GARP and Mahalanobis distance models were better predictors (Tsoar et al. 2007). Some of these types of models have been shown to be effective for building models using data from museum and herbarium records (Soberón et al. 2000; Graham et al. 2004; Elith et al. 2006).

An appealing feature of distance metrics is that they incorporate the correlations among habitat variables that often characterize ecological data sets, and thus are closely linked to the concept of the niche. Presence-only models do not allow for model comparison or evaluation based on statistical tests, thus it can be difficult to determine whether the parameters included in a model are significant or not.

11.2.1.2 Tree-Based Models

A second class of models – classification and regression trees (CART) – presents a powerful tool that allows for the use of multiple variables that may not have common effects across the sample (Breiman et al. 1984, De'ath and Fabricius 2000, Vayssiéres et al. 2000). Additionally, CART models have been shown to be better predictors at geographically removed locations than other models (i.e., discriminant function models; Dettmers et al. 2002). CART models partition data recursively into subsets based on constraints rather than correlates to build a binary decision tree that partitions "habitat" and "non-habitat" (Breiman et al. 1984). CART models are optimized via the recursive minimization of within-node misclassifications. However, CART models tend to be over-fitted for this same reason; thus, *post-hoc* pruning is necessary (Breiman et al. 1984). CART models may also bias toward covariates that have many possible split values; thus, applying unbiased selection of covariates based on conditional distributions has recently been suggested to reduce this bias generated by the data (Hothorn et al. 2006). More recently, bagging and boosting has been suggested as an alternative set of algorithms for classification (Borra and DiCiaccio 2002); two common tools include multivariate adaptive regression splines (MARS; Friedman 1991) and boosted regression tress (BRT; Friedman et al. 2000). A further method to come out of CART models is the machine-learning approach, Random Forests (Breiman 2001a, see also Chap. 8 in this text).

11.2.1.3 Regression-Based Models

The most common applied statistical model across all three sub-fields (Fig. 11.5) are regression-based models (e.g., GLM, GAM, logistic models). In general, the methodology here is to develop a suite of possible models using available predictor data (e.g., environmental, habitat and resource data layers). The challenge for predictive modeling is to select the best-plausible model. Model selection is usually performed using a step-wise approach (e.g., forward, backward, or both) or an information-theoretic (I-T) approach. In some cases, Akaike Information Criterion (AIC) values are used in a step-wise fashion (Fig. 11.6), although the literature strongly recommends against this practice as it violates the spirit of the I-T approach (Anderson and Burnham 2002; Burnham and Anderson 2002). Rather AIC should be applied to a suite of models that are based on a priori hypotheses that have underlying biological/ecological mechanisms to support them (Anderson and Burnham 2002)

The difficulty with using logistic models for predicting habitat is deciding what the probability threshold for determining habitat/non-habitat should be. In the absence of information, a value of $p = 0.5$ is often set with probability greater than

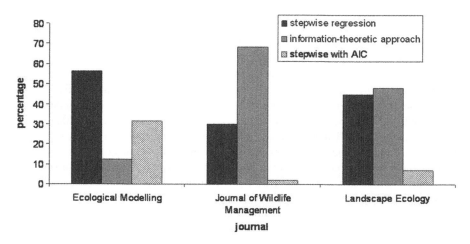

Fig. 11.6 Methods of model selection (as a percentage of total number of articles) used to select from a suite of regression-based models in predictive modelling papers in the three journals from 1998 to 2007. Sample size in each journal: *Ecological Modelling* (16), *Journal of Wildlife Management* (60), *Landscape Ecology* (29)

0.5 representing potential habitat, and values less than 0.5 representing non-habitat. This implicitly assumes that the prior probabilities are equal and the "cost" of a false positive is the same as that for a false negative. With the availability of testing data, receiver operating characteristic (ROC) curves can be used to "tune" the model and determine an optimum threshold that maximizes model sensitivity and specificity (Fielding and Bell 1997, Pearce and Ferrier 2000). In addition, ROC curves give a measure of the area under curve (AUC) when sensitivity is plotted against 1-specificity. AUC values range from 0 to 1 and can be used to interpret how well the model is performing (Hanley and McNeil 1982). The AUC value indicates the percent of occasions that a random selection from the positive predictions would have a greater classification score than a random selection from the negative predictions. An AUC of 0.5 indicates that the classification is no more effective than a coin toss. Generally an AUC >0.9 is considered "excellent," >0.8 is considered "very good," and >0.7 is considered "good" (Hanley and McNeil 1982).

11.2.2 Model Considerations

The *Journal of Wildlife Management* is particularly notable for its inclusion of a series of reflective/opinion papers that debate the merits of frequentist (e.g., step-wise regression) versus I-T approaches to model selection (Johnson 1999; Anderson et al. 2000; Robinson and Wainer 2002; Anderson and Burnham 2002; Guthrey et al. 2005; Steidl 2006, 2007; Sleep et al. 2007). The crux of the debate hinges on perspectives of hypothesis testing. Frequentist statistics are applied to null hypothesis testing, whereas I-T approaches examine multiple working hypotheses in *sensu* Chamberlain (1965). Moreover, frequentist approaches focus on *statistical* hypotheses, rather than *biological* hypotheses (Johnson 2002; Anderson and Burnham 2002; Guthrey et al. 2005; Sleep et al. 2007). Critics of stepwise regression argue that reliance on the stepwise approach cannot be considered hypothesis testing, and rather should be viewed as data-dredging which can lead to spurious conclusions (Breiman 2001b; Sleep et al. 2007) (Johnson 2002; Sleep et al. 2007). Proponents of frequentist approaches argue that there is a risk that important information or "best models" are missed when the information-theoretic approach is applied (Steidl 2006, 2007).

The value of the I-T approach is that it allows one to evaluate a suite of *a priori* hypotheses. Rather than select models based on a statistical comparison of the model with and without a particular parameter (essentially testing the null of "no difference" with and without the parameter within a given probability, usually a *p* value of 0.05), the I-T approach gives a relative weighting of all plausible models. This has particular value for generating predictions because it allows for approximate model averaging (Anderson and Burnham 2002; Sleep et al. 2007). This is not to say that frequentist statistics do not have their place. Indeed, Robinson and Wainer (2002) make compelling arguments for the role of null hypothesis testing in wildlife science; however, their comments are more applicable to manipulative experiments and not as germane to predictive habitat models. Keating and Cherry

(2004) discuss how to properly apply logistic regression models in wildlife research within a frequency-based approach. However, the I-T approach is not immune to abuse. Anderson and Burnham (2002) and Guthrey et al. (2005) present cogent critiques of how the I-T approach can be misused. AIC has also been shown to bias toward overparametrized models in some cases (Link and Barker 2006). A common misuse is the application of a combination of frequentist and I-T approaches to model selection (Anderson and Burnham 2002). This practice was observed in journals surveyed, although it was least frequent in the *Journal of Wildlife Management* (Fig. 11.6).

It is interesting to note that research papers in both the *Journal of Wildlife Management* and *Landscape Ecology* made extensive use of AIC (the most common application of the I-T approach), yet *Landscape Ecology* carried no reflective/critique type papers on the topic of model selection during the same period. An examination of those papers using I-T for model selection in *Landscape Ecology* showed that nearly 50% (6 of 14) used count data as the source for building models. Anderson and Burnham (2002) advise applying QAIC (a modification of the AIC that accounts for overdispersion) for model selection in this case, because of the tendency for count data to be overdispersed. In contrast, only 7% of the papers in the *Journal of Wildlife Management* that used count data for model building used AIC when QAIC might have been more appropriate. Anderson et al. (2001) and Anderson and Burnham (2002) present clear guidelines for information that should be reported when applying the I-T approach; 30 of 41 papers in *Journal of Wildlife Management* and 13 of 14 papers in *Landscape Ecology* failed to include complete information (e.g., log-likelihood, number of parameters, AIC (or comparable information criterion), differences (Δ_i) or Akaike weights (w_i)) in their results section. Fifty percent of the papers in *Landscape Ecology*, and three of the papers in *Journal of Wildlife Management* contained no summary data on model comparisons; the others failed to include some key information, usually log-likelihood and raw AIC values; most papers in both journals that included some results of model comparisons reported delta AIC and Akaike weights, although this information was sometimes missing. It is important to report more than just the fact that the model with the lowest AIC was selected, because models with a difference in AIC value of <2 are all considered plausible. It is also instructive to see the "best" model weighted against the other models. This is helpful for making biological inferences about underlying mechanisms. The strength of the I-T approach is that it allows comparison across models so as to assess their plausibility relative to one another. A frequentist-based approach only provides the statistically "best" model, with no comment on how alternative models may have performed in predicting data. Further information can be generated by expanded the I-T approach to include the Baysian Information Criterion (BIC; Burnham and Anderson 2002; Link and Barker 2006). A common theme emerging from the debate between frequentist and I-T based approaches is the importance of considering underlying ecological theories when developing models (Anderson and Burnham 2002; Burnham and Anderson 2002; Johnson 2002; Guthrey et al. 2005), which is one of the main themes we hope will emerge from this volume as a whole.

11.3 Model and Model Validation Techniques

Different statistical models can yield very different predictions of where suitable habitats may be located and this can have important implications for landscape management. When predictive habitat models are based only on one statistical model, the risk for identifying unsuitable areas as suitable, and vice versa, arises. In the literature survey, only two papers in the *Journal of Wildlife Management* used more than one statistical model. Interestingly, there were a high number of papers ($n = 13$) in *Ecological Modelling* that made use of more than one statistical model. A comparison of the papers' findings may be found in Table 11.2. The meta-analysis of predictive habitat models constructed using different statistical models illustrates that there can sometimes, though but not always, be significant variation in the parameters chosen for each model. For example, in the paper by deFrutos et al. (2007), the final model generated using a step-wise GLM yielded three predictors for the summer range of the least kestrel, while hierarchical partitioning suggested only one variable was significant. This variable (autocorrelation to the index of relative kestrel density within an eight-neighbor block) also explained 42% of the variation in the GLM; however, the GLM also suggested a significant negative relationship of distance to roost and distance to nearest colony with more than ten breeding pairs. Thus, using a GLM, management for kestrel would suggest a focus on roosting habitat and maintenance of colonies with greater than ten pairs. However, if hierarchical partitioning had been the only model applied, these variables might not have been accounted for, which could result in negative consequences for kestrel if the other predictors are indeed significant. In terms of prioritization of habitat for protection or intensive management, the resulting predictive maps can also vary between models. Garzon et al. (2006) show three predictive maps for the potential habitat of *Pinus sylvestris*; these differ significantly from each other and from the real distribution of the species. In other cases, however, the predictors chosen between different statistical models were identical (Jelaska et al. 2003; Holloway and Malcolm 2006), or very similar (e.g., Olivier and Wotherspoon 2005; Yang et al. 2006; Lippitt et al. 2008).

It is clear that, from a management perspective, decisions made based on one model might vary if a different statistical model were applied. Thus, managers should be cautious when interpreting habitat models and should first and foremost consider the underlying ecological model. They should also be aware of the more common misapplications of particular statistical techniques, particularly with regard to the data the model(s) were built upon. Ideally, modelers should apply more than one statistical model in order to check for consistency. We greatly value the software support and kind cooperation with Salford Systems, Dan Steinberg and his team. If multiple models yield similar predictions, then managers can be more confident in their accuracy. However, time and resources do not always allow for this. When multiple statistical models are not applied, managers should carefully consider the validity of the model used and whether it was appropriate to the data available. A close examination of the assumptions inherent in the models (see Chap. 5 by Laurent et al.) and issues of data accuracy (see Chap. 4

Table 11.2 Summary of papers that used more than one statistical model to develop predictive habitat models

Paper	Models used	Main findings
deFrutos et al. (2007)	GLM	Three variables were included in the final model after stepwise selection
	Hierarchical partitioning	Only one variable was shown to be significant in final model; this variable explained 42% of the variance in the GLM
Dzeroski and Drumm (2003)	Regression trees Linear regression Instance-based learning	All three models had similar predictive power (r), but instance-based learning had highest mean absolute error. Parameter coefficients for regression trees conform better to expert knowledge than coefficients for linear regression. Most important coefficients in both linear regression and regression trees were similar; coefficients not calculated for IBL
Garzon et al. (2006)	CART Neural networks RF	All three models had similar accuracy measures, as measured by AUC and Kappa scores. However, the three models varied in the parameters included, as well as in the total amount of habitat predicted to be suitable. Random forest model was deemed most accurate
Holloway and Malcolm (2006)	Stepwise linear/ logistic regression Linear/logistic and AIC	Model selected via stepwise regression was generally one of best models selected based on AIC
Jelaska et al. (2003)	GLM Discriminant analysis CART	Predictive models were not significantly different from each other. Models were not pruned to select only most significant predictors
Lippitt et al. (2008)	Back propagation neural networks Logistic regression Multi-criteria evaluation	BPN provided the highest accuracy as measured by AUC scores, followed by MCE and logistic regression. MCE overpredicted below 90% true positive threshold, and thus would be best model if unlimited training data were available. There was similarity in variables included in each model, though variation in relative importance of the predictors
Maisonneuve et al. (2006)	Stepwise logistic regression CART	Variables selected in stepwise regression and CART were identical; overall classification accuracy was slightly higher with CART model
Miller and Franklin (2002)	GLM CART	Both statistical models had generally high AUC for four sets of models for different vegetation alliances, classification trees had slightly higher accuracy on the training data than the GLMs, but were less robust against the testing data
Moisen et al. (2006)	CART GAM Stochastic gradient boosting	Stochastic gradient boosting had higher specificity, whereas GAMs had higher sensitivity for the majority of species. Thresholds varied for each model. Model performance varied independently of species prevalence

(continued)

Table 11.2 (continued)

Paper	Models used	Main findings
Olivier and Wotherspoon (2005)	CART GLM GAM	In general, GLM classification accuracy (presence and absence) was higher than CART, across scales and cutoff points, for both training and testing data. CART and GLM models had 86% overlap in the total area predicted to be suitable habitat
Pittman et al. (2007)	Mmultiple linear regression Neural networks CART	CART outperformed neural networks in terms of predicting species richness; both were "good" models compared with multiple linear regression models. Ability to predict spatial patterns of species richness varied; neural networks perfomed well for areas with low species richness and poor for areas of high species richness, multiple linear regression performed the opposite
Yang et al. (2006)	Logistic regression Expert knowledge	Expert knowledge produced a slightly more accurate predictive map (but not statistically significant). They have identical sensitivity and the expert model had higher specificity
Yen et al. (2004)	GLM CART Multiple adaptive regression Artificial neural networks	CART had the lowest difference between observe and predicted values; other models tended to overestimate abundance. Across locations, however, models vary in their predictive ability, lending support for the use of more than one model across a species range or a study area
Zaniewski et al. (2002)	GAM ENFA	GAMs had the highest correlation between observe and predicted presence/absence

by Zuckerberg and Huettman) should also be carried out before decisions are made. Most importantly, the final model(s) chosen should be evaluated within the context of the species' biology/ecology.

11.4 Data Availability and Suitability

Key data challenges in predictive habitat modeling have always been the dominance of presence-only data. Many models rely on museum or herbarium specimens, which may not have been collected according to rigorous sampling design guidelines. However, new modeling techniques (such as MARS) appear to be well-suited for these type of data (Elith and Leathwick 2007; see also Chap. 4) because they require only presence data and can model multiple responses.

In addition to considerations regarding sample size, recent work has empirically analyzed data characteristics to infer consequences of different data types. For example, Graham et al. (2008) modeled uncertainty in georeferenced locations of species' to assess how such error may affect model predictions. They showed that predictive error caused by location errors was minimal when boosted regression trees were used and that predictive models built using maximum entropy approaches were not affected by these errors. Guisan et al. (2007a) examined the effects of data, statistical techniques, and species models in predictive accuracy. They found model performance varied more within techniques across species versus techniques within species. Overall, the newer modeling methods (e.g., BRT, MAXENT, GAM, and MARS) had the highest predictive power. In general, species characteristics appeared to be the best prediction of model accuracy. Those models for species with narrow ecological niches and limited geographic distribution performed best. Species that are rarely dominant and/or are early successional species did not model well under any statistical technique (Guisan et al. 2007a). In a similar study, Guisan et al. (2007b) examined the effect of grain size on model performance. In tests on 50 species in five regions, they showed a tenfold change in grain size did not severely affect predictive habitat models. Again, BRT generally performed best across all variations in grain.

11.5 Past, Current, and Future Applications

One of the most comprehensive exercises in model comparison was carried out by Elith et al. (2006) who compared 16 modeling methods using presence-only data for 226 species from around the world. Elith et al. (2006) used some of the models discussed above – including GLM and regression trees – as well as envelope-style models, machine learning models, and generalized dissimilarity models. All models were tested with independent presence-absence data for model specificity and sensitivity (i.e., ability to accurately predict presences and absences). There was a wide degree of variation in model performance, but in general the newer sets of models (e.g., MARS community models, boosted regression trees, generalized dissimilarity models and maximum entropy models) performed best (Elith et al. 2006).

MARS models have not been widely applied to date, but are part of the family of generalized linear models that accommodate nonlinear responses. MARS community models allow modelers to assess community data (i.e., occurrence of all species at once) (Elith and Leathwick 2007). Boosted regression trees and maximum entropy models are based on machine-learning approaches and appear to work well with presence-only data. Boosted regression trees use two algorithms – a boosting algorithm and a regression-tree algorithm. This method can overcome inaccuracies in single tree models (Elith et al. 2008). Maximum entropy models find the most uniform distribution subject to constraints in each environmental variable (Phillips et al. 2006). Generalized dissimilarity models (GDM) are based on combining matrix regression and GLM to model spatial turnover in community composition as a function of environmental differences between pairs of sites (Ferrier et al. 2002). Elith et al. (2006) advocate the adoption of these community-based models (e.g., MARS-COMM and GDM) when considering rare species, which have always been difficult to model accurately.

In general, the comparison performed by Elith et al. (2006) yielded some general recommendations about statistical models. First, specialist species generally have higher AUC scores (i.e., higher predictive accuracy) than generalist species, and second, sample size was not related to modeling success. However, in a follow-up study that explicitly compared the effect of sample-size across a range of data and statistical models, Wisz et al. (2008) found that most models showed decreasing accuracy with smaller sample size. They found that maximum entropy models were less sensitive to sample size, and that genetic algorithms performed best in low sample sizes. However, all models performed poorly when the sample size fell below 30 (Wisz et al. 2008).

In general, it appears that the field of predictive habitat modeling is continually developing and evolving. The recent literature includes a number of reviews and critiques, authored by established and new practitioners. The tone of these is largely constructive, and new statistical techniques have been developed based on earlier models (e.g., BRT are improvements on CART models; MARS is an alternative GLM). Preliminary evaluation and testing suggests that these models may be of great use in when uncertain data are being used in predictive habitat modeling. Nonetheless, modelers and managers should not simply apply the latest statistical model because it appears to be the current trend. The fact that the predictive habitat models will be affected by the statistical model chosen should always be kept in mind, particularly when results are being interpreted. Even with more sophisticated statistical techniques that appear to be robust to errors and uncertainty in data, it is still preferred to construct multiple statistical models to enable a full evaluation and comparison of predictors. When evaluating significance and values of predictors between models, it is also important to keep ecological models in mind, and to chose based on known ecological and/or biological relationships.

Acknowledgments F. Huettmann, M. Hooten, T. Lookingbill and two anonymous reviewers provided helpful comments on an earlier draft of this chapter. Also thanks to N. Laite for assistance with compilation of journal articles for the meta-analysis.

References

Anderson DR, Burnham KP, Thompson WL (2000) Null hypothesis testing: problems, prevalence, and an alternative. J Wildl Manag 64:912–923.

Anderson DR, Link WA, Johnston DJ, Burnham KP (2001) Suggestions for presenting the results of data analysis. J Wildl Manag 65:373–378.

Anderson DR, Burnham KP (2002) Avoiding pitfalls when using information-theoretic methods. J Wildl Manag 66:912–918.

Austin MP (1985) Continuum concept, ordination methods, and niche theory. Annu Rev Ecol Syst 16:39–61.

Borra S, DiCiaccio A (2002) Improving nonparametric regression methods by bagging and boosting. Comput Stat Data Anal 38:407–420.

Boyce MS, Vernier PR, Nielsen SE, Schmiegelow FKA (2002) Evaluating resource selection functions. Ecol Modell 157:281–300.

Breiman L, Friedman JH, Olshen RA, Stone CJ (1984) Classification and regression trees. Wadsworth International Group. Belmont, California.

Breiman L (2001a) Random forests. Mach Learn 45:5–32.
Breiman L (2001b) Statistical modeling: the two cultures. Stat Sci 16: 199–231.
Burnham KP, Anderson DR (2002) Model selection and multimodel inference: a practical information-theoretic approach. 2nd edition. Springer, New York.
Busby JR (1991) BIOCLIM – a bioclimate analysis and prediction system. In: Margules CR, Austin MR (eds) Nature conservation: cost effective biological surveys and data analysis. CSIRO, Melbourne.
Carpenter G, Gillison AN, Winter J (1993) DOMAIN: a flexible modelling procedure for mapping potential distributions of plants and animals. Biodivers Conserv 2:667–680.
Clark JD, Dunn JE, Smith KG (1993) A multivariate model of female black bear habitat use for a geographic information system. J Wildl Manag 7:519–526.
Chamberlin TC (1965) The method of multiple working hypotheses. Science 148:754–759.
De'ath G, Fabricius KE (2000) Classification and regression trees: a powerful yet simple technique for ecological data analysis. Ecol 81:3178–3192.
deFrutos A, Olea PP, Vera R (2007) Analyzing and modelling spatial distribution of summering lesser kestrel: the role of spatial autocorrelation. Ecol Modell 200:33–44.
Dettmers R, Buehler DA, Bartlett JB (2002) A test and comparison of wildlife-habitat modeling techniques for predicting bird occurrence at a regional scale. Pages 607–615 In: Scott JM, Heglund PJ, Morrison ML, Haufler JB, Raphael MG, W. A. Wall WA, Samson FB (eds) Predicting species occurrences: issues of accuracy and scale. Island Press, Washington, DC.
Dzeroski S, Drumm D (2003) Using regression trees to identify the habitat preference of the sea cucumber (*Holothuria leucospilota*) on Rarotonga, Cook Islands. Ecol Modell 170:219–226.
Elith J, Graham CH, Anderson RP, Dudík M, Ferrier S, Guisan A, Hijmans RJ, Huettmann F, Leathwick JR, Lehmann A, Li J, Lohmann LG, Loiselle BA, Manion G, Moritz C, Nakamura M, Nakazawa Y, Overton JMC, Peterson AT, Philips SJ, Richardson K, Scachetti-Pereira R, Schapire RE, Soberón J, Williams S, Wisz MS, Zimmerman NE (2006) Novel methods improve prediction of species' distributions from occurrence data. Ecography 29:129–151.
Elith J, Leathwick J (2007) Predicting species distributions from museum and herbarium records using multiresponse models fitted with multivariate adaptive regression splines. Divers Distrib 13:265–275.
Elith J, Leathwick JR, Hastie T (2008) A working guide to boosted regression trees. J Anim Ecol 77:802–813.
Fielding AH, Bell JF (1997) A review of methods for the assessment of prediction errors in conservation presence/absence models. Environ Conserv 24:38–49.
Ferrier S, Drielsma M, Manion G, Watson G (2002) Extended statistical approaches to modelling spatial pattern in biodiversity: the northeast New South Wales experience II. Community level modelling. Biodivers Conserv 11:2309–2338.
Friedman JH (1991) Multivariate adaptive regression splines (with discussion). Ann Stat 19:1–141.
Friedman JH, Hastie T, Tibshirani R (2000) Additive logistic regression: a statistical view of boosting. Ann Stat 28:337–407.
Garzon MB, Blazek R, Neteler M, Sanchez de Dios R, Ollero HS, Furlanello C (2006) Predicting habitat suitability with machine learning models: the potential area of *Pinus sylvestris* L. in the Iberian Peninsula. Ecol Modell 197:383–393.
Graham CH, Ferrier S, Huettman F, Mortiz C, Peterson AT (2004) New developments in museum-based informatics and applications in biodiversity analysis. Trends Ecol Evol 19:497–503.
Graham CH, Elith J, Hijmans RJ, Guisan A, Peterson AT, Loiselle BA, NCEAS PSDWG (2008) The influence of spatial errors in species occurrence data used in distribution models. J Appl Ecol 45:239–247.
Guisan A, Zimmermann NE (2000) Predictive habitat distribution models in ecology. Ecol Modell 135:147–186.
Guisan A, Zimmermann NE, Elith J, Graham CH, Phillips S, Peterson AT (2007a) What matters for predicting the occurrences of trees: techniques, data or species' characteristics? Ecol Monogr 77:615–630.

Guisan A, Graham CH, Elith J, Huettmann F, NCEAS SDMG (2007b) Sensitivity of predictive species distribution models to change in grain size. Divers Distrib 13:332–340.
Guthrey FS, Brennan LA, Peterson MJ, Lusk JJ (2005) Information theory in wildlife science: critique and viewpoint. J Wildl Manag 69:457–465.
Hanley JA, McNeil BJ (1982) The meaning and use of the area under a receiver operating characteristic (ROC) curve. Radiology 143:29–36.
Hirzel AH, Arlettaz R (2003) Modeling habitat suitability for complex species distributions by environmental-distance geometric mean. Environ Manag 32:614–623.
Holloway GL, Malcolm JR (2006) Sciurid habitat relationships in forests managed under selection and shelterwood silviculture in Ontario. J Wildl Manag 70:1735–1745.
Hothorn T, Hornik K, Zeileis A (2006) Unbiased recursive partitioning: a conditional inference framework. J Comput Graph Stat 15:651–674.
Jelaska SD, Antoni O, Nikoli T, Hrsak V, Plazibat M, Krizan J (2003) Estimating plant species occurrence in MTB/64 quadrants as a function of DEM-based variables-a case study for Medvednica Nature Park, Croatia. Ecol Modell 170:333–343.
Johnson DH (1999) The insignificance of statistical significance testing. J Wildl Manag 63:763–772.
Johnson DH (2002) The role of hypothesis testing in wildlife science. J Wildl Manag 66:272–286.
Keating KA, Cherry S (2004) Use and interpretation of logistic regression models in habitat selection studies. J Wildl Manag 68:774–789.
Li J, Hilbert DW (2008) LIVES: a new habitat modelling technique for predicting the distribution of species' occurrences using presence-only data based on limiting factor theory. Biodivers Conserv 17:3079–3095.
Link WA, Barker RJ (2006) Model weights and the foundations of multimodel inference. Ecology 87:2626–2635.
Lippitt CD, Rogan J, Toledana J, Sangermano F, Eastman JR, Mastro V, Sawyer A (2008) Incorporating anthropogenic variables into a species distribution model to map gypsy moth risk. Ecol Modell 210:339–350.
MacKenzie DI, Nichols JD, Royle JA, Pollock KH, Bailey LL, Hines JE (2006) Occupancy estimation and modeling: inferring patterns and dynamics of species occurrence. Academic Press, Burlington, MA.
Maisonneuve C, Belanger L, Bordage D, Jobin B, Grenier M, Beauliu J, Gabor S, Filion B (2006) American black duck and mallard duck breeding distribution and habitat relationships along a forest-agricultural gradient in southern Quebec. J Wildl Manag 70:450–459.
Manly BJF (1985) Measuring selectivity from multiple choice feeding-preference experiments. Biometrics 5:709–715.
Manly BJF, McDonald LL, Thomas DL (1993) Resource selection by animals: statistical design and analysis for field studies. 1st edition. Chapman and Hall, London.
Manly BJF, McDonald LL, Thomas DL, McDonald TL, Erickson WP (2003) Resource selection by animals: statistical design and analysis for field studies. 2nd edition. Kluwer Academic Publishers, Dordrecht, NL.
Miller J, Franklin J (2002) Modeling the distribution of four vegetation alliances using generalized linear models and classification trees with spatial dependence. Ecol Modell 157:227–247.
Moisen GG, Freeman EA, Blackard JA, Frescino TS, Zimmermann NE, Edwards TC (2006) Predicting tree species presence and basal area in Utah: a comparison of stochastic gradient boosting, generalized additive models, and tree-based methods. Ecol Modell 199:176–187.
Naugle DE, Higgins KF, Nusser SM, Johnson WC (1999) Scale-dependent habitat -use in three species of prairies wetland birds. Landsc Ecol 14:267–276.
Olivier F, Wotherspoon SJ (2005) GIS-based application of resource selection functions to the prediction of snow petrel distribution and abundance in East Antarctica: comparing models at multiple scales. Ecol Modell 189:105–129.
Pearce J, Ferrier S (2000) Evaluating the predictive performance of habitat models developed using logistic regression. Ecol Modell 133:225–245.

Phillips SJ, Anderson RP, Schapire RE (2006) Maximum entropy modeling of species geographic distributions. Ecol Modell 190:231–259.

Pittman SJ, Chistensen JD, Caldow C, Menza C, Monaco ME (2007) Predictive mapping of fish species richness across shallow-water seascapes in the Caribbean. Ecol Modell 204:9–21.

Ribe R, Morganti R, Hulse D, Shull R (1998) A management driven investigation of landscape patterns of northern spotted owl nesting territories in the high Cascades of Oregon. Landsc Ecol 13:1–13.

Robinson DH, Wainer H (2002) On the past and future of null hypothesis significance testing. J Wildl Manag 66:263–271.

Rotenberry JT, Knick ST, Dunn JE (2002) A minimalist approach to mapping species' habitat: Pearson's planes of closest fit. Pages 281–289 In: Scott JM, Heglund PJ, Morrison ML, Haufler JB, Raphael MG, Wall WA, Samson FB (eds) Predicting species occurrences: issues of accuracy and scale. Island Press, Washington, DC, USA.

Saarenmaa H, Stone ND, Folse LJ, Packard JM, Grant WE, Makela ME, Coulson RN (1988) An artificial intelligence modelling approach to simulating animal/habitat interactions. Ecol Modell 44:125–141.

Sleep DJH, Drever MC, Nudds TD (2007) Statistical versus biological hypothesis testing: response to Steidl. J Wildl Manag 71:2120–2121.

Soberón JM, Llorente JB, Onate L (2000) The use of specimen-label databases for conservation purposes: an example using Mexican Papilionid and Pierid butterflies. Biodivers Conserv 9:1441–1466.

Steidl RJ (2006) Model selection, hypothesis testing, and risks of condemning analytical tools. J Wildl Manag 70:1497–1498.

Steidl RJ (2007) Limits of data analysis in scientific inference: reply to Sleep et al. J Wildl Manag 71:2122–2124.

Stockwell DRB, Peters AT (1999) The GARP modelling system: problems and solutions to automated spatial prediction. Int J Geogr Inf Sci 13: 143–158.

Tsoar A, Allouch O, Steinitz O, Rotem D, Kadmon R (2007) A comparative evaluation of preesence-only methods for modelling species distribution. Divers Distrib 13:397–405.

Vayssiéres MP, Plant RE, Allen-Diaz BH (2000) Classification trees: an alternative non-parametric approach for predicting species distributions. J Veg Sci 11:679–694.

Whittaker GA, McCuen RH (1976) A proposed methodology for assessing the quality of wildlife habitat. Ecol Modell 2:251–272.

Wisz MS, Hijmans RJ, Li J, Peterson AT, Graham CH, Guisan A, NCEAS PSDWG (2008) Effects of sample size on the performance of species distribution models. Divers Distrib 14:736–773.

Yang X, Skidmore AK, Melick DR, Zhou Z, Xu J (2006) Mapping non-wood forest product (matsutake mushrooms) using logistic regression and a GIS expert system. Ecol Modell 198:208–218.

Yen PPW, Huettmann F, Cooke F (2004) A large-scale model for the at-sea distribution and abundance of Marbeled Murrelets (*Brachyramphus marmoratus*) during the breeding season in coastal British Columbia, Canada. Ecol Modell 171:395–413.

Zaniewski AE, Lehmann A, Overton JMC (2002) Predicting species spatial distributions using presence-only data: a case study of native New Zealand ferns. Ecol Modell 157:261–280.

Chapter 12
Expert Knowledge as a Basis for Landscape Ecological Predictive Models

C. Ashton Drew and Ajith H. Perera

12.1 Introduction

Defining an appropriate role for expert knowledge in science can lead to contentious debate. The professional experience of ecologists, elicited as expert judgment, plays an essential role in many aspects of landscape ecological science. Experts may be asked to judge the relevance of competing research or management questions, the quality and suitability of available data, the best balance of complexity and parsimony, and the appropriate application of model output. Even the initial decision to pursue modeling follows expert judgment regarding the cost and benefits of a model relative to data collection and the suitability of alternative modeling approaches for the specific application. Increasingly, however, professionals are asked to provide expertise to complement or even substitute for scarce data in landscape ecological models, by quantifying their personal experiences and anecdotal observations. In such cases, the professional is asked to reference their knowledge against geospatial data or landscape metrics derived from such data. We offer our chapter to raise awareness and promote discussion of this particular development within landscape ecological modeling. We draw examples from cases where expertise is provided as data in support of the predictive species-habitat models used to inform conservation planning objectives and strategies.

Most of the chapters in this book describe modeling approaches that are data intense. However, few taxa and few regions of the globe offer high-quality spatial data. Although protocols for sampling populations and habitat at landscape scales have advanced rapidly, few species or habitats have yet been systematically monitored over long temporal and broad spatial scales. Therefore, empirical data limitations are common both when setting broad-scale national population and habitat objectives, and when debating local management decisions to meet these objectives. The trend to substitute expertise for empirical data is most evident where models inform decisions

---

C.A. Drew (✉)
North Carolina Fish and Wildlife Cooperative Research Unit, Department of Biology, North Carolina State University, Raleigh, NC 27695, USA
e-mail: cadrew@ncsu.edu

regarding the risk or utility of natural resource management alternatives, where ecological systems are complex and poorly defined, and where resource management questions are deemed too urgent to await the empirical information from rigorous experiments or field surveys. To some, turning to expert-based modeling opens a Pandora's Box of potential prejudice and error into what should be a carefully controlled, unbiased, repeatable, and transparent process. Others, however, are confident that careful attention to methodology and application can produce expert-based models of the same rigorous standards as their data-driven counterparts.

The growing popularity of expert-based models in applied landscape ecology reflects several trends in scientific knowledge, social values, and resource conflicts. Expert knowledge is viewed as subjective and sometimes associated with high uncertainty and bias. Yet, experts are often the most accessible and cost-effective source of immediate ecological information. Increasing urgency of conservation in the face of such threats as global climate change, habitat fragmentation and alteration by land use, and invasive species proliferation promotes the use of expert knowledge in conservation planning. Not only are such threats expected to intensify in the future (Balmford and Cowling 2006), but conservation managers are also challenged to consider potential impacts and trade-offs over broader spatial scales and longer temporal horizons than have been customary in the past. Expert-based landscape ecological models now regularly support natural resource management decisions by projecting how land cover is expected to change and where different types of habitat will be lost and gained, predicting the likely quality of present and future habitat patches, and evaluating the potential conservation value of future landscapes (Marcot et al. 2001; Pearce et al. 2001; Store and Kangas 2001; Petit et al. 2003; Yamada et al. 2003; Martin et al. 2005; MacMillan and Marshall 2006; Bashari et al. 2009; Murray et al. 2009; Doyon et al. 2010; Rothlisberger et al. 2010; Teck et al. 2010; Perera et al. in press). Unlike decision models designed to assess the quality or value of local resources in the present, predictive models define likely outcomes given complex and uncertain future conditions. The case is made that management decisions cannot await further data across such a broad range of species, habitats, and issues, so experts are called upon to fill ecological knowledge gaps. However, in contexts where management plans must meet "science-based" criteria, it is often implied that expert knowledge must gradually be tested and replaced by data gathered through monitoring in an adaptive management framework.

Given local and international attention to complex, landscape-relevant challenges such as climate change, alternative energy development, water resource conflicts, and population growth and migration, we expect expert knowledge to play an increasingly important role in landscape ecological applications. Therefore, for those landscape ecologists and land managers that find themselves data-limited and facing difficult decisions, we offer this introduction to expert-based modeling procedures for landscape-scale prediction of species and habitat distribution patterns. In this chapter, we discuss how expert knowledge is currently elicited and applied within predictive models, identify common traits that distinguish successful models from their unsuccessful counterparts, and review how expert judgment is typically quantified and compared prior to being incorporated into predictive models. We conclude with summary recommendations, drawn from the literature and our own experience in expert-based modeling, by which we hope to improve the rigor, appropriate use, and acceptance of expert-based modeling approaches.

12.2 Who Is An Expert and What Is An Expert-Based Model?

In this chapter, we define an "expert" as anyone who has special knowledge, gained through a combination of formal training, direct personal experience, and reflection of the ecological pattern or process under investigation. Within the scope of this definition, the specific vocation of experts targeted for knowledge elicitation can be variable and goal-specific. It certainly includes professional ecologists and resource managers, but could, in some instances, also include non-professional members of the public. Hunters or wildlife photographers, for example, can provide valuable knowledge of certain species and habitats (Yamada et al. 2003; Gilchrist et al. 2005). Members of traditional cultures can also offer knowledge of historical and present day species-habitat associations and population dynamics (Huntington 2000).

Predictive landscape-scale species–habitat modeling approaches that incorporate expert knowledge include simple rule-based models (Petit et al. 2003; Drescher and Perera 2010), multi-attribute value functions (Store and Kangas 2001; Geneletti 2005), and Bayesian approaches. For this discussion, we focus on a subset of expert-based models: those that, in the absence or limited availability of empirical data to define statistical probability of a future event, call upon experts to substitute their own subjective estimates of probability via a formalized elicitation process. Much of the recent growth in this area of expert-based modeling for conservation applications builds upon the development of probabilistic modeling approaches that incorporate Bayesian statistics (Nyberg et al. 2006; Uusitalo 2007; O'Leary et al. 2008; Bashari et al. 2009; Low Choy et al. 2009; James et al. 2010). Most Bayesian procedures use Bayes rule to assimilate incomplete knowledge formalized as the subjective prior probability distribution updated with data and their corresponding likelihood function. The scaled product of the prior and likelihood result in a posterior probability distribution, which formalizes the updated uncertainty of a future event given the data and model. In the absence of prior information, expert elicitation can be used to generate the prior probability distributions. The opportunity to begin with a model based upon expert knowledge and then iteratively update both the model and the uncertainty through the addition of new information is an attractive feature of Bayesian approaches, because it fits well with the current paradigm of adaptive monitoring and management (Williams 2003; Williams et al. 2009). Bayesian models offer the opportunity to meet criteria of employing "best available science" in support of complex decisions, while also using the model to develop monitoring protocols that will gather the data necessary to test and improve the model for the next round of decisions.

Although we focus here on recent applications of expert knowledge in Bayesian models for landscape-scale species–habitat conservation, we do not wish to imply that the role of expert knowledge in natural resource planning is a novel development.

There exists a rich history of theoretical and methodological advances for eliciting and using expert knowledge within natural resource management research and planning (Coulson et al. 1987; Rykiel 1989; Starfield and Bleloch 1991; Giles 1998). Historically, when eliciting knowledge, the primary objective was to develop expert systems that could reliably reproduce expert judgments by codifying expert knowledge and reasoning (e.g., artificial intelligence). In contrast, the elicitation methods discussed here aim not to define and replace expert reasoning, but to provide the best possible representation of the existing knowledge base to support expert reasoning. However, while the fundamental objectives of eliciting knowledge may differ, many of the ideas presented within this chapter benefited from and build upon the earlier efforts to improve elicitation methods.

12.3 Evidence of Expert-Based Model Strengths and Weaknesses

Long-term success or failure in applying expert-based predictive models is not evident within landscape ecological literature. The absence of quantitative tests reflects both the relative short history of available techniques and the failure to implement monitoring programs to validate or update the original models. Most assessments of expert-based model performance have focused on the internal consistency of expert judgment either when an individual's knowledge is elicited by multiple methods (Yamada et al. 2003; Low Choy et al. 2009) or when multiple experts' knowledge is elicited by a single method (Yamada et al. 2003). Narrow credible intervals (the Bayesian approximation of confidence intervals) that result from expert agreement are interpreted as evidence that, in the absence of empirical data, using "expert information as prior in ecological models is a cost-effective way of making more confident predictions about the effect of management" (Martin et al. 2005). Studies have shown expert-based models to be highly sensitive to variation in expert opinion (Yamada et al. 2003; Drescher et al. 2008; Aspinall 2010). This suggests a need for caution when only one or few experts' knowledge can be gathered and supports a strong recommendation that all models be subjected to thorough uncertainty and sensitivity analyses, as well as stringent external review, prior to application in management decisions (Yamada et al. 2003; Johnson and Gillingham 2004). Unfortunately, while measures of expert consensus and consistency provide information on model precision, they provide no data on model accuracy.

Publications that compare results from expert-based versus data-based models typically do so for a single spatial and temporal extent. For example, distribution patterns of 93 faunal species in New South Wales Australia were predicted more accurately (based on comparison of model output with independent survey data) by logistic regression models fit to data than by expert-based models constructed from the elicited knowledge of three regional experts. (Pearce et al. 2001). Another multi-taxa study using expert-based rules to predict species occurrence concluded that poor model performance was possibly a result of low species prevalence, as models for insects performed much better than those for plants and birds (Petit et al. 2003). However, direct comparisons of expert-based with data-based models in data-rich settings may not address the correct questions to truly evaluate potential strengths and weaknesses of predictive models as applied in landscape conservation contexts. In such cases, when managers begin with little or no data, expert-based models are developed to summarize the only available relevant data – that held within the experts' personal experiences. If the applied alternative to an expert-based model is no model, a fairer test of the value of expert knowledge may be to compare expert predictions to random or null model predictions. When expert-based models are a starting point for an adaptive management process (Marcot et al. 2006; Nyberg et al. 2006), the implication is that new data will be added through time to gradually improve the model and move toward data-based prediction. It would therefore be useful to test whether expert-based models, through the gradual incremental addition of new data, do converge toward the same conclusions as data-based models. Furthermore, it would be insightful to test whether an expert-based model or a data-based model fitted to a very small dataset (e.g., simulating a delay in management action to conduct a one year baseline study) would more rapidly converge towards the conclusions of a data-based model fitted with a large dataset.

12.4 Strategies for Sampling Expertise

It is perhaps helpful to initially think of "sampling" expertise in the same sense as sampling any other landscape ecological phenomenon. For example, experts who observe species in the field generate informal, hypothesized species–habitat relationships. Elicitation seeks to acquire predictive value from these past observations. During elicitation, one or more experts may be asked to identify important ecological variables or processes, to categorize or rank variables, to define response curves with confidence intervals, and possibly even to classify their own level of expertise. Methods range from informal interviews to complex, customized graphical programs (Al-Awadhi and Garthwaite 2006; Drescher et al. in press), from mail-in survey questionnaires to large gatherings run by professional facilitators (Moody and Grand in press; Silbernagel et al. in press). Different methods provide different types of data, which in turn support different modeling objectives and applications. Therefore, as in any other ecological study, a clearly defined objective and study design (both for the study itself and the role of experts within the study) is essential to later interpret and apply the data within the predictive model. Basic questions that must be addressed at the onset include: (1) what is the purpose of the model? (2) where and when will the model be applied? And (3) how will the model be calibrated, evaluated, and validated? Only after these initial questions have been answered can individuals with the relevant expertise be identified and can a method be designed that will elicit the desired information. Furthermore, by defining the extent and resolution of the knowledge required, these questions clarify the bounds of the elicited knowledge and facilitate appropriate use and effective testing of the model products.

12.4.1 Identifying and Calibrating Experts

Developing a suitable method to identify individuals with the requisite expertise can be challenging, and is easily compromised as participation defaults to those willing to offer their time and knowledge. It is easy to visualize a scale of expertise – ranging from low to high – where expert utility varies from serving as a temporary substitute for point data to quantifying parameter rates or hypothesizing cause–effect relationships. However, in practice it can be quite difficult to assess an individual's level of expertise, especially when multiple experts with diverse experiences offer conflicting judgments. In general, the utility of knowledge from individuals attributed with "high" levels expertise will be of highest value to the scientific community. These are individuals that not only have direct experience relevant to the modeling objectives, but are also known for their ability to synthesize, extrapolate, and generalize from their experiences. They are also the same individuals whose professional judgment is frequently sought to review and validate data-driven models. There is greater dispute on the value and appropriate use of experts with very localized (in space or time) experience. Such experts may mistakenly be viewed as inexperienced and may simultaneously be more likely to undervalue their own knowledge. However, if the resolution of available data or high-level expert knowledge is more coarse than the scale of the proposed model and management application, local experts with low-level expertise may still be valuable to set global knowledge within the local context. (Drescher et al. 2008).

In reviewing the qualifications of an expert, it is not adequate to simply ask them to quantify their own expertise. Expert self-confidence can vary by gender, age, and personality type as much as by any ecologically relevant criteria. It is therefore important to consider multiple factors, including where, when, and how an expert gained their experience and how they have since filtered these experiences through conversation, reading, and reflection (Yamada et al. 2003; Battisti et al. 2008; Drew and Collazo in press). Most experts have built their knowledge base through a variety of sources, including academic course work, field research, primary literature, workshops or symposia, conversations with colleagues, and anecdotal/recreational observations.

12.4.2 Where Did Experts Acquire Knowledge?

Every expert has a "home range" from which they have gained their experience. Ideally (but unrealistically) the home-range of an expert would perfectly overlap the extent of the proposed species–habitat model and they would have visited all regions within their home range regularly. This would provide systematic knowledge of available habitats against which to reference their observations. However, expert observation, just like empirical data, never perfectly or describes a species' ecological niche. The knowledge of each expert reflects the perspective they have gained from a unique spatial and temporal setting. Careful consideration of the home range where an expert gained their knowledge and experience can suggest when differences in judgment reflect differences in landscape, rather than, or in addition to, different levels of observational skill and experience. For example, experts from coastal North Carolina and Virginia marshes work in conservation lands that vary greatly in their landscape and microhabitat characteristics. Some regions offer only narrow fringing marshes whereas others offer island and mainland marsh patches of varying size and isolation. In the absence of data to characterize the home range of the expert, their judgment regarding the relevance of patch size, shape, and isolation could look wildly discordant, as each tended to discount the relative importance of factors outside their experience (Drew and Collazo in press). Similarly, experts from different regions in Australia offered conflicting opinions regarding the influence of elevation and geology on brush-tailed wallabies (Murray et al. 2009). Their lack of consensus also reflected real geological differences influencing species–habitat associations. In boreal Canada, expert knowledge of forest succession is biased by the spatial configurations, as well as temporal periods, of the experts range of familiarity (Drescher et al. 2008). These examples suggest that although the ability of experts to extrapolate beyond their region of knowledge is highly variable and sometimes very poor (Murray et al. 2009), it may be possible to assess such ability through comparison of the experts' home ranges and the extent of the modeling landscape.

Distinguishing when differences among experts reflect true ecological insight from unique landscapes, rather than simply differences in their respective levels of knowledge or ability, is especially important when deciding whether (and if so how) to aggregate elicited knowledge or force consensus. A common assumption has been that multi-expert elicitations are better than single-expert elicitations because they reduce effects of individual bias (Yamada et al. 2003; MacMillan and Marshall 2006). However, if individual experts provide locally precise and accurate information within a geographically large and environmentally diverse region, the value of using multiple experts is not to remove bias so much as to ensure the diversity of the region is fully sampled (Drescher et al. 2008; Murray et al. 2009). This implies a need to attend to expert selection to recognize potential problems of spatial bias, autocorrelation, and uneven sampling of environmental conditions, just as when designing sampling protocols of species or habitats.

12.4.3 When Did Experts Acquire Knowledge?

The time period over which experts gained their knowledge also provides an important context for eliciting and applying their judgment. This is especially true if elicitation methods will require experts to use mapped data or aerial photography, which offer temporal snapshots. Processes such as succession, erosion, and development can rapidly and significantly change landscapes and alter species geographic distribution. Locations that an expert visited in the past may have experienced significant change and may no longer reflect conditions recalled by the expert. Orienting experts to landscape conditions, as represented in the map or aerial imagery allows experts to identify regions that may have changed since their knowledge acquisition. Without this orientation, accurate knowledge of past species–habitat associations could easily be inaccurately applied to present landscapes, introducing an avoidable source of model error. The timing of knowledge acquisition is also important as it relates to extreme or episodic environmental conditions that affect species population dynamics (Stenseth et al. 2002; Battisti et al. 2008). For example, observations made during El Niño versus La Niña climate periods could lead to very different, but equally valid, conclusions about species–habitat associations and distribution patterns (Davis 2000; Kim et al. 2008). Exposure to extreme or rare events can also introduce bias in probability estimates through the heuristic termed "availability", if sensational or unusual events are recalled more easily or in greater detail than common events (Tversky and Kahneman 1974; Kynn 2008).

12.4.4 How Did Experts Acquire Knowledge?

As in any other landscape ecological study design, scale plays an important role in determining the relevance of available information. Elicitation of species–habitat associations will be confounded if experts do not clearly understand the relationship between the scale of their own knowledge relative to the scale of the proposed model (King and Perera 2006). For example, most field biologists or hunters do not consider species–habitat associations using the landscape metrics commonly used to define landscape patterns. Such metrics have simply not played a major role in their training (unless recently graduated) or experience, and so have not served as reference points for the synthesis of their knowledge. Furthermore, unless population surveys are conducted aerially, most empirical observations offer insights closer to the microhabitat perspective of an individual organism than to the coarser landscape perspective of geographic information systems (GIS) (Yamada et al. 2003; Drew and Collazo in press). While this would not present a challenge to the creation of site level habitat suitability indices or community level associations (e.g., matching species to forest types within a vegetation classification systems), it does confound efforts to construct accurate GIS-based landscape ecological models from expert knowledge. During elicitation and model development, it often becomes necessary to associate elicited fine-scale knowledge to coarse-scale mapped features through the use of indirect proxy variables. For example, if experts provide predictions based on forest structure, LIDAR data might provide a useful landscape-scale proxy for field measurements of forest structure (Lefsky et al. 2002). More subtle perhaps, and of unknown effect for expert-based models, are the potential consequences of translating their elicited knowledge to a GIS when proxy models seem unnecessary. For example, experience may lead an expert to predict a marsh species has a strong association with edges near open water. In a raster representation of marsh and water cover, locations classified as edge will be subject to the grain (e.g., pixel cell size) of the data. If the raster represents features using a 30-m grid, then any water feature within the marsh that is smaller than 30 m × 30 m would be mapped as marsh rather than water; edge habitat of small water features would not appear in the map data. If an expert assessment of a species' dependence on edge habitat stems from observations near water features smaller than this threshold, then predicted suitability or occupancy of any given marsh pixel and the landscape as a whole would potentially be underestimated. It is therefore essential to elicit the relevant scale of the expert's experience and observations, and ensure that participating experts understand the scale relationships between their knowledge and the proposed application of their knowledge.

It is also valuable to know the observation methods used by experts in the field. In field research, gear selection and survey method can have a strong impact on estimates of population size, vital rates, and habitat associations. Therefore, knowing how their knowledge was acquired would provide insight into differences in judgments. If the relative efficiency of two observation methods is known, it may be possible (and desirable) to correct for differences attributed to method prior to contrasting or aggregating elicited judgments.

12.4.5 Challenges Unique to Sampling Expertise

12.4.5.1 Miscommunication During Elicitation

Sampling expertise differs from sampling other ecological phenomenon in that communication between the elicitor and experts presents a unique source of potential confusion, uncertainty, and error (Ray and Burgman 2006). Effective communication requires early and frequent interaction among all participants to ensure transparency and to establish a common lexicon. This includes not only those involved in the elicitation and synthesis of expert knowledge, but also those who will have responsibility for model development, delivery, and application. Due to confusion in terminology, especially across fields, it is essential to ensure that the language of the elicitation matches that of the expert being queried (Johnson and Gillingham 2004). Many problems relative to eliciting and comparing expert knowledge arise through the use of vague terms and concepts (Elith et al. 2002). If knowledge is elicited without direct interaction (e.g., only through mail-in survey), it can be very difficult to assess how the questions were interpreted. Imprecise language, the use of different terms in parallel disciplines, and vague concepts and terminology compound knowledge uncertainties (Elith et al. 2002). For example, categorical descriptors of gradients such as wet/dry, large/small, steep/dry have specific local meanings and will confound an elicitation and misinform models unless defined and expressed in unequivocal terms. Without an effective communication plan, individual experts may also hold different assumptions about what information is desired.

The elicitation questions provide the sampling "quadrat" applied to expert knowledge. While few ecologists would haphazardly use different quadrat sizes during a single vegetation survey, elicitations are sometimes conducted without clearly defining and consistently using terminology that matches the objectives. For example, suppose expert knowledge is sought to generate a model of predicted species distribution. Then, throughout the elicitation, experts are asked questions about where species have been detected, with little thought to the fact that detections often under-represent true occupancy and the two quantities can vary independently in relation to habitat characteristics (MacKenzie et al. 2003). The risk of miscommunication may be greatest in informal elicitations where the modelers more easily interchange terms that are synonymous in casual conversation, but which have strictly different meaning in the context of predicting species-habitat associations (e.g., Where have you seen the species? Where have you detected the species? Where does the species live? What habitats are suitable for the species?). Similar semantic confusion has been documented for the terms use, selection, and preference in reference to species–habitat associations (Jones 2001).

12.4.5.2 Cognitive Limitations and Group Dynamics

Several choices are made during the selection and development of an elicitation approach. Foremost among these decisions are: (1) whether to conduct group or individual elicitation, and (2) if using multiple experts, whether to seek consensus. The social science and psychology literature provide valuable insight regarding how individual biases and group dynamics influence the interpretation and summary of experience during elicitation procedures (Tversky and Kahneman 1974; Kynn 2008). Work assessing the influence of such factors on expert-based ecological models used to support conservation management decisions is just beginning (Anderson 1998; Perera et al. in press). However, these same authors illustrate that human cognitive limitations and social dynamics, which can generate bias in elicited data, remain even after the elicitation terminology and questions are clearly defined.

The knowledge of any one individual will reflect the sum of their experience gained through direct observation, peer interaction, literature review, and personal reflection and synthesis of these combined inputs. In an elicitation to support Bayesian models, individuals are asked to summarize past experience to generate probability estimates to define Bayesian priors or conditional probability tables. People make predictable cognitive errors during probabilistic reasoning and commonly provide answers that defy the laws of probability (Anderson 1998; Baddeley et al. 2004; Kynn 2008) due to over-dependence upon heuristic rules-of-thumb. The use of heuristics to process information has been linked to predictable errors (Tversky and Kahneman 1974; but see Kynn 2008), such as over-estimating the probability of a rare event, if such an event has been recently experienced or is easily recalled (i.e., the bias of availability). Also, although direct observations are personal experiences, this information can quickly disperse through the expert community either formally through published literature or informally through conversation. Knowledge sharing reduces the independence of expert knowledge as data, as individual experts qualify their personal experiences in light of group knowledge and opinion (i.e., the bias of anchoring and adjustment). Fortunately, good elicitation design can reduce the potentially significant impact of these biases and other cognitive errors (Meyer and Booker 2001). For example, experts tend to more accurately estimate the frequency of events, than the probability of a single event (Anderson 1998).

Common cognitive and motivational biases of individuals can be magnified or unduly suppressed through group dynamics. Baddeley et al. (2004) identified a herding trend among experts asked to provide judgments in contexts offering little data, but the opportunity to apply heuristics. Without skillful mediation of the elicitation process, such group dynamics can lead to strong consensus without any foundation in objective reality and lead to over-confidence in the group's conclusions (Baddeley et al. 2004).

12.5 Methods and Tools for Eliciting Knowledge

12.5.1 Elicitation Framework

An elicitation framework is the step-by-step process by which expert knowledge is obtained, summarized, and documented. Framework details will be specific to each project, given different objectives and available resources (experts, time, and money). However, guidelines for a successful elicitation framework share a number of recommendations. These include bringing the experts together at least once; clearly defining the issues, objectives, and terms; explaining the methods and application; providing some training in probability assessment and common biases; eliciting and documenting disagreement as well as consensus; and performing a dry run to test communication strategies (Cooke 1991). Meyer and Booker (2001, Chap. 7) outline six components of an elicitation framework: elicitation techniques (verbal or ethnographic), modes of communication (mail, telephone, or in person), elicitation situations (individual, interactive group, or Delphi), response modes (e.g., quantities, probabilities, ranks), aggregation schemes (mathematical or behavioral), and documentation. Their guidelines clarify when each component is necessary and highlight critical methodological choices made within each component. We refer readers to their work, but also here outline and expand upon their discussion of situation and response mode choices.

When the judgments of multiple experts are sought, these may be elicited independently and compared, elicited independently and combined, or elicited together for group consensus. Individual face-to-face interviews eliminate bias due to group dynamics, but also eliminate the possibility of synergistic effects from inter-expert discussion (Meyer and Booker 2001). Two common and versatile approaches for multi-expert elicitation in the natural resource and environmental assessment literature are Delphi or modified-Delphi surveys (Marcot et al. 2001; Meyer and Booker 2001; Hess and King 2002; MacMillan and Marshall 2006) and structured decision-making workshops (Ralls and Starfield 1995). The traditional Delphi process uses an iterative series of questionnaires to build consensus among survey participants who remain anonymous to one another. After each survey round, experts are returned their own answers along with summary statistics defining the mean and variance of all participants' responses. Experts then have the opportunity to revise their own response or, if choosing not to alter an outlier response, to provide a written justification of their unusual observation or prediction. Modified Delphi surveys typically place less emphasis on anonymity and, in bringing experts together for group elicitation, allow more opportunity to explore differences among expert judgment. Structured decision-making (Lyons et al. 2008) is typically performed in a workshop setting (Moody and Grand in press). The method requires participants to define three aspects of a management scenario: (1) the objective, (2) a set of alternative potential actions, and (3) the expected consequences of each action, stated in terms of the objective (Lyons et al. 2008). The elicitation of expert knowledge as data occurs in step three, when experts can define the probability of how each alternative action will impact the modeled system.

Within the structure of the elicitation framework, the modeler must pose clearly stated questions to obtain the necessary knowledge in the necessary format to support the modeling objective (Cooke 1991; Meyer and Booker 2001). Questions may be posed to elicit knowledge as physical quantities, such as, "What is the minimum patch size that would support this animal's home range?" Alternatively, they may be posed to elicit knowledge as probabilities or frequencies: "What is the probability that this animal would establish a territory in a 20 hectare patch?" or "In surveys of 100, 20-hectare patches, how many would contain an active territory of this animal?" Questions of physical quantity are direct and easily interpreted, so experts are often most comfortable with this approach. Eliciting probabilities also provides a direct means of expressing each individuals' uncertainty within the model, although combining elicited quantities from multiple experts could provide a measure of group uncertainty (Aspinall 2010). As estimates of probabilities are prone to various cognitive and motivational biases (Tversky and Kahneman 1974; Kynn 2008), it is essential that the framework include some means to assess or calibrate experts.

It is helpful if the elicitation includes opportunities for experts to visualize their responses in multiple formats so they can seek internal consistency (Cooke 1991; Meyer and Booker 2001; Yamada et al. 2003; Denham and Mengersen 2007; James et al. 2010). In landscape ecological applications, complementary visualizations usually include GIS maps and graphs (e.g., response curves). These graphs typically illustrate the relationship between a univariate predictor variable and the response variable (e.g., suitability, probability of occupancy) with some depiction of variation or uncertainty. In a GIS, the expert can interactively view species point or range data, or simply locate familiar sites, in relation to the available spatial data layers. In the past, maps or graphs of expert knowledge would be produced after the elicitation and then returned to the experts for their review and feedback. However, an increasing number of tools are available to supply these graphics interactively throughout the elicitation process (Yamada et al. 2003; Al-Awadhi and Garthwaite 2006; James et al. 2010). Where data are elicited and presented within graphs and maps simultaneously, both the participating experts and the modelers conducting the research claim that the final quantified judgments better reflect true beliefs (Yamada et al. 2003).

12.5.2 Combining Knowledge of Multiple Experts

The issue of how many experts to interview and how (or whether) to combine their elicited data deserves careful consideration. The knowledge of multiple experts may be combined by requiring all experts to reach consensus together or by combining results of individual elicitations (Morris 1977; Cooke 1991). Where a single consensus or merged result is the objective, there is an implicit assumption that combining the knowledge of multiple experts will produce more reliable results than any one individual (MacMillan and Marshall 2006). Tests of this assumption have received mixed reviews (Cooke 1991, Aspinall 2010). Ultimately, the value of a group versus individual judgment depends on the time and space scale of each individual's experience in relation to the proposed application of the elicited knowledge (Drescher et al. 2006).

Again, it is helpful to consider elicitation in light of traditional research study design. Certainly, as in experimental studies, more sample points offer potentially greater analytical power to discern pattern and infer process. However, to offer this benefit, sample locations must be carefully selected to fully represent variability in the factors of interest while controlling for variability that would confound interpretation of results. Similarly, it is critical to consider and control for consistency and representativeness in the pool of expertise that is sampled. Divergent personal experience (e.g., where, when, and how experience was gained) can lead equally qualified experts to provide widely divergent species-habitat predictions (e.g., Yamada et al. 2003; Murray et al. 2009). If the purpose is to construct a broadly applicable model predicting habitat associations or climate change effects over a large area, then a sample of experts from across the country may contribute useful information and consensus may provide a balanced view of how species respond to different environmental gradients. If instead the purpose is to construct a model for a very localized system, it would be wise to carefully consider the potential value of an expert familiar with the system generally, but not familiar with the local landscape. As a separate example, if experts are asked to characterize species response to a given gradient, then consensus may avoid problems of poor calibration (see below). However, if experts are asked to develop a single importance ranking for each environmental gradient to be applied over a broad spatial scale, then this may be at odds with ecological theory stating that limiting factors for species can vary within their range (Root 1988; Brown et al. 1996). In this case, there may sometimes be more information in the conflicting judgment of experts rather than in a single final consensus estimate.

If the knowledge of multiple experts is elicited on an individual basis, these data must somehow be combined to provide the required probability and uncertainty estimates for modeling (unless specifically intended to serve as alternative hypotheses). The simplest approach is by averaging the elicited values. An unweighted average assigns all experts equal weight and generally assumes that each expert represents an independent and unbiased sample. However, there are a variety of reasons that an expert's knowledge might be highly correlated, poorly calibrated, or strongly biased. Therefore it is generally more common to calculate a weighted average. An expert's weight maybe based on their rank as determined either through a self-assessment of their own degree of confidence, or a metric of expertness (e.g., years experience). Alternatively, experts may be weighted by their ability to accurately estimate known values and judge uncertainty (Cooke 1991; Aspinall 2010; Rothlisberger et al. 2010; Teck et al. 2010). If a suitable test can be developed, such a performance-weighted approach offers the most quantitative and defensible method of weighting experts (Aspinall 2010).

12.6 Landscape Ecological Theory in Expert-Based Models of Conservation

Landscape ecologists have clearly demonstrated that landscape context, extent, resolution, hierarchy, and spatial geometry matter in conservation planning. It is also now a firmly established working hypothesis that landscape structural heterogeneity is important to maintain healthy populations and functioning ecosystems. However, variability in space and time limits our ability to transfer lessons from data and models describing one place and time to applications set in another, new place or time. Furthermore, ecological theory continues to move away from an equilibrium, steady-state worldview toward one that embraces complexity (Holling 2001) and acknowledges the emergent properties of ecological systems. Landscape ecological principles of hierarchical structure and function provide a framework for scaling ecological investigations and mediating resource conflicts, but novel methods and tools take time to permeate into conservation and management practice.

An expert's exposure to landscape ecological theory influences their knowledge of ecological pattern and processes relevant to species distribution patterns. Experts who regularly peruse and discuss the primary literature may have very different perspectives on species distribution dynamics than those that primarily interact with agency technical reports and empirical data. There is, not surprisingly, a substantial delay between the first proposal of ecological theory and the establishment of management guidelines based on that theory, as ideas are tested and debated in diverse contexts. Furthermore, nuances in the theoretical literature are often lost in translation to the heuristics used to guide management decisions. Thus, the theoretical assumptions and understanding that shape expert judgment may differ greatly depending on their exposure to recent advancements in the field of ecology. It is not uncommon for applied ecology literature to generate guidelines that, unintentionally, become established as conservation gospel. One example would be the idea that habitat patches connected by corridors offer higher conservation value than isolated patches (Beier and Noss 1998). Although experts rarely hold absolutely to such simplistic assumptions, bias is introduced when these heuristics serve as anchors or reference points in their interpretation and synthesis of personal experiences.

Although most landscape ecology textbooks highlight the link between research and applied landscape management for conservation (e.g., Gutzwiller 2002; Liu and Taylor 2002; Bissonette and Storch 2003; Millspaugh and Thompson 2009), significant gaps remain between theory and practice (Perera et al. 2006). Interviews of individuals active in the management of conservation lands revealed a scale mismatch between the individuals' expertise derived from experience and expertise derived from literature (Drew and Collazo in press). Despite broad acceptance of the proposition that landscape structure likely influences species occurrence in a landscape, experts were more confident and consistent when defining a species association with either a very fine-scale microhabitat feature or a vegetation community class, than with various metrics of landscape structure. This discrepancy appears to arise because whereas microhabitat and vegetative community observations are easily noted in the field even for informal species observations, landscape structure associations can only be made through purposeful sampling design or after-the-fact plotting of the observation coordinates onto a map.

The expectation that wildlife professionals gain increased familiarity with their landscapes requires that relevant ecological knowledge be actively managed and reviewed in a spatially-explicit format. As GIS and their associated map products become more commonplace, landscape ecological theory should become increasingly accessible and applicable to wildlife conservation practitioners. Just as the term landscape ecology arose from exposure to early aerial imagery and contemplation of the potential ecological significance of this newly revealed perspective (Troll 1939) – so too the illustration of species data on land cover maps can prompt experts to more easily consider their observations within a framework of landscape context, connectivity, and patchiness.

12.7 Conclusions and Recommendations

As expert-based models become more prevalent in landscape ecology applications, more attention must be focused on developing methods to utilize expert knowledge in a manner that is rigorous, transparent, repeatable, and unbiased. These criteria concern every aspect of expert modeling, but especially the stages of: (1) designing and implementing the elicitation procedure, (2) conveying multiple possibly discordant responses, and (3) quantifying uncertainty. Collaboration among ecologists, statisticians, and social scientists to define best practices for expert-based ecological modeling is an active area of research (e.g., Cleaves 1994; Ralls and Starfield 1995; O'Leary et al. 2008). Awareness of potential individual and group biases has led to improved elicitation and weighting techniques (Aspinall 2010). Bayesian statistical methods have provided a foundation for integrating expert knowledge and other data sources within complex ecological models, by providing a means to formalize expert knowledge within a statistical framework (e.g., Garthwaite et al. 2005; McCarthy 2007; Low Choy et al. 2009). New software packages developed or customized for wildlife and natural resource applications to seamlessly integrate elicitation and modeling processes are increasingly accessible to the research and practitioner community (James et al. 2010). Much of the work to improve elicitation of landscape ecological expertise draws upon the well-developed literature on decision risk analysis and uncertainty assessment (Cooke 1991; O'Hagan 1998; Ayyub 2001; Meyer and Booker 2001; O'Hagan 2006).

The process of eliciting expert knowledge requires the same level of planning and attention to detail as does the collection of empirical data. In the field, experience and training ensure that ecologists are alert to potential sources of error and uncertainty in their data. The skills and knowledge necessary to recognize and either avoid or account for error and uncertainty in expert knowledge, however, are more typically taught to social scientists than to ecologists. Therefore, we conclude with several practical recommendations to landscape ecologists considering an expert-based modeling approach.

1. Spend some time reviewing expert elicitation literature, focusing particularly on the methods used to identify and correct expert bias, develop and apply good survey design and interview technique strategies, and compare or combine knowledge from multiple experts.
2. Seek advice from social scientist colleagues, possibly even asking such a colleague to observe and review a practice elicitation session. Training in interview or group facilitation techniques could also be helpful.
3. Allow time initially in every elicitation to define or clarify project objectives, to review key terminology, and orient experts to common heuristics of probabilistic reasoning and their associated biases. This will help to ensure clear communication and facilitate translation of elicitation responses to the model. Be aware that individuals' understanding and use of terminology will vary greatly, as will their ability to translate their personal experiences to probabilistic estimates.
4. Clearly communicate to experts the scale (grain and extent) at which their knowledge will be applied and explicitly define the scale of each expert's knowledge. Both spatial and temporal scale mismatches can introduce unnecessary model error and bias. Where scale mismatches cannot be directly reconciled (e.g., microhabitats not captured in remote imagery), proxy variables could be proposed for expert consideration.
5. Quantify the expertise of participating experts, preferably by testing their ability to accurately assess probability and uncertainty using independent data.
6. Critically explore disagreements among experts rather than simply forcing consensus. If different experts respond based upon different scale or contextual perspectives, their divergent responses may reflect important variation within the focal landscape or alternate hypotheses regarding ecological processes.
7. Provide clear documentation of the entire elicitation process. "Recording the reasoning underlying the experts' professional judgment, and their beliefs in the best or most feasible outcome is essential to keep track of how the evaluation was made and to promote clarity and trust in the model." (Kontic 2000) As expert subjectivity and bias can never be fully eliminated, the interviews or survey materials themselves must be accessible to those that review or use the models. Along these same lines, experts should never be anonymous in cases where their knowledge will be used as data, to ensure maximum accountability and credibility.
8. Always provide an assessment of model quality. At minimum, define a scale of expertness and quantify uncertainty of both expert knowledge and model output. Sensitivity analysis can provide valuable insight through identifying which elicited model parameters most contribute to model uncertainty.

Acknowledgments We thank J. Collazo and two anonymous reviewers for helpful comments provided on an earlier draft of this chapter.

References

Al-Awadhi SA, Garthwaite PH (2006) Quantifying expert opinion for modelling fauna habitat distributions. Comp Stat 21:121–140.

Anderson JL (1998) Embracing uncertainty: the interface of Bayesian statistics and cognitive psychology. Ecol Soc 2 [online] http://www.consecol.org/vol2/iss1/art2.

Aspinall W (2010) A route to more tractable expert advice. Nature 463:294–295.

Ayyub BM (2001) Elicitation of expert opinions for uncertainty and risks. CRC Press, Boca Raton, Florida.

Baddeley MC, Curtis A, Wood RA (2004) An introduction to prior information derived from probabilistic judgements: elicitation of knowledge, cognitive bias and herding. In: Curtis A, Wood R (eds) Geological prior information: informing science and engineering. Special Publications 239, Geological Society, London.

Balmford A, Cowling M (2006) Fusion or failure? The future of conservation. Conserv Biol 20:692–695.

Bashari H, Smith C, Bosch OJH (2009) Developing decision support tools for rangeland management by combining state and transition models and Bayesian belief networks. Agri Sys 99:23–34.

Battisti C, Luiselli L, Pantano D, Teofili C (2008) On threats analysis approach applied to a Mediterranean remnant wetland: is the assessment of human-induced threats related to different level of expertise of respondents? Biodiv Conserv 17:1529–1542.

Beier P, Noss RF (1998) Do habitat corridors provide connectivity? Conserv Biol 12:1241–1252.

Bissonette JA, Storch I (eds) (2003) Landscape ecology and resource management: linking theory with practice. Island Press, Washington DC.

Brown JH, Stevens GC, Kaufman DM (1996) The geographic range: size, shape, boundaries, and internal structure. Ann Rev Ecol Syst 27:597–623.

Cleaves DA (1994) Assessing uncertainty in expert judgments about natural resources. General Technical Report so-1 10, USDA Forest Service, Southern Forest Experimental Station, New Orleans, Louisiana.

Cooke RM (1991) Experts in uncertainty: opinion and subjective probability in science. Oxford University Press, New York.

Coulson RN, Folse LJ, Loh DK (1987) Artificial intelligence and natural resource management. Science 237:262–267.

Davis JLD (2000) Changes in tidepool fish assemblages on two scales of environmental variation: seasonal and El Niño Southern Oscillation. Limnol Oceanogr 45:1368–1379.

Denham R, Mengersen KL (2007) Geographically assisted elicitation of expert opinion for regression models. Bayes Anal 2:99–136.

Doyon F, Sturtevant BR, Papaik M, Fall A, Messier C, Kneeshaw D (2010) A comparison of landscape dynamics derived from expert knowledge-based succession models and process-based landscape models. In: Perera AH, Drew CA, Johnson C (eds) Expert knowledge and landscape ecological applications. Springer, New York.

Drescher M, Perera AH (2010) Comparing two steps of forest cover change knowledge used in forest landscape management planning. J Environ Plan Manag. DOI: 10.1080.10964056100372710.

Drescher M, Perera AH, Buse LJ, Ride K, Vasiliauskas S (2006) Identifying uncertainty in practitioner knowledge of boreal forest succession in Ontario through a workshop approach. Forest Research Report 165, Ontario Ministry of Natural Resources, Ontario Forest Research Institute, Canada.

Drescher M, Perera AH, Buse LJ, Ride K, Vasiliauskas S (2008) Uncertainty in expert knowledge of forest succession: a case study from boreal Ontario. Forest Chron 84:194–209.

Drescher M, Buse LJ, Perera AH, Ouellette MR (in press) Eliciting and formalizing expert knowledge of forest succession supported by a software tool. In: Perera AH, Drew CA, Johnson C (eds) Expert knowledge and landscape ecological applications. Springer, New York.

Drew CA, Collazo JC (in press) Expert knowledge as a foundation for management of rare or secretive species and their habitat. In: Perera AH, Drew CA, Johnson C (eds) Expert knowledge and landscape ecological applications. Springer, New York.

Elith J, Burgman MA, Regan HM (2002) Mapping epistemic uncertainties and vague concepts in predictions of species distribution. Ecol Model 157:313–329.

Garthwaite PH, Kadane JB, O'Hagan A (2005) Statistical methods for eliciting probability distributions. J Am Stat Assoc 100:680–701.

Geneletti D (2005) Formalising expert opinion through multi-attribute value functions: an application in landscape ecology. J Environ Manag 76:255–262.

Giles Jr, RH (1998) Natural resource management tomorrow: four currents. Wild Soc Bull 26:51–55.

Gilchrist G, Mallory M, Merkel F (2005) Can local ecological knowledge contribute to wildlife management? Case studies of migratory birds. Ecol Soc 10 [online] URL: http://www.ecologyandsociety.org/vol10/iss1/art20.

Gutzwiller KJ (ed) (2002) Applying landscape ecology in biological conservation. Springer, New York.

Hess GR, King TJ (2002) Planning open spaces for wildlife I. Selecting focal species using a Delphi survey approach. Landsc Urban Plan 58:25–40.

Holling CS (2001) Understanding the complexity of economic, ecological, and social systems. Ecosystems 4:390–405.

Huntington HP (2000) Using traditional ecological knowledge in science: methods and applications. Ecol App 10:1270–1274.

James A, Low Choy S, Mengersen KL (2010) Elicitator: an expert elicitation tool for regression in ecology. Environ Model Softw 25:129–145.

Johnson CJ, Gillingham MP (2004) Mapping uncertainty: sensitivity of wildlife habitat ratings to expert opinion. J App Ecol 41:1032–1041.

Jones J (2001) Habitat selection studies in avian ecology: a critical review. Auk 118:557–562.

Kim DH, Slack RD, Chavez-Ramirez F (2008) Impacts of El Niño-Southern Oscillation events on the distribution of wintering raptors. J Wildl Manag 72:231–239.

King AW, Perera AH (2006) Transfer and extension of forest landscape ecology: a matter of models and scale. In: Perera AH, Buse LJ, Crow TR (eds) Forest landscape ecology: transferring knowledge to practice. Springer, New York.

Kontic B (2000) Why are some experts more credible than others? Environ Impact Assess Rev 20:427–434.

Kynn M (2008) The 'heuristics and biases' bias in expert elicitation. J R Stat Soc A: Stat Soc 171:239–264.

Lefsky MA, Cohen WB, Parker GG, Harding DJ (2002) Lidar remote sensing for ecosystem studies. BioScience 52:19–30.

Liu J, Taylor WW (eds) (2002) Integrating landscape ecology into natural resource management. Cambridge University Press, New York.

Low Choy SL, O'Leary R, Mengersen, KL (2009) Elicitation by design in ecology: using expert opinion to inform priors for Bayesian statistical models. Ecology 90:265–277.

Lyons JE, Runge MC, Lasowski HP, Kendall WL (2008) Monitoring in the context of structured decision making and adaptive management. J Wildl Manag 72:1683–1692.

MacKenzie DI, Nichols JD, Hines JE, Knutson MG, Franklin AB (2003) Estimating site occupancy, colonization, and extinction when a species is detected imperfectly. Ecology 84:2200–2207.

MacMillan DC, Marshall K (2006) The Delphi process – an expert-based approach to ecological modeling in data-poor environments. Anim Conserv 9:11–19.

Marcot BG, Holthausen RS Raphael MG, Rowland MM, Wisdom MJ (2001) Using Bayesian belief networks to evaluate fish and wildlife population viability under land management alternatives from an environmental impact statement. Forest Ecol Manag 153:29–42.

Marcot BG, Steventon JD, Sutherland GD, McCann RK (2006) Guidelines for developing and updating Bayesian belief networks applied to ecological modeling and conservation. Can J For Res 36:3063–3074.

Martin TG, Kuhnert PM, Mengersen K, Possingham HP (2005) The power of expert opinion in ecological models using Bayesian methods: impact of grazing on birds. Ecol App 15:266–280.

McCarthy MA (2007) Bayesian methods in ecology. Cambridge University Press, New York.

Meyer, MA, Booker JM (2001) Eliciting and analyzing expert judgment: a practical guide. Society for Industrial and Applied Mathematics, Philadelphia, Pennsylvania.

Millspaugh JJ, Thompson III FR (eds) (2009) Models for planning wildlife conservation in large landscapes. Academic Press, Massachusetts.

Moody AT, Grand JB (in press) Incorporating expert knowledge in decision support models for bird conservation. In: Perera AH, Drew CA, Johnson C (eds) Expert knowledge and landscape ecological applications. Springer, New York.

Morris PA (1977) Combining expert judgements: a Bayesian approach. Manag Sci 23:679–693.

Murray JV, Goldizen AW, O'Leary RA, McAlpine CA, Possingham HP, Choy SL (2009) How useful is expert opinion for predicting the distribution of a species within and beyond the region of expertise? A case study using brush-tailed rock-wallabies *Petrogale penicillata*. J App Ecol 46: 842–851.

Nyberg JB, Marcot BG, Sulyma R (2006) Using Bayesian belief networks in adaptive management. Can J For Res 36:3104–3116.

O'Hagan A (1998) Eliciting expert beliefs in substantial practical applications. J R Stat Soc Ser D: the Statistician 47:21–35 (with discussion, pp. 55–68).

O'Hagan A (2006) Research in elicitation. In: Upadhyay SK, Singh U, Dey DK (eds) Bayesian statistics and its applications. Anamaya, New Delhi.

O'Leary RA, Murray JV, Low Choy SJ, Mengersen KL (2008) Expert elicitation for Bayesian classification trees. J App Prob Stat 3:95–106.

Pearce JL, Cherry K, Drielsma M, Ferrier S, Whish G (2001) Incorporating expert opinion and fine-scale vegetation mapping into statistical models of faunal distribution. J App Ecol 38:412–424.

Perera AH, Buse LJ, Crow TR (eds) (2006) Forest landscape ecology: transferring knowledge to practice. Springer, New York.

Perera AH, Drew CA, Johnson C (eds) (in press) Expert knowledge and ecological applications. Springer, New York.

Petit S, Chamberlain D, Haysom K, Pywell R, Vickery J, Warman L, Allen D, Firbank L (2003) Knowledge-based models for predicting species occurrence in arable conditions. Ecography 26:626–640.

Ralls K, Starfield AM (1995) Choosing a management strategy: two structured decision making methods for evaluating the predictions of stochastic simulation models. Conserv Biol 9:175–181.

Ray N, Burgman MA (2006) Subjective uncertainties in habitat suitability maps. Ecol Model 195:172–186.

Root T (1988) Environmental factors associated with avian distributional boundaries. J Biogeogr 15:489–505.

Rothlisberger JD, Lodge DM, Cooke RM, Finnoff DC (2010) Future declines of the binational Laurentian Great Lakes fisheries: the importance of environmental and cultural change. Front Ecol Environ 8: 239–244.

Rykiel Jr, EJ (1989) Artificial intelligence and expert systems in ecology and natural resource management. Ecol Model 46:3–8.

Silbernagel JM, Price J, Miller N, Swaty R, White M (in press) An iterative, interactive elicitation process sheds light into black box of forest conservation scenarios. In: Perera AH, Drew CA, Johnson C (eds) Expert knowledge and landscape ecological applications. Springer, New York.

Starfield A, Bleloch AL (1991) Building models for conservation and wildlife management. Second edition, The Burgess Press, Edina, Minnesota.

Stenseth NC, Mysterud A, Ottersen G, Hurrell JW, Chan KS, Lima M (2002) Ecological effects of climate fluctuations. Science 297:1292–1296.

Store R, Kangas J (2001) Integrating spatial multi-criteria evaluation and expert knowledge for GIS-based habitat suitability modeling. Landsc Urban Plan 55:79–93.

Teck SJ, Halpern BS, Kappel CV, Micheli F, Selkoe KA, Crain CM, Martone R, Shearer C, Arvai J, Fischhoff B, Murray G, Neslo R, Cooke R (2010) Using expert judgment to estimate marine ecosystem vulnerability in the California Current. Ecol App. DOI: 10.1890/09-1173.

Tversky A, Kahneman D (1974) Judgement under uncertainty: heuristics and biases. Science 185:1124–1131.

Troll C (1939) Luftbildplan und ökologische Bodenforschung. Zeitschrift der Gesellschaft für Erdkunde, Berlin, pp 241–298.

Uusitalo L (2007) Advantages and challenges of Bayesian networks in environmental modeling. Ecol Model 203:312–318.

Williams BK (2003) Policy, research, and adaptive management in avian conservation. Auk 120:212–217.

Williams BK, Szaro RC, Shapiro CD (2009) Adaptive management: the US Department of the Interior technical guide. Adaptive Management Working Group, US Department of the Interior, Washington, DC.

Yamada K, Elith J, McCarthy M, Zerger A (2003) Eliciting and integrating expert knowledge for wildlife habitat modelling. Ecol Model 165:251–264.

Part IV
Designing Models for Increased Utility

Chapter 13
Choices and Strategies for Using a Resource Inventory Database to Support Local Wildlife Habitat Monitoring

L. Jay Roberts, Brian A. Maurer, and Michael Donovan

13.1 Introduction

Wildlife habitat models are a necessary component of ecosystem management and play a critical role in determining conservation priorities and making management decisions. They are vital to managers who must perform conservation activities with very limited information. As habitat modeling and other conservation projects are implemented there are a multitude of decisions that must be made as to the specific components that will be included in the models, not to mention the sources of these data. First, managers must evaluate to what purpose the models will be applied. For example, if the goal is to assess the coarse-scale distribution of a habitat across a region, then correspondingly coarse environmental data – such as classified land-coverage from satellite imagery – may be suitable. However, if managers wish to predict which forest stands are likely to provide suitable habitat for a species, or to assess the number of acres of suitable habitat on a landscape, then more detail (meaning both the spatial resolution and number of vegetation measurements) is needed to assess which acres are suitable for a species and which are not, as well as whether the spatial arrangement of those areas is appropriate.

Such decisions go hand in hand with the limitations provided by research budgets and the difficulty (cost) of acquiring more detailed and accurate data. This chapter describes the process by which a local (fine-scale) wildlife habitat modeling project can be developed from a forest resource database and highlights some of the typical choices that managers must entertain when constructing models and the consequences of those choices. We will do this by showing the relative tradeoffs in accuracy versus effort that result from the inclusion of more detailed forest inventory information in the construction of forest bird habitat models. We hope to help answer questions such as: Do you have to go out and measure vegetation, or can you rely on models (i.e., classified satellite imagery) that interpolate/extrapolate existing

L.J. Roberts (✉)
PRBO Conservation Science, 3820 Cypress Drive #11, Petaluma, CA, 94952, USA
e-mail: ljroberts@prbo.org

data?; How much vegetation data do you need, and at what scale should it be measured?; and What relative accuracy levels can you expect for habitat generalists versus specialists, or abundant versus rare species? We will compare the potential strengths and weaknesses of a data-rich forest inventory database with satellite-derived vegetation/land cover for use in local monitoring and decision-making.

Commonly available environmental data (the independent variables in wildlife habitat models) may include categorical habitat classes, vegetation or substrate measurements, and climate or other abiotic features. These can be generated in any number of ways, from classified satellite imagery to intensive field samples, to interpolation of measurements at weather stations. These data are often either: (1) coarse resolution and broad in extent, (2) medium resolution and narrow in extent, or (3) fine resolution and sparsely sampled, and are correspondingly ranked in terms of cost (per unit area) to develop. Dependent variables typically consist of species occurrence records, and these data also require many methodological choices that involve tradeoffs in accuracy, sample size, information content, and acquisition cost. Alternatively, when species occurrence data are not available, expert opinions can be used to draw the links between species and their associated habitats (see Chap. 12).

Much work has been done to improve the quality of wildlife habitat models through a wide variety of means, as exemplified by the chapters in this volume. That is because the success of many conservation activities is closely tied to the quality and proper application of wildlife habitat models (Fahrig 2001; Jetz et al. 2008). The choice of inputs (both the dependent and independent variables) for wildlife habitat models should be taken very seriously in the project design, and managers and scientists alike often must weigh the costs of generating new data (vs. incorporating existing data) against the incremental benefits that may result from inclusion of more information into models. The successful application of models can be disrupted by poor-quality input data, a lack of understanding of species ecology that leads to incorrect model design, or by inappropriate conclusions from model results (Austin 2007). For example, the same model may not be appropriate for abundant and rare species, or the same input data for habitat generalists and specialists. In a similar sense, models built for use as a coarse-filter description of potential habitat distribution (i.e., GAP) are sometimes applied as a prediction of species occurrence (Scott et al. 1993). When models perform poorly in one particular context they are sometimes assumed to be not useful or simply wrong, despite the fact that they may be useful in another context; their value is thus discounted. A clear need exists to refine and evaluate our understanding of the ecological correlates between wildlife habitat model accuracy and species traits such as prevalence, habitat specificity, and detectability (Seoane et al. 2005a). A better understanding of the inherent relationships between species ecological traits and model performance would provide a basis for implementing wildlife habitat models for use in monitoring and decision-making (McPherson and Jetz 2007).

As inventory technology and data resources continue to advance, the evaluation and refinement of existing assets such as forest resource inventories – which often are the product of activities not focused on conservation – can ensure that they are efficiently translated for conservation research and management applications. By evaluating local and regional database resources, modelers and managers can leverage the need for detailed vegetation inventory with the need to monitor and conserve important wildlife habitats. Correctly implementing forest resource data, which can be rich in vegetation measurements (e.g., plant species composition and structure), necessitates a solid understanding of the ecological relationship between species occurrence and vegetation communities (Gottschalk et al. 2005; Austin 2007). Furthermore, where inventory data are regularly maintained and easily accessed (this may be rare), there exists the opportunity to link wildlife population and habitat data to obtain near real-time updates in support of adaptive management decisions. Though we illustrate this approach using a forest inventory database, similar strategies could be used for rangeland, agricultural lands, stream networks, coastal wetlands, or any matrix of habitats for which detailed resource measurements have been developed.

13.2 Management Challenge, Ecological Theory, and Statistical Framework

There are many examples that illustrate the relationship between species ecology and the ability to predict distribution of habitats and species occurrence. Rarer species can be more difficult to model than abundant species, in part because of sampling issues (Karl et al. 2002), and in part due to ecological reasons such as the higher frequency of local extinctions associated with metapopulation dynamics (Storch and Sizling 2002). Species that have greater specialization on measureable environmental characteristics are usually more accurately modeled than generalists because statistical models are able to discriminate between used and unused sites (Seoane et al. 2005a; Tsoar et al. 2007). It is generally assumed that the more environmental variables included, the better the model performance, but care should be taken to avoid spurious relationships due to chance. Ideally, these variables are chosen to reflect specific habitat cues that are important for the species included in the study. It is hard to know which variables are important and difficult to avoid inclusion of arbitrary characteristics – even expert opinion may not provide useful information (Seoane et al. 2004, 2005b).

The major challenge in building wildlife habitat models lies in the ability of modelers to identify environmental variables that account for the variability in species' occurrence across a landscape. A species' habitat associations can vary across its range, and its absence in a location can be due to many factors (in addition to habitat associations) including food availability, competition (Whittaker and Levin 1975; Herzog and Kessler 2006), population abundance or conservation status (Linder et al. 2000; Hepinstall et al. 2002), dispersal and site fidelity (Knick and Rotenberry 2000; Pulliam 2000; Mörtberg 2001), and more (Guisan and Thuiller 2005). Species location data can be expensive to generate (i.e., funding a crew of field surveyors) and as a result the availability of such data can be restricted to small areas or spread sparsely over the landscape of interest. The lack of extensive species occurrence data can lead to a high rate of sampling error and the inability of statistical methods to fit wildlife habitat models to these sparse data (Araújo and Guisan 2006). This would support the idea that for conservation projects that include many species, both abundant and rare, expert-based descriptions of habitat associations would be a necessary strategy (see Chap. 12). However, in order to tailor expert-based or statistical habitat models to each species, practitioners must first have a general understanding of the inherent types of error associated with species traits, such as prevalence (proportion of sites where a species is present) and habitat specificity (variety of habitats used by a species).

Most wildlife habitat models estimate a probability of occurrence (or presence, abundance, habitat suitability, or detection) for each species to each site (or groups of sites) in a sample. When groups of sites (in this case, corresponding to different "habitat types") are ranked and plotted against the probability of occurrence, the result is a declining function representing the likelihood of the species' presence on each habitat type (Fig. 13.1). The shape of this function differs for each species, and in reality it is unknown. The purpose of wildlife habitat relationship models is to estimate this unknown function. For common species, there may be a high probability of occurrence across a large portion of the habitats (as in Fig. 13.1a). Less common species will show a lower probability of occurrence on a smaller proportion of the habitats (Fig. 13.1b). Habitat specialists will be associated with a narrower band of habitats than generalists.

Wildlife habitat relationship models attempt to fit this function as closely as possible. The simplest models are binary (i.e., GAP). The shaded areas on Fig. 13.1 represent the set of habitats that are identified by the model as appropriate for that species. The area of the plot where the prediction surface (shaded area) overlaps the occurrence function reveals the correct presences, while the area above the curve reveals the incorrect presence predictions (commission errors). The breadth of the model prediction surface has a large effect on the proportion of omission and commission errors. If a threshold in probability of occurrence is chosen, this defines the boundary of the prediction surface and can have a large effect on the number of omission and commission errors. Comparing Fig. 13.1a, c the narrower prediction surface (defined by the probability of occurrence threshold of 0.5) has the result of decreasing the number of commission errors, but increasing the omission errors. Figure 13.1b, d show a similar situation for a rare species. In this case, the use of the 0.5 threshold results in a very large proportion of all the habitats occupied by this species to be predicted as absent. Despite being a very common default value, 0.5 is rarely appropriate for use as a threshold value for a large set of species with varying prevalence and habitat specificity. A threshold tied to each species' prevalence is a better approach (Freeman and Moisen 2008). Choice of this kind of threshold value is an example of how model use and interpretation (even the calculation of accuracy [see below]) can have major implications as to the results and conclusions drawn from a set of models. For example, one can use a relatively low threshold to identify all the areas of potential habitat, but a higher threshold to identify only the most appropriate habitats.

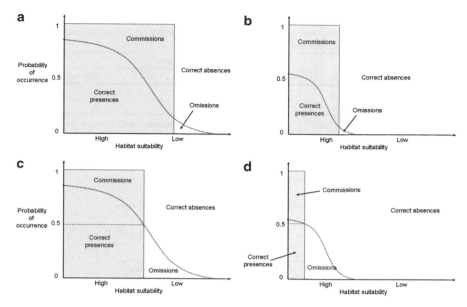

Fig. 13.1 Overlay of species probability of occurrence functions with predictive surfaces. The *vertical axis* refers to the actual probability that a species will be recorded in a field survey, the *horizontal axis* represents a gradient of habitats. The *sigmoid shaped functions* represent a declining probability of occurrence function as habitats become less and less suitable for that species. Finally, the *shaded areas* represent the areas where a model predicts that species will be present, the *unshaded areas* are predicted absences. Figures modified from Karl et al. (2000). (**a**) A binary model (e.g., GAP potential habitat) for an abundant habitat generalist includes nearly all of the habitats that are potentially used by this species. Omission errors are relatively low because a large proportion of all the sites are included, and commission errors are relatively low because this species is present on most of the habitats in the study area. (**b**) A binary model (e.g., GAP potential habitat) for a rare habitat specialist includes nearly all of the habitats that are potentially used by this species. Omission errors are relatively low because a large proportion of all the habitats used by this species are included, and commission errors are high because this species is present on only a few of the habitats in the study area. (**c**) A binary model for an abundant habitat generalist with a threshold for probability of occurrence set at 0.5. The predicted presence area includes most of the habitats that are potentially used by this species, but less than in Fig. 13.2a. Omission errors have been increased, but commission errors are lower. (**d**) A binary model for a rare habitat specialist with a threshold for probability of occurrence set at 0.5. The predicted presence area includes a relatively small portion of the habitats that are potentially used by this species because the required probability of occurrence (0.5) is met at only a few of the sites. Omission errors are high because a large proportion of all the habitats used by this species are not included in the predicted presence set, and yet commission errors are still high because this species is present on only a few of the habitats in the study area

13.3 Model and Model Validation Techniques

As noted previously, most wildlife habitat models give an ordinal probability value to each site or habitat type, instead of a binary one as shown in Fig. 13.1. These data can be very useful, for example as an index of habitat suitability (Guisan and

Table 13.1 Error matrix used to calculate Kappa, omission and commission error rates, sensitivity and specificity, and other accuracy measures (but not ROC/AUC)

Predictions	Observations	
	Presence	Absence
Presence	a	b
Absence	c	d

Cells "a" and "d" are the number of correct presence and absence predictions, respectively. Cell "b" is the number of incorrect presence predictions, and cell "c" is the number of incorrect absence predictions

Accuracy measure equations:

(Total number of samples $= n = a + b + c + d$)

$$\text{Kappa} = \frac{\left(\frac{(a+d)}{n}\right) - \left(\frac{(a+b)(a+c)+(c+d)(b+d)}{n^2}\right)}{1 - \frac{((a+b)(a+c)+(c+d)(d+b))}{n^2}}$$

$$\text{Commission error} = \frac{b}{a+b}$$

$$\text{Omission error} = \frac{c}{a+c}$$

$$\text{Percent correctly classified (PCC)} = \frac{a+d}{n}$$

$$\text{User's accuracy} = 1 - \text{commission error} = \frac{a}{a+b}$$

$$\text{Producer's accuracy} = 1 - \text{omission error} = \frac{a}{a+c} = \text{sensitivity}$$

Thuiller 2005). But the use of a probability threshold (to define what is "suitable habitat" and what is not) is still necessary to calculate many model accuracy measurements (such as percent correctly classified (PCC), Kappa, omission and commission error, sensitivity, and specificity) that are calculated from an error matrix (Table 13.1). Kappa and some other metrics [such as the true skill statistic (Allouche et al. 2006)], are designed to correct for the bias resulting from large differences in the number of presences and absences between species, but these measures are still very sensitive to the choice of a threshold in probability of occurrence that defines the boundary between presence and absence. For this reason, threshold independent measures, like area under the receiver-operator characteristic curve (ROC/AUC) have become popular, but still are susceptible to problems (Lobo et al. 2008).

No accuracy measure is best in all situations. For a successful conservation project, wildlife habitat model users need to be aware of not only the properties of each accuracy measurement, but also of the expected performance of models built for species with particular ecological traits. In the following sections, we will show examples of how wildlife habitat model performance is influenced by the specificity of species-habitat associations, and the prevalence of species in a sample. The results should help modelers know what to expect, in terms of model quality and accuracy, from their particular data.

We are using results from a field bird survey project on state forest lands in the lower peninsula of Michigan (Roberts 2009, unpublished data). At each survey location we sampled the presence of all birds seen or heard within 100 m as well as vegetation characteristics within a 50 m radius plot. At each of these locations we also recorded the stand-level habitat from a statewide forest resource database (MDNR 2005). We built habitat relationship models with both the plot-level field vegetation samples and the stand-level inventory measurements, and compared both of these with the potential habitat distribution maps generated from the Michigan Gap Analysis project (MIGAP; Donovan et al. 2004). For each model, all samples were included in the training dataset and we have not validated the models. As a result, all of these comparisons are of model fit, not accuracy in the sense of many other modeling projects that use a separate dataset to test models. As the sample size is small (393 field sites) and two of the species included in the analyses were present at fewer than 15% of the sites, subsetting the data would likely result in unreliable validation. Furthermore, models are not being applied to generate predictions of species habitat distribution, so validation in this context is of secondary importance. The purpose of the model comparisons is to generate relevant conclusions about the relative utility of each dataset and to show the potential improvements that can be obtained with increased investment into gathering more detailed species occurrence and vegetation measurement data.

We built a database of habitat conditions targeted toward deciduous, coniferous, and mixed forest bird species (Table 13.2) at each field survey site. The same set of variables was calculated from both the plot-level field surveys and stand-level vegetation measurements, and these data were compared in this study to examine the effect that resolution has on the habitat association models. Because very similar methods were used to take the vegetation measurements, the main difference between these two samples is the scale at which the data are aggregated (50 m plots vs. 1–200+ acre stands). The choice of variables to include in this dataset was large. In this particular forest resource database, each measurement unit included the size and cover of each canopy species; the size, density, and height of sub-canopy species; dominant ground cover; and stand-level variables such as basal area, presence of slash, overall size, land cover type/vegetation cover class, management type (plantation, even or uneven aged), upland or lowland, and canopy closure. The list of available habitat variables may differ with other systems, so careful consideration

Table 13.2 List of habitat variables included in the recursive partitioning models

Habitat variable	Vegetation measurements
Vegetation cover class	Stand data: 20 classes (8 forest) Plot data: 19 classes (9 forest)
Basal area	Average of three measurements per stand
Diameter at breast height	Proportional average for all species in stand
Canopy closure	Visual estimate (four 25% categories)
Proportion of deciduous canopy cover	Sum of deciduous cover divided by total
Canopy species richness	Count of canopy species
Canopy species diversity	Simpson's (1/P) diversity of canopy species cover
Subcanopy cover	Sum of individual species cover
Proportion of deciduous subcanopy cover	Sum of deciduous cover divided by total
Subcanopy richness	Count of subcanopy species
Subcanopy species diversity	Simpson's (1/P) diversity of subcanopy species cover
Overall size of canopy trees	Average size of dominant trees (sap, pole, log)
Upland or lowland	Binary
Plantation	Binary
Location	North/South

The number and detail of vegetation cover classes are comparable to Level 3 in the hierarchical ecological classification system developed by Anderson et al. (1976)

of the species included in the project should drive variable selection and extraneous variables should be avoided (depending on type of model [e.g., GLMs], Thompson et al. 2000; Johnson and Gillingham 2005).

The list of species included in this analysis was reduced to include only those species that are likely to be observed in field surveys (i.e., eliminating nocturnal and non-vocal birds), and abundant enough to produce statistical habitat models (Table 13.3). These species represent a variety of upland, lowland, forest, and forest-edge habitats. We simplified the recorded abundance of each species at each site to presence/absence. Wildlife habitat models were generated with a statistical algorithm known as recursive partitioning (Feldesman 2002) using the RPART module (Atkinson and Therneau 2000) in R. We used recursive partitioning to predict each species' probability of presence at each sample location, and compared these predictions with the field observations. A more detailed description of the model construction methods is covered in Roberts (2009).

Each model constructed was evaluated using multiple statistical criteria (Table 13.1). We show the results for omission/commission error, Kappa, and area under the curve of the receiver-operator characteristic plot (ROC/AUC). All the accuracy measures shown in Table 13.1 (but not ROC/AUC) are calculated using a 2×2 error matrix (actual presence/absence vs. predicted presence/absence). As mentioned previously, the construction of error matrices requires that a response value threshold (probability level that separates presence from absence) be set so that sites were classified into the binary presence/absence categories. We used two thresholds in order to illustrate the effect that this choice can have in the interpretation of model results (our results are in terms of model fit, not accuracy *per se*).

Table 13.3 List of bird species included in models

Common name	Scientific name	Prevalence	Habitat	Specificity
Ovenbird	*Seiurus aurocapillus*	0.55	Forest	Specialist
Red-eyed Vireo	*Vireo olivaceus*	0.55	Forest	Specialist
Blue Jay	*Cyanocitta cristata*	0.42	Forest	Specialist
American Goldfinch	*Carduelis tristis*	0.41	Grassland	Specialist
Black-capped Chickadee	*Poecile atricapillus*	0.40	Forest	Generalist
American Robin	*Turdus migratorius*	0.34	Mixed	Generalist
Common Yellowthroat	*Geothlypis trichas*	0.32	Wetland	Specialist
Rose-breasted Grosbeak	*Pheucticus ludovicianus*	0.31	Forest	Specialist
Eastern Wood-Pewee	*Contopus virens*	0.31	Forest	Specialist
Chipping Sparrow	*Spizella passerina*	0.30	Mixed	Generalist
Veery	*Catharus fuscescens*	0.25	Forest	Specialist
Gray Catbird	*Dumetella carolinensis*	0.25	Mixed	Generalist
Scarlet Tanager	*Piranga olivacea*	0.24	Forest	Specialist
Indigo Bunting	*Passerina cyanea*	0.24	Mixed	Generalist
American Redstart	*Setophaga ruticilla*	0.22	Forest	Specialist
Great Crested Flycatcher	*Myiarchus crinitus*	0.21	Forest	Specialist
Red-winged Blackbird	*Agelaius phoeniceus*	0.20	Wetland	Specialist
Eastern Tufted Titmouse	*Baeolophus bicolor*	0.19	Mixed	Generalist
Wood Thrush	*Hylocichla mustelina*	0.19	Forest	Specialist
Eastern Towhee	*Pipilo erythrophthalmus*	0.18	Forest	Specialist
Field Sparrow	*Spizella pusilla*	0.17	Mixed	Generalist
Northern Flicker	*Colaptes auratus*	0.16	Mixed	Generalist
White-breasted Nuthatch	*Sitta carolinensis*	0.15	Forest	Specialist
Hermit Thrush	*Catharus guttatus*	0.13	Forest	Specialist
Cedar Waxwing	*Bombycilla cedrorum*	0.13	Mixed	Generalist
Yellow-billed Cuckoo	*Coccyzus americanus*	0.13	Mixed	Generalist
Pine Warbler	*Dendroica pinus*	0.09	Forest	Specialist
Nashville Warbler	*Vermivora ruficapilla*	0.07	Forest	Specialist
Black-throated Green Warbler	*Dendroica virens*	0.07	Forest	Specialist
Alder Flycatcher	*Empidonax alnorum*	0.05	Wetland	Specialist

Prevalence lists the proportion of survey sites at which each species was present (out of 393 total). Most (17) of the species are associated with forest habitats, some (9) are associated with mixed (edge) habitats, and fewer are wetland (3) and grassland (1) species (Peterjohn and Sauer 1993)

The first sets the threshold equal to each species' prevalence; the second makes the predicted prevalence of the recursive partitioning model equal to the observed prevalence for each species. This second method is supported by Freeman and Moisen (2008). Kappa accounts for large differences in the number of sites in the present and absent categories (Karl et al. 2000; Manel et al. 2001) and reflects the improvement over a random distribution among the categories. To provide a threshold independent measure of model fit we used ROC/AUC (Fielding and Bell 1997; McPherson et al. 2004). In general, Kappa and ROC/AUC are highly correlated, but ROC/AUC is more apt to represent the fit of models built for less prevalent species (Allouche et al. 2006).

The models included in this analysis represent not only different levels of scale and detail of the input data, but also a progression of effort required to implement them. The MIGAP models use a statewide land cover map and simple expert opinion models to generate potential habitat distributions. The plot-level field sample models represent perhaps the largest input of effort and cost to develop, requiring intense fieldwork to gather vegetation and species occurrence samples. The stand-level cover type only models would be considered intermediate in this collection of wildlife habitat models. Stand-level inventory models will vary in cost depending on whether other organizations have developed them for other uses (e.g., tracking timber resources), and if not then they would be much more costly than field samples.

The fit of the models was assessed against field survey data using the statistical criteria described above. In the case of the MIGAP models, we are assessing them as predicted species distributions and validating with field samples of species occurrence. In the case of the other models, we are assessing the fit of models parameterized with these same field samples of species occurrence. The model fit measures were averaged over all species to evaluate overall patterns, and within groups of species based on life-history traits and prevalence. The purpose of this approach was to fully describe the strengths and weaknesses of the available datasets, particularly in relation to different species characteristics (e.g., common vs. rare, and specificity of habitat associations), and the choice of threshold for binning predictions into a binary response. While we understand that it would be almost trivial to assert that more detailed habitat data will yield better model results, we do feel that putting the results into the context of the effort (and cost) required to generate these additional habitat details will show in which cases it is warranted.

We also show detailed model results for four individual species that differ in prevalence and habitat specificity (Table 13.3). Ovenbird is a very prevalent forest habitat specialist, American Robin is a prevalent bird that is a habitat generalist, Black-throated Green Warbler is a low-prevalence bird in this sample that has specific habitat associations, and Yellow-billed Cuckoo is a low-prevalence bird associated with edge habitats. For each of these species we list detailed model results, and link the statistical models to a conceptual diagram that relates each species' prevalence with habitat suitability.

13.3.1 Model Results

The overall fit of the models ranked in the order of increased vegetation detail and effort (Figs. 13.2 and 13.3). The Kappa values (Fig. 13.2) averaged over all 30 species for the MIGAP models (Kappa=0.09) are lower than for recursive partitioning stand-level cover type only models (Kappa=0.29), and when the composition and structure data are included, the average Kappa values increase (Kappa=0.39 and 0.40, respectively for stand and plot-level models). ROC/AUC results show a similar difference in fit between the vegetation class only models and the composition and structure models (Fig. 13.3).

13 Choices and Strategies for Using a Resource Inventory Database 261

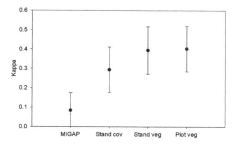

Fig. 13.2 Model accuracy (kappa, scale -1 to +1) by model type, results averaged for all 30 species. The accuracy of GAP models is much lower than for vegetation inventory models. There is an increase in accuracy between the stand cover-type only ('Stand cov') model and the stand vegetation inventory ('Stand veg') model (key difference is the addition of vegetation structure and composition information), but the differences between the plot-level ('Plot veg') and stand-level vegetation inventory models is small. Error bars show 1 standard deviation. The threshold used to calculate kappa is where predicted prevalence = actual prevalence

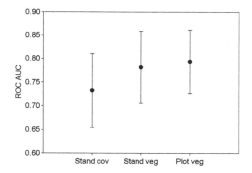

Fig. 13.3 Model accuracy (ROC/AUC, scale 0.5 to 1.0) by model type, results averaged for all 30 species. The difference between the stand cover type model and the full vegetation measurements models is similar with ROC/AUC and Kappa. The GAP models are not included because they are binary so there is no way to calculate ROC/AUC. Error bars show 1 standard deviation

The majority of the species models in each set (22/30 for stand-level and 27/30 for the plot-level data) achieved a good model fit (Kappa=0.3 or better). The only species that showed a large difference between the stand and plot-level models (defined as a difference between the models of Kappa≥0.15) were mixed habitat species, Field Sparrow, and Northern Flicker. ROC/AUC supports this result, the largest differences in AUC values between stand and plot-level models were for Cedar Waxwing (mixed habitat guild) and Northern Flicker. For all of these species the plot-level models were more accurate. Model fit (both Kappa and AUC) tended to be higher for habitat specialists than for generalists for both the stand and plot-level models (Table 12.4). Note that model fit for generalists (Kappa~0.30–0.35, Table 13.4) is only slightly better for models built with the full suite of vegetation composition and structure variables as it is for cover type only models (Fig. 13.2).

The choice of threshold in predicted presence probability has a relatively small, but important, effect on these results (Table 13.4). The results above were calculated with the threshold that sets predicted prevalence equal to actual prevalence. The overall difference in Kappa between stand- and plot-level models is slightly larger when threshold is equal to each species' prevalence, and there are more species that show a large difference between the stand and plot-level models, but again most of these were mixed habitat birds: Yellow-billed Cuckoo (mixed habitat guild), Nashville Warbler (forest habitat guild, but is also found in shrub habitats), and Gray Catbird (mixed habitat guild), in addition to Northern Flicker and Field Sparrow.

These two threshold choices appear to differ mainly in the balance between omission and commission errors (Table 13.4, Fig. 13.4). Remember that both omission and commission errors contribute to lowering accuracy, but changing the threshold does not necessarily result in a 1:1 tradeoff between omissions and commissions. When the threshold is set where predicted prevalence = actual prevalence for each species, commission and omission error rates are relatively even (Fig. 13.4a). There is a noticeable, but not significant, decline in both commission and omission error rates between the vegetation cover class only models and the full vegetation measurements models. When threshold = prevalence, commission error rates are much higher than omission error rates. MIGAP models show very high commissions and relatively low omissions in comparison to the other models. The MIGAP models predict a binary response value, so choice of threshold has no effect on the results for these models. Despite these large differences in error rates, threshold choice has only a small effect on kappa values. ROC/AUC is independent of threshold values. For a more detailed analysis of the relationships between model accuracy and species prevalence (see Freeman and Moisen 2008).

Looking at Kappa and ROC/AUC, the models for Ovenbird and Black-throated Green Warbler had relatively good fit for both the stand level and plot level vegetation data, whereas American Robin and Yellow-billed Cuckoo had relatively poor fitting models (Table 13.4). This result agreed with the averaged fit for generalist versus specialist groups. Specialists (like Ovenbird and Black-throated Green Warbler) showed a significantly better fit with both Kappa and ROC/AUC, for both the stand and plot-level vegetation measurements. Recursive partitioning (RPART) plot-level models for each of the four species we chose as examples are represented in Fig. 13.5a–d in a manner similar to Fig. 13.1. Each vertical box shows one habitat class generated by the RPART model (each of these boxes corresponds to an end-node of the partition tree), and the width of each box represents the number of samples included in that class. The classes are ranked along the horizontal axis according to their predicted presence value (analogous to habitat suitability), shown on the vertical axis. Dark shaded portions of each box represent the proportion of presences observed in the data, and light shaded areas represent absences. Both threshold values for generating the error matrix are shown in this figure, the first set at the prevalence of each species (along the vertical axis, dotted line), and the

Table 13.4 Results of model accuracy measurements for the four species targeted for detailed examination and averaged for the 10 habitat specialists and 20 habitat generalists included in this analysis

	Threshold independent		Threshold = prevalence				Threshold where predicted prevalence = actual			
	AUC		Kappa		Commissions/ ommissions (%)		Kappa		Commissions/ ommissions (%)	
	Stand	Plot	Stand	Plot	Stand	Plot	Stand	Plot	Stand	Plot
Ovenbird	0.86	0.84	0.61	0.61	16.3/19.8	17.2/16.0	0.60	0.61	18.9/16.1	17.2/16.0
American Robin	0.72	0.70	0.35	0.33	43.5/41.8	48.7/35.1	0.35	0.33	43.5/41.8	48.7/35.1
Yellow-billed Cuckoo	0.78	0.76	0.14	0.36	80.6/2.0	61.4/44.3	0.31	0.32	59.6/62.0	59.4/57.4
Black-throated Green Warbler	0.89	0.85	0.54	0.49	52.3/25.0	56.8/29.0	0.59	0.59	42.4/32.1	6.7/54.8
Generalists	0.80*	0.81	0.30*	0.35**	56.1/28.2	56.8/29.6	0.31*	0.34*	38.0/59.7	44.0/50.5
Specialists	0.74*	0.77	0.40*	0.41**	51.4/25.8	50.6/26.1	0.44*	0.43*	41.4/42.1	37.9/45.7

*$p=0.05$
**The average values for generalists and specialists are significantly different from each other at $p=0.1$
This table shows kappa and commission/omission error calculated with the threshold = prevalence, and the threshold where predicted prevalence = actual prevalence. Significance calculated with an independent groups T-test

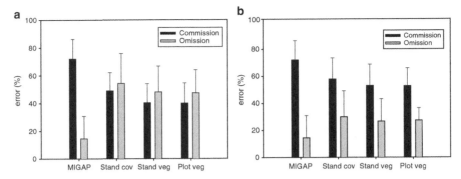

Fig. 13.4 Model accuracy (commission and omission error) by model type, results averaged for all 30 species. Commission errors are those species that were predicted to be present but are absent in field surveys, omission errors are those species that were predicted to be absent but are present in field surveys. (**a**) Omission and commission error rates when the threshold that sets predicted prevalence = actual prevalence for each species is used. In comparison to the statistical models (Phases 2 and 3) MIGAP models show relatively high errors of commission, while keeping omission errors lower. This threshold has the effect of keeping commission and omission errors relatively even. Error bars show 1 standard deviation. (**b**) Omission and commission error rates when the threshold = prevalence for each species is used. In comparison to the statistical models MIGAP models show relatively high errors of commission, while keeping omission errors lower. This threshold has the effect of favoring low omission error rates at the expense of relatively high commission error. Error bars show 1 standard deviation

second at the point where the models predicted prevalence equals the observed prevalence for each species (along the top horizontal, dashed lines).

The more accurate models (Ovenbird [Fig. 13.5a] and Black-throated Green Warbler [Fig. 13.5d]) have a smaller proportion of their habitat classes near the intermediate (0.5) values in predicted presence probability, while the less accurate models (American Robin [Fig. 13.5b] and Yellow-billed Cuckoo [Fig. 13.5c]) have a relatively large proportion of predictions at intermediate values. The two thresholds result in identical cutoffs for Ovenbird and American Robin (the more prevalent species) and thus the model fit does not change depending on threshold used (Table 13.4). The two thresholds result in different cutoffs for Yellow-billed Cuckoo and Black-throated Green Warbler (the less prevalent species), and thus the measurements of model differ between the two thresholds (Table 13.4). The second method of selecting threshold values (predicted prevalence = actual) appears to set the number of omissions and commissions very close to equal, while the first method (threshold = prevalence) tends to minimize omissions. With the first threshold, the increase in commission error rates with less prevalent species (Fig. 13.4a, b) can be explained by the fact that there are many fewer predicted presences for rare species (i.e., the denominator in the equation for commission error is smaller; Table 13.1). Likewise omission error rates decline or stay the same because there are many more observed absences (i.e., the denominator in the equation for omission error is larger; Table 13.1).

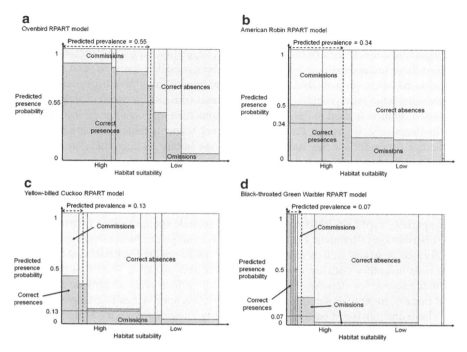

Fig. 13.5 Graphical representation of single species recursive partitioning models. The height of the *dark shaded boxes* represent the predicted presence probability for a group of sites, the width represents the proportion of all sites that fall into that group. The threshold values for calculating accuracy measures (like kappa) are shown by the dotted and dashed lines (see text for details). (**a**) Ovenbird, (**b**) American Robin, (**c**) Yellow-billed Cuckoo, (**d**) Black-throated Green Warbler

13.4 Data Availability and Suitability

When designing a new wildlife habitat-monitoring project, it can be helpful to evaluate the ecological characteristics of the target wildlife species. Then, depending on whether those species are habitat generalists or specialists, and prevalent or rare, one can assess the type (and level of detail) of vegetation measurements that are likely to be important resources to those species and thereby needed to build accurate models. Finally, the sources of available data, as well as those that can be potentially developed, can be assessed along with the cost (money and time) to acquire each dataset, and note the expected increase in accuracy that would result with each additional dataset. For example, for a 2,000–3,000 acre study area (like the ones in this study) there may be a land cover dataset available at no cost. However, if there is not, or the available data are outdated or of poor quality, then expect to fund a team of between two and five people $10,000–$100,000 or more to purchase and classify aerial photography or satellite imagery to complete a relatively simple vegetation/land cover dataset (i.e., no vegetation composition or structure measurements). For the same amount of money, the same field crew could likely complete from 50 to

100 detailed vegetation plot-level surveys, including vegetation composition and structure, but only cover perhaps 1% of the areal coverage of the study area. Spatially explicit forest inventories (which contain both vegetation type and structure/composition) therefore are extremely valuable to habitat modeling projects, and should be considered whenever available. These costs are in addition to wildlife species occurrence data, which can potentially double the costs (time and money) of fieldwork, or if expert opinion is used to build the associations then that cost can be minimized.

In this study, we found that the stand-level forest inventory database has the potential to model species habitat associations nearly as accurately as intensive plot-level field vegetation sampling (with the possible exception of edge and mixed habitat associated species). Therefore the extra efforts required to build plot-level habitat association models are perhaps not warranted given that the fit of stand and plot-level models are very close. In addition, because these data are typically stored in a spatially explicit GIS format, models can include spatial pattern metrics to map the distribution of habitats. Furthermore, for habitat generalists, cover-type models performed nearly as well as models built with more detailed vegetation data. When resource inventory data are available, however, this work suggests that they are a valuable source for step-down habitat modeling projects and can provide an important improvement over satellite-derived cover class approaches like Gap Analysis.

There are consistent patterns of model utility emerging from the wide variety of wildlife habitat modeling projects throughout the world, and we have summarized some of these patterns with specific examples. For most forest bird species it is clearly necessary to include detailed vegetation measurements (the exception being some generalists), and this is likely to be the case for many other wildlife species. Wildlife habitat models for habitat specialists will inherently be more accurate than generalists because statistics can more easily define habitat classes that clearly delineate appropriate from poor habitats. This pattern has been supported in our results. The type of environmental input data also plays a large role in the ability to accurately predict species locations. Stand-level vegetation measurements (in comparison with plot-level measurements) will not be as appropriate for describing edge and mixed habitat associations, but tend to be well suited to other species and have the added advantage of providing the possibility of calculating spatial pattern information (not included in this study).

The correlation between model accuracy and prevalence tends to be positive (e.g., Vaughan and Ormerod 2005) because rarer species are simply less likely to be present on any given location of appropriate habitat (Manel et al. 2001). Accuracy measures rely on the models to predict appropriate habitat sites with a high probability of presence value, but rare species actually have a low probability of being detected on even the best sites. Most common accuracy measures fail to take this into account and therefore are not reflecting the actual quality of the model. Therefore, expectations of model accuracy should typically be relaxed for habitat generalists and species with low prevalence. When a model performs poorly, it can be due to many factors, including the fact that the species has a low prevalence across its range/niche and is not likely to have a high probability of occurrence at any location. When this is the case, a high-quality model can still result in very low accuracy. Other factors result in a model having both low quality and poor accuracy, for example the modelers may have missed important cues for species habitat associations (the wrong vegetation and environmental measurements were included in model construction).

Many common accuracy measures, including Kappa when a 0.5 threshold is used, are relatively inflexible to the species' ecological characteristics. Kappa, however, can be used effectively when the choice of threshold in probability of occurrence is flexible, and tied to each species' prevalence. Our results support the conclusions of Freeman and Moisen (2008). The relationship between model accuracy and prevalence was weak (Kappa) or even negative (ROC/AUC) (Roberts 2009). For any statistical wildlife–habitat relationship model, there is a tradeoff between sensitivity and specificity (Allouche et al. 2006). Sensitivity is the probability of correctly classifying a presence, whereas specificity is the probability of correctly classifying an absence (Table 13.1). By definition, when one is increased the other declines. ROC/AUC assesses model accuracy across all values of sensitivity/specificity, but Kappa can change drastically (Allouche et al. 2006; Freeman and Moisen 2008) at any particular value. The proper choice of threshold values can optimize the specificity versus sensitivity tradeoff, even when using Kappa (but see Manel et al. 2001). The two threshold values used in this study had relatively small effects on Kappa but large effects on commission and omission error rates (Table 13.4).

All of these models, except for species with very specific habitat associations, are prone to inherently large commission errors. As the ultimate goal of all the work described in this chapter is the conservation of wildlife habitat, it may be desirable to minimize the omission error rate even at the expense of increasing the commission error rate. The reason for this would be to preserve as much potential habitat for each species as possible, as any increase in omission errors associated with wildlife habitat models will lead to neglecting potential habitat for that species (Wilson et al. 2005). Another way to say this is to favor sensitivity over specificity (Table 13.1). If this is a desirable condition of a wildlife habitat-modeling project for a large set of species, then the threshold for considering a location as appropriate habitat should be equal to the prevalence of the species (the second threshold used in this study). When this is the case commission error rates were relatively high, but omission errors were low across all species (Fig. 13.4b). This pattern may not be desirable, however, if the purpose of the project is to target only the most appropriate habitats for protection, for example when conservation resources and funds are limited. When the focus of a conservation project is to identify only the highest quality locations for protection, then using the threshold that sets predicted prevalence=actual for each species may be appropriate (Fig. 13.4a).

13.5 Past, Current, and Future Applications

Natural resource inventory databases have been designed for the purposes of monitoring the stocks and sustainable management of important assets, but traditionally do not incorporate wildlife among those assets. We suggest that wildlife habitat-resource modules can successfully be implemented into natural resource inventory decision support tools. If these tools are combined in the future, it could be possible to track changes in wildlife habitat in real time with each resource management action. Clearly, forest resource inventory data (e.g., Welsh et al. 2006) are amenable to this task, but rangeland, wetland, stream network, and other natural resources should be as useful. Wildlife habitat monitoring is increasingly becoming more feasible due to sharing of species occurrence information, as evidenced by the growing number of open-access and cooperative data repositories and organizations dedicated to managing and distributing quality ecological data (for birds these include: Breeding Bird Survey, Avian Knowledge Network, Partners in Flight, Ornithological Information System, eBird, and others).

Forest resource inventories could augment these wildlife occurrence database efforts where appropriate, and provide a valuable resource towards developing dynamic, predictive models in support of management decisions. There are significant challenges to overcome in achieving these goals, the most obvious of which is filling in the gaps where wildlife species occurrence data or vegetation inventory data do not exist. When appropriate wildlife survey data from which to fit statistical models do not exist, we advise creating expert-based models for each species, in a manner similar to the GAP method, based on published habitat accounts and local habitat associations. Remote sensing has the potential to map many vegetation structural and composition attributes, and can do so across jurisdictional boundaries (Wulder and Franklin 2003), and provides the most cost-efficient means to assess wildlife habitat on a per-area basis.

Acknowledgments We would like to thank the Michigan Department of Natural Resources IFMAP (Integrated Forest Monitoring, Assessment, and Prescription) program for funding the fieldwork described in this chapter, as well as the Maurer Lab students who participated in the project, including; J Skillen, J Nesslage, S Damania, A Axel, J Karl, E Mize, M Cook, and D Lipp. We would also like to thank the editors of this book and two anonymous reviewers for many helpful edits.

References

Allouche O, Tsoar A, Kadmon R (2006) Assessing the accuracy of species distribution models: prevalence, kappa and the true skill statistic (TSS). J Appl Ecol 43:1223–1232.
Anderson JR, Hardy EE, Roach JT, Witmer RE (1976) A land use and land cover classification system for use with remote sensor data. Professional Paper 964. U.S. Geological Survey, Washington, DC.
Araújo MB, Guisan A (2006) Five (or so) challenges for species distribution modelling. J Biogeogr 33:1677–1688.
Atkinson EJ, Therneau TM (2000) An introduction to recursive partitioning using RPART routines. Mayo Foundation, Rochester MN.
Austin M (2007) Species distribution models and ecological theory: a critical assessment and some possible new approaches. Ecol Model 200:1–19.
Donovan ML, Nesslage GM, Skillen JJ, Maurer BA (2004) The Michigan Gap Analysis Project final report. Michigan Department of Natural Resources – Wildlife Division, Lansing, MI.
Fahrig L (2001) How much habitat is enough? Biol Conserv 100:65–74.
Feldesman MR (2002) Classification trees as an alternative to linear discriminant analysis. Am J Phys Anthropol 119:257–275.

Fielding AH, Bell JF (1997) A review of methods for the assessment of prediction errors in conservation presence/absence models. Environ Conserv 24:38–49.

Freeman EA, Moisen GG (2008) A comparison of the performance of threshold criteria for binary classification in terms of predicted prevalence and kappa. Ecol Model 217:48–58.

Gottschalk TK, Huettmann F, Ehlers M (2005) Thirty years of analysing and modelling avian habitat relationships using satellite imagery data: a review. Int J Remote Sens 26:2631–2656.

Guisan A, Thuiller W (2005) Predicting species distribution: offering more than simple habitat models. Ecol Lett 8:993–1009.

Hepinstall JA, Krohn WB, Sader SA (2002) Effects of niche width on the performance and agreement of avian habitat models. In: Scott JM, Heglund PJ, Morrison ML (eds) Predicting species occurrences: issues of accuracy and scale. Island Press, Washington.

Herzog SK, Kessler M (2006) Local vs. regional control on species richness: a new approach to test for competitive exclusion at the community level. Glob Ecol Biogeogr 15:163–172.

Jetz W, Sekercioglu CH, Watson JEM (2008) Ecological correlates and conservation implications of overestimating species geographic ranges. Conserv Biol 22:110–119.

Johnson CJ, Gillingham MP (2005) An evaluation of mapped species distribution models used for conservation planning. Environ Conserv 32:117–128.

Karl JW, Heglund PJ, Garton EO, Scott JM, Wright NM, Hutto RL (2000) Sensitivity of species habitat-relationship model performance to factors of scale. Ecol Appl 10:1690–1705.

Karl JW, Svancara LK, Heglund PJ, Wright NM, Scott JM. 2002. Species commonness and the accuracy of habitat-relationship models. In: Scott JM, Heglund PJ, Morrison ML (eds) Predicting species occurrences: issues of accuracy and scale. Island Press, Washington.

Knick ST, Rotenberry JT (2000) Ghosts of habitats past: contribution of landscape change to current habitats used by shrubland birds. Ecology 81:220–227.

Linder ET, Villard MA, Maurer BA, Schmidt EV (2000) Geographic range structure in North American landbirds: variation with migratory strategy, trophic level, and breeding habitat. Ecography 23:678–686.

Lobo JM, Jiménez-Valverde A, Real R (2008) AUC: a misleading measure of the performance of predictive distribution models. Glob Ecol Biogeogr 17:145–151.

Manel S, Ceri Williams H, Ormerod SJ (2001) Evaluating presence-absence models in ecology: the need to account for prevalence. J Appl Ecol 38:921–931.

McPherson JM, Jetz W (2007) Effects of species' ecology on the accuracy of distribution models. Ecography 30:135–151.

McPherson JM, Jetz W, Rogers DJ (2004) The effects of species' range sizes on the accuracy of distribution models: ecological phenomenon or statistical artefact? J Appl Ecol 41:811–823.

MDNR (2005) IFMAP field manual. Field Manual Michigan Department of Natural Resources, Lansing, MI.

Mörtberg UM (2001) Resident bird species in urban forest remnants; landscape and habitat perspectives. Landsc Ecol 16:193–203.

Peterjohn BG, Sauer JR (1993) North American Breeding Bird Survey annual summary 1990–1991. Bird Popul 1:1–15.

Pulliam HR (2000) On the relationship between niche and distribution. Ecol Lett 3:349–361.

Roberts LJ (2009) Improving wildlife habitat model performance: sensitivity to the scale and detail of vegetation measurements. PhD Dissertation. Michigan State University, East Lansing, MI.

Scott JM, Davis F, Csuti B, Noss R, Butterfield B, Groves C, Anderson H, Caicco S, D'Erchia F, Edwards TC Jr, Ulliman J, Wright RG (1993) Gap analysis – a geographic approach to protection of biological diversity. Wildl Monogr 123:1–41.

Seoane J, Bustamante J, Díaz-Delgado R (2004) Competing roles for landscape, vegetation, topography and climate in predictive models of bird distribution. Ecol Model 171:209–222.

Seoane J, Carrascal LM, Alonso CL, Palomino D (2005a) Species-specific traits associated to prediction errors in bird habitat suitability modelling. Ecol Model 185:299–308.

Seoane J, Bustamante J, Díaz-Delgado R (2005b). Effect of expert opinion on the predictive ability of environmental models of bird distribution. Conserv Biol 19:512–522.

Storch D, Sizling AL (2002) Patterns of commonness and rarity in central European birds: reliability of the core-satellite hypothesis within a large scale. Ecography 25:405–416.

Thompson FR III, Brawn JD, Robinson S, Faaborg J, Clawson RL (2000) Approaches to investigate effects of forest management on birds in eastern deciduous forests: how reliable is our knowledge? Wildl Soc Bull 28:1111–1122.

Tsoar A, Allouche O, Steinitz O, Rotem D, Kadmon R (2007) A comparative evaluation of presence-only methods for modelling species distribution. Divers Distrib 13:397–405.

Vaughan IP, Ormerod SJ (2005) The continuing challenges of testing species distribution models. J Appl Ecol 42:720–730.

Welsh HH Jr, Dunk JR, Zielinski WJ (2006) Developing and applying habitat models using forest inventory data: an example using a terrestrial salamander. J Wildl Manage 70:671–681.

Whittaker RH, Levin SA (eds) (1975) Niche: theory and application. Dowden, Hutchinson and Ross, Stroudsburg, Pennsylvania.

Wilson KA, Westphal MI, Possingham HP, Elith J (2005) Sensitivity of conservation planning to different approaches to using predicted species distribution data. Biol Conserv 122:99–112.

Wulder MA, Franklin SE (eds) (2003) Remote sensing of forest environments: concepts and case studies. Kluwer Academic Publishers, Boston.

Chapter 14
Using Species Distribution Models for Conservation Planning and Ecological Forecasting

Josh J. Lawler, Yolanda F. Wiersma, and Falk Huettmann

14.1 Introduction

Conservation practitioners and resource managers must often work with limited data to answer critical, time-sensitive questions. In many regions of the world, even the most basic information about the distribution of species is lacking. Knowing the geographic extent of a given species or ecological system is the first step in planning for its management or conservation. The sustainable management of fish stocks, timber, waterfowl populations, and biodiversity in general requires high quality spatial data on species distributions. Selecting preserves or easements to protect plants and wildlife, for instance, requires detailed knowledge of where different species are on the landscape. Such information is the foundation of science-based management and is necessary for assessing the risks of land-use actions, management scenarios, or other human activities to plant and wildlife populations (Huettmann et al. 2005).

Species distribution models provide one tool for addressing the lack of species distribution data (Boyce and McDonald 1999). These models can be used to fill gaps in our knowledge by projecting habitat suitability in areas with few or no occurrence records. These models can also be used to forecast the effects of changes in environmental conditions on species distributions. Given the immense threat that climate change, land-use change, and invasive species pose to ecosystem services and to rare species in particular (Wilcove et al. 1998; Sala et al. 2000), having the ability to forecast potential future impacts allows practitioners to assess alternative policies and actions and to plan for change (Huettmann et al. 2005; Nielsen et al. 2008). Thus, forecasting will likely play an even more important role in conservation as forecasting tools become more accurate and more accessible.

Many different types of models have been applied to these problems. These have included highly mechanistic models that simulate population dynamics, species interactions, and dispersal (Bugmann 1996; Schumaker et al. 2004; Battin et al.

J.J. Lawler (✉)
School of Forest Resources, University of Washington, Seattle, WA 98195, USA
e-mail: jlawler@u.washington.edu

2007), empirical models that rely on correlative relationships between known distributions and environmental conditions (Guisan and Zimmermann 2000), and models that combine mechanistic and empirical approaches (Iverson et al. 2004). Here, we explore how species distribution models can be applied to two types of conservation applications: conservation planning and ecological forecasting. We begin with a brief description of these conservation applications, concentrating on some of the specific questions that species distribution models have been used to help answer. We then discuss the types of models that are available for these applications and whether, and how, those models have been validated. We conclude with two case studies in which we describe specific applications of species-distribution modeling to reserve selection and to forecasting climate impacts.

14.2 Management Challenge, Ecological Theory, and Statistical Framework

14.2.1 Conservation Planning

Arguably, the most effective way to conserve biodiversity is through the use of nature reserves (i.e., strictly legislated protected areas such as parks or wilderness reserves). In both Canada and the United States, the first protected areas were either national parks set aside primarily for recreation and tourism purposes (Runte 1987; Sellars 1997; MacEachern 2001), or wildlife and migratory bird sanctuaries for the purpose of conserving particular species (Wiersma et al. 2005). Similar motivations drove much of the protected-area planning in other parts of the world. In the late nineteenth and early twentieth centuries, the areas identified as suitable sites for protection were chosen largely for their scenic beauty. Areas were often chosen opportunistically from the portions of the landscape that were less suitable for agriculture or human development. Globally, this has resulted in a disproportionate representation of high-altitude, low-biodiversity ("rock and ice") parks, and relatively little representation of low-elevation, high biodiversity areas (Pressey 1994; Khan et al. 1997; Rodrigues et al. 1999; Howard et al. 2000; Scott et al. 2001; Brooks et al. 2004).

In the last two decades, increases in computing power, and the widespread availability of geographic information systems (GIS) and spatially referenced data have aided a move away from ad hoc planning for protected areas (Ardron et al. 2008). Conservation biologists developed a wide array of tools for systematically identifying protected area networks explicitly designed to protect biodiversity (Kirkpatrick 1983; Margules et al. 1988; Pressey and Nicholls 1989; Pressey and Cowling 2001). These new, more analytical approaches emphasize efficiency by identifying sites that protect a maximum amount of biodiversity with a minimum amount of space or for the lowest cost. The tools themselves include heuristic algorithms (Pressey and Nicholls 1989), stochastic optimization approaches (McDonnell et al. 2002), integer programming (Underhill 1994), and metrics such as irreplaceability (Pressey et al. 1993). Many of these tools have been incorporated into software packages such as SITES/MARXAN (Andelman et al. 1999), C-PLAN (Pressey et al. 2005), PORTFOLIO (Urban 2002), and Zonation (Moilanen and Kujala 2008) that have been used by conservation practitioners around the globe.

All of the tools for systematic conservation planning rely heavily on spatially referenced biodiversity data. Because conservation planning occurs at broad spatial extents (often within provinces, territories, or large ecologically-defined regions), detailed data on species occurrences are typically not available. Instead, these algorithms frequently employ data on the extent of species' occurrences, which may be taken from taxonomic atlases or range maps. Given the coarse resolution of atlases and range maps, these sources tend to overestimate species occurrences by ignoring the porous nature of species ranges (van Jaarsveld et al. 1998a). Consequently, species distribution models have been used to provide finer resolution estimates of species occurrences for the design of protected areas networks (Margules and Stein 1989; Scott et al. 1993; Jennings 2000; Cabeza et al. 2004; Wilson et al. 2005)

14.2.2 Forecasting

Climate change, land-use change, and invasive species are likely to have some of the largest and most pervasive effects on biodiversity in the coming century (Sala et al. 2000). Understanding how these forces will change the distribution of biota and how managers and conservation planners can respond to these changes can be greatly facilitated by predictive species distribution models.

Global average temperatures have risen 0.6°C over the last century and are projected to rise between 1.1 and 6.4°C over the next 100 years (IPCC 2007a). Both the paleoecological record and recent ecological evidence indicate that species distributions shift in response to climate change (IPCC 2007b). Over the past century, plants and animals have experienced geographic range shifts, both at rates and in directions that are consistent with recent increases in temperature (Parmesan and Yohe 2003; Root et al. 2003; Parmesan 2006). Many species are responding to increasing temperatures by moving upward in elevation (Lenoir et al. 2008) or poleward in latitude (Parmesan et al. 1999; Thomas and Lennon 1999).

A whole field of research has evolved that focuses on projecting species range shifts in response to climate change. The models used to project climate-driven range shifts are often referred to as climate–envelope models, bioclimatic models, or ecological niche models. These models are built using known relationships between current climatic conditions and current species ranges. Projected future climate data are then used as inputs to these models to project potential future species distributions. Such shifts in distributions have been used to assess potential climate-driven extinction rates (Thomas et al. 2004), where and how much flora or fauna will likely change in response to climate change (Thuiller et al. 2005; Araújo et al. 2006; Lawler et al. 2009), and whether current protected-area networks will adequately protect biodiversity in a changing climate (Araújo et al. 2004; Hannah et al. 2007). Additionally, species distribution models have been used to examine how human disease vectors may respond to projected changes in climate.

Despite the imminent and pervasive nature of climate change, land-use change potentially poses an even greater threat to biodiversity over the coming century (Jetz et al. 2007). Deforestation and land-conversion for agriculture and residential development currently drive much of the landscape change on the Earth's surface. The replacement of natural landscapes with human dominated ones often results in the loss of habitat and subsequent shifts in species distributions. Understanding how species will respond to particular types and patterns of land-use change allows land-use and conservation planners to design landscapes to better accommodate both human and non-human resource needs (Forman 1995; Onyeahialam et al. 2005).

Forecasting the effects of potential future changes in land-use on species distributions has been carried out at many different scales for several different systems, using a wide variety of approaches. In general, models are built that relate species occurrences or abundances to measured habitat attributes such as the composition and structure of vegetation, landscape patterns, soil types, and topography. Projected changes in land-cover, landscape pattern, and vegetation structure are then used as inputs to the models to project changes in species distributions or in habitat suitability. Predictive models have been used to assess the effects of different forest-management practices (Spies et al. 2007; Nielsen et al. 2008) tax incentives (Nelson et al. 2008), and changes in land use (White et al. 1997; Schumaker et al. 2004; Huettmann et al. 2005) on species distributions.

Invasive species are often cited as the second or third leading threat to biodiversity (Wilcove et al. 1998), yet they tend to be less well studied than many other threats (Lawler et al. 2006a). Once they become established in a system, most invasive species are incredibly difficult and costly to eradicate. For example, invasive species incur an economic cost of $138 billion in the United States annually (Pimentel et al. 2000). To prevent invasions, managers and policy makers need to know where a given species is likely to invade and how likely it is to be able to persist and outcompete native species (Vander Zanden and Olden 2008). Forecasting has been used to describe potential future invasions and to identify areas that are particularly susceptible to the spread of invasive species (Peterson 2003). Predictive models have been used to project ant invasions (Steiner et al. 2008), potential new interactions between Spotted (*Strix occidentalis caurina*) and Barred Owls (*S. varia*) (Peterson and Robins 2003), and potential invasions of plant species (Higgins et al. 1999).

14.3 Modeling and Model Validation Techniques

In general, species distribution models can be classified as either empirical or mechanistic. Empirical models use observed data to build correlative relationships between species occurrences or abundances and environmental factors. Mechanistic models use theoretical or observed relationships to simulate some subset of the processes that determine species distributions. Here, we discuss some of the modeling techniques that have been used for conservation planning and ecological forecasting as well as some of the approaches that have been used to validate these models.

14.3.1 Empirical Models

Most of the models that have been used to project species distributions for conservation planning, ecological forecasting, and risk assessment are empirical models. The most common among these are statistical approaches such as generalized linear models (e.g., linear and logistic regression) (Higgins et al. 1999), generalized additive models (GAMS) (Guisan et al. 2002), and discriminant analysis (Manel et al. 1999). In addition to these more traditional approaches, ecologists now have several machine-learning-based tools at their disposal (Olden et al. 2008). These tools include, but are not limited to, classification and regression trees (CART), artificial neural networks (ANN) (Olden and Jackson 2001), genetic algorithms (Peterson and Vieglais 2001), and maximum entropy models (Phillips et al. 2006). Additional model-averaging approaches such as random forests (Cutler et al. 2007) that make use of these statistical and machine-learning tools may improve accuracy in some cases (Prasad et al. 2006). Wiersma (see Chap. 12) provides a more detailed discussion of the implications of choosing one statistical approach over another. All of these approaches are used to model species occurrences or abundances as a function of one or more independent environmental variables.

There have been many comparisons of the performance of different empirical approaches with modeling species distributions (Olden and Jackson 2002; Elith et al. 2006; Lawler et al. 2006b; Pearson et al. 2006; Prasad et al. 2006; Wiersma Chap. 12). In general, these comparisons have demonstrated that the more flexible and more complex approaches tend to be the most accurate for the largest number of species (Elith et al. 2006). One common conclusion from these studies is that model selection should be based on detailed knowledge about the suspected relationships between the species in questions and its environment (Guisan and Zimmermann 2000; Guisan et al. 2006) as well as on the quality of the assessment data.

Empirical models have several limitations. The main limitation with any correlative model is that correlation does not imply causation. For example, in any correlative model, it is possible that the variables in the model are merely surrogates for the causal factors that actually determine the distribution of a species. If these variables play the same surrogate role in the same way in the new region or new time period over which the model is being applied for prediction, then the model may still represent the distribution of the species. If, however, the relationships between the surrogate variables and the causal factors change, or the causal factors themselves change, the empirical model will be unable to describe the species distribution in the new landscape or the new time period. Empirical models also assume that the species being modeled is in equilibrium with its environment (i.e., at the very least, that the species occupies all of the habitat available to it). This may not be the case for many species, particularly for invasive species that are still expanding into unoccupied portions of their range, or when climate change alters the factors that determine a species' distribution.

Additionally, empirical models are static and thus cannot directly account for dispersal or evolution. Therefore, using empirical models to forecast changes in species distributions over long time periods requires making assumptions about these two factors. Most commonly, when empirical models are used for forecasting, researchers assume unlimited or conversely, no dispersal. Occasionally, more complex methods are used to assess the ability of a species to move through a changing landscape (Peterson et al. 2002). Empirical forecasts assume no evolution. Furthermore, empirical models generally do not take interspecific interactions into account. At coarse spatial resolutions, these factors likely have less of an impact on species occurrences, but at finer resolutions, competition and predation may play important roles in determining occurrence or abundance.

Although it is not necessarily a limitation, spatial scale is another important issue that must be considered in the modeling process. Both grain and extent (sensu Wiens 1989) can affect the fitting of correlative models. The data used for model building and prediction are often mapped to a spatial grid (see also discussion of data issues in Zuckerberg et al., Chap. 4). To provide the most accurate models, the resolution (grain) of the grid should match the underlying data. Coarser resolution grids used to summarize individual point data can overestimate the distribution of species by portraying presences across large areas in which the species may occur in only limited locations (Rahbek and Graves 2001; Rahbek 2005). Alternatively, using a relatively fine resolution grid to portray course resolution range-map data can overestimate presences implying a false degree of accuracy. The spatial extent of the data used to build a model can also influence model fit and model predictions (Thuiller et al. 2004). For example, models built with data covering only a portion of a species' range may not capture the full range of suitable environmental conditions for the species and hence may provide poor predictions.

Closely related to the issue of spatial scale is the issue of spatial autocorrelation. Spatial autocorrelation occurs when observations are more or less similar as a result of the distance between them in Euclidean space. These similarities and differences may be driven by the environmental patterns that are being represented in the model being developed (e.g., temperature, soil moisture, elevation), or by some other process (e.g., population dynamics at individual sites). Spatial autocorrelation that is not fully accounted for by the environmental variables used as predictors in a model can result in pattern in the model residuals, thus violating the assumptions of several modeling approaches and potentially biasing parameter estimates and increasing Type I error rates. There are, however, several methods available for addressing spatial autocorrelation in species distribution models (Dormann et al. 2007; but see clarification in Betts et al. 2009).

One of the greatest strengths of empirical models is that they can be built rather easily with relatively little data on the specific biology of a given species. This means that these models can be used to relatively quickly project distributions for large numbers of species. Thus, these models are useful tools for conservation planning and forecasting when many species must be modeled with little biological data. When interpreting the predictions of these models, as with any models, the user must fully understand the model limitations and their implications for particular applications of the results.

14.3.2 Mechanistic Models

Far fewer mechanistic or process-based models have been used to project species distributions for conservation planning and ecological forecasting. Many of the mechanistic models that have been used tend to be population models. Spatially explicit population models have been used to investigate the potential effects of both climate change and land-use change on populations and species distributions (Carroll 2007). Much simpler dispersal models have also been used in conjunction with empirical approaches to assess species responses to climate change (Iverson et al. 2004). By directly simulating many of the mechanisms that drive species distributions, mechanistic models overcome several of the limitations of empirical models. However, building a mechanistic model that captures enough of the ecological processes that determine the distribution of a species will require more natural history knowledge than we presently have for most species. Even when enough data can be found to build and parameterize such a model, there is no guarantee that a more complex mechanistic model will out-perform an empirical model (Robertson et al. 2003).

14.3.3 Model Validation

Assessing the accuracy of species distribution models is critical for both conservation planning and ecological forecasting. For both applications, the different types of errors these models produce have different ramifications. For conservation planning, whether errors of commission are more costly in terms of biodiversity protection than errors of omission will depend on the specific situation. For example, if one can select three sites to protect a given species, and the species distribution model has predicted the species should be present at thirty sites when it is really only present at five sites, there is a good chance that the three sites selected from the thirty will provide no protection for the species. Conversely, if the species occurs at ten sites, but the species distribution model predicts the species will be present at only five of those sites, the species will at least be found at all three of the five sites selected to protect the species. Alternatively, if one has unlimited funds and plans to protect the entire habitat of a given species, errors of omission may result in only a subset of the actual habitat being protected. Errors of commission, on the other hand, will result in more land than is needed being protected, but more of the actual habitat will likely be protected. For forecasting the effects of climate change on species distributions, errors of commission will result in overpredictions of species range expansions and errors of omission will produce exaggerated estimates of species range contractions or will underestimate the degree of range expansion.

In general, model validation is lacking in many studies involving species distribution modeling (Araújo et al. 2005a). Independent data sources for validation are often difficult to come by. Models are often evaluated solely on the data with which they were constructed. This amounts to a non-independent assessment and provides relatively unreliable information on the accuracy of the model. The most common form of model validation involves splitting the data set in two and reserving one set of data for model assessment and using the other (often larger) set of data for model building. This approach provides a semi-independent assessment, as the data are still likely to be spatially correlated with the data used to build the model. The most independent assessments are conducted with data from a different region or a different time period, although the data sets from a different region may still be temporally correlated and the data from a different time period may still be spatially correlated.

Several different metrics have been applied to evaluate model performance. These include simple estimates of the percentage of presences and/or absences correctly predicted, Cohen's Kappa (Stehman 1997), and the area under the receiver operator curve (AUC). The latter two measures provide a synthetic estimate of model error that includes both the ability to predict presence and absences. Because many species distribution models predict probabilities, one often has to select a threshold for categorizing predictions as presences or absences. AUC is a metric that assesses model accuracy across a range of potential thresholds. Although AUC is widely used to assess model performance, it can be a highly misleading measure and its use has been discouraged by some for multiple reasons (Boyce et al. 2002; Lobo et al. 2008). Simpler measures of the percentage of presences and absences correctly classified by a model for a given threshold (or even for multiple thresholds) provides a more meaningful and useful assessment of model performance, particularly if that model will be applied using a given presence–absence threshold.

14.4 Past, Current, and Future Applications

14.4.1 Systematic Reserve Selection Based on Species Distribution Modeling

The vast majority of reserve-selection studies have made use of atlas or range-map data. Because these two types of distribution data generally provide relatively coarse-resolution representations of the extent of occurrence of a species, there are scale-dependent limitations to their utility (Van Jaarsveld et al. 1998b; Habib et al. 2003; Hurlbert and White 2005; Graham and Hijmans 2006). Range maps, for example, are often coarsely defined polygons that encompass known locations of a given species. Atlas occurrences, on the other hand, are represented on large spatial grids. Grid cells highlighted as having occurrences have at least one known occurrence, but do not necessarily have individuals distributed throughout the cell. Consequently, both of these types of occurrence data overestimate the distribution of species'. Unless the spatial scale of reserves is selected to match that of the range-map or atlas data, reserves selected based on these data will likely under-represent some species.

Species distribution models can be used to refine range-map or atlas data to provide finer scaled estimates of species occurrences. However, only a few studies have integrated predictive habitat models with reserve design (e.g., Alidina et al. 2008), largely because most reserve-design studies assess representation requirements for a wide range of species, and the labor involved in developing predictive distribution models for so many species is usually prohibitive. Below, we describe a few projects that have attempted to integrate distribution models with reserve selection or reserve evaluation.

There is a relative long history of integration of predictive distribution models with reserve design within the United States Gap Analysis Program (Scott et al. 1993; Jennings 2000). GAP analysis seeks to identify the "gaps" in conservation protection based on predicted occurrence of key species and mapped land cover. The earliest GAP analyses used simple wildlife–habitat rules to predict occurrences, and not the types of empirical or mechanistic models discussed above. However, since its earliest application in the 1990s, the GAP program in the United States has grown. The GAP program website (http://www.gapanalysis.nbii.gov) now lists hundreds of projects, some of which have used predictive distribution models such as those described above to evaluate which species and natural features are under-represented in protected area networks. In addition, the GAP program has focused primarily on evaluating existing reserves. However, there has been some effort to use the GAP data to prioritize areas for reserve selection (Kiester et al. 1996).

Other studies have also used species distribution models to assess current reserves and/or locate new reserves. Margules and Stein (1989) applied a simple systematic reserve–design analysis to predictive occurrence data for 32 tree species in south-eastern Australia. They compared the predicted occurrences of these species with the distribution of current reserves, as well as to a suite of simulated reserves. They found that the simulated reserves that followed an east–west orientation tended to capture more of the predicted distribution of tree species than the existing suite of reserves, or than simulated reserves along a north–south gradient. They concluded that many tree species were not well represented in existing reserves, and that reserves should be large and follow rainfall and elevation gradients (Margules and Stein 1989). The validity of these conclusions, of course, relies on the accuracy of the underlying species distribution models (Margules and Stein 1989).

Wilson et al. (2005) used five different sets of predicted future distributions for four Australian plant species to investigate the potential impact of model accuracy on resulting reserve networks. The distribution models they used predicted a probability of occurrence for each 1-hectare grid cell for each species. Three of the different predicted distributions were attained using different methods for classifying the predicted probabilities into presences and absences. These three methods included: (1) an arbitrary threshold of $p=0.5$, (2) a threshold that achieved a predefined level of sensitivity and specificity, and (3) a threshold based on a receiver–operator characteristic curve that took an assumed relative cost of false positive and false negative errors into account. The two other approaches they tested involved using different reserve–selection strategies based on the raw predicted probabilities of occurrence. They then selected sets of reserves to protect the four plant species based on these different predicted distributions.

The results of Wilson et al. (2005) demonstrated that selected reserve networks can vary greatly based on seemingly small differences in the underlying predicted species distributions. The reserves varied in the amount of land, and number of planning units selected, and their location. The reserves selected based on predictive data with a presence–absence threshold of 0.5 occupied the least total area. More risk-averse predictive models (e.g., presence–absence classifications that incorporated tradeoffs between specificity and sensitivity or that considered the relative costs of different types of model errors) required a larger area of land to represent the four species. However, these did not generally allow for the original conservation targets to be met (15% of pre-European distribution to be conserved within reserves) because the predictive thresholds were such that only a small amount of the region was actually predicted to contain the species (Wilson et al. 2005). Although this study only included four plant species, it showed that reserve–design algorithms are highly sensitive to the uncertainty inherent in predictive species distribution models. Planners who are risk-averse would seek more stringent models, but might sacrifice efficiency in the final reserve network. Networks based on the more stringent predictive models were also less flexible (Wilson et al. 2005). Thus, uncertainty in predictive habitat models carries direct, predictable implications for how reserve networks are designed.

Two simple steps can be taken to improve the validity of reserve–selection analyses when they are based on projected species distributions (e.g., Alidina et al. 2008). First, it is important to match the spatial resolution of the predicted distribution data with that of the reserves. For example, if the grain of the distribution data is much larger than the reserves that comprise the reserve network, the fact that a reserve overlaps an area (pixel) within a species distribution, will not necessarily mean that the species is likely to be found in the reserve. Second, it is critical to assess the accuracy of the predicted distribution data. Although these models can reduce the overestimation of species distributions inherent in range-map or atlas data, they can still be relatively inaccurate, resulting in inefficient, or worse, ineffective reserve networks.

14.4.2 Forecasting Climate-Induced Range Shifts for Three North American Vertebrates

Climate change can greatly alter ecological systems. As species shift their distributions in response to changing temperatures and precipitation regimes, communities are disassembled. Linkages within ecosystems are broken and new linkages are formed. Changes in phenology will likely decouple critical relationships, driving some populations extinct and potentially forging new bonds between others. Furthermore, climate change will drive changes in hydrology, fire regimes, the distribution and impact of pathogens, as well as the distribution and cultures of human populations. Managing forests, wildlife refuges, and other natural resources in a changing climate requires some understanding of how these processes and relationships will be altered.

Here, we describe forecasts of the potential effects of climate change on the ranges of three North American birds: the Blackburnian Warbler (*Dendroica fusca*), Lark Bunting (*Calamospiza melanocorys*), and Clark's Nutcracker (*Nucigrafa Columbiana*). These forecasts were part of a larger study designed to assess the effects of climate change on the distribution of birds, mammals, and amphibians in the western hemisphere (Lawler et al. 2009). We chose to focus on these three bird species here because their ranges cover different parts of the continent and the species distribution models project that climate change will affect their distributions in markedly different ways.

Data for building the models came from several sources. Species distribution data were derived from digital range maps of the birds breeding ranges (Ridgely et al. 2003). These data were mapped to a 50-km by 50-km grid resulting in a record of presence or absence for each species in each grid cell. Current climate data used to build the models were based on two historic data sets, the 30-min Climatic Research Unit (CRU) CL 1.0 (New et al. 1999) data set and the 10-min CRU CL 2.0 data set (New et al. 2002). These data were downscaled to the 50-km grid and averaged for a period between 1961 and 1990. From the temperature, precipitation, and sunshine data derived from these data sets, a set of 37 bioclimatic variables was calculated. These variables included measures such as potential evapotranspiration and total annual snowfall and were selected to represent the biological mechanisms likely to influence the ranges of plants and animals and thus best predict species distributions. See Lawler et al. (2009) for a full description of the downscaling methods and the bioclimatic variables.

Lawler et al. (2009) used empirical models to model species distributions as a function of current climate. The models were built using random-forest predictors (Breiman 2001; Cutler et al. 2007, see also Evans et al., Chap. 9, this volume). Random forests are a model-averaging approach based on classification or regression trees – our models were based on classification trees. This machine-learning-based approach has been shown to perform as well as, if not better than, other approaches for modeling species distributions (Lawler et al. 2006b; Prasad et al. 2006; Cutler et al. 2007; Magness et al. 2008). The models were built using 80%

of the data, reserving 20% of the data for a semi-independent model validation. The models were validated through assessment of the percentage of test-data-set presences and test-data-set absences that were correctly predicted.

Projected future climate data from ten different climate simulations were used to project potential changes in the ranges of the three species. These simulations were derived from ten different general circulation models run for a mid-high (SRES A2) greenhouse-gas emissions scenario. The future climate projections were obtained from the World Climate Research Programme's Coupled Model Intercomparison Project phase 3 multi-model archive (http://www-pcmdi.llnl.gov/ipcc/about_ipcc.php). As for the historic climate data, the projected future temperature anomalies were downscaled to the 50-km grid. These anomalies were added to the current climate variables and the same 37 bioclimatic variables were calculated. The future climate data were calculated as a 30-year average for the period between 2071 and 2099.

The models for all three species were relatively accurate, predicting between 88% and 93% of all presences and 99% of the absences in the test-data sets correctly. The projections for the Blackburnian Warbler were typical of the poleward shift in species ranges that has been observed for many species over the last half century (Fig. 14.1). For this species, the model projected contractions of the potential range along the current southern boundary and expansions in the potential range at the northern edge. There was substantial agreement across the climate projections ($\geq 80\%$) that the potential range of the warbler would shift out of most of the Appalachian mountain range and that the range would expand to the north. Although the model built for the Lark Bunting projected contractions at the southern edges of the species' range, it projected only modest expansions of the potential range at the northern boundary (Fig. 14.1). Instead, the model projected an expansion of the potential range to the east. There was, however, far less agreement across model projections for the eastern expansion than there was for the contractions at the southern range boundary. Finally, the model for Clark's Nutcracker projected contractions across much of the species' current range and very few areas of potential range expansion (Fig. 14.1). Although there was more agreement in the projected range contractions at the southern range boundary and scattered areas throughout the range, in general, there was relatively little agreement for the contractions projected across much of the species' range.

The projected potential future ranges for these three species allow us to draw several conclusions. First, as they have in the past (Davis and Shaw 2001), species will respond independently to climate change. It is likely that communities will disassemble and new communities – in some cases without current analogs – will be formed. Second, not all species will experience poleward or upward range shifts, some may respond to changes in moisture regimes resulting in longitudinal, or more complex range shifts. Third, despite variability in climate-change projections, species distribution models can be used to discern potential range shifts. Although different general circulation models vary in their projections of the magnitude of warming and sometimes both the direction and magnitude of changes in precipitation, there were still areas in which at least eight of ten climate-change scenarios

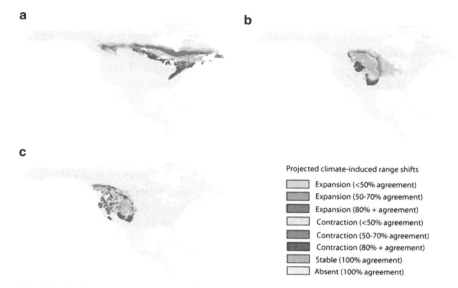

Fig. 14.1 Projected climate-driven shifts in the potential ranges of the Blackburnian Warbler (a), Lark Bunting (b), and Clark's Nutcracker (c). Maps show agreement across ten different climate-change scenarios for a period averaged over the years 2071-2100

resulted in potential range contractions or expansions for these three species. The degree of agreement in model projections, based on different climate scenarios, varied by species. As discussed above, correlative distribution models have their limitations. More detailed explorations of the limitations of these models – specifically with respect to modeling potential climate-driven shifts in species distributions – can be found in Pearson and Dawson (2003), Heikkinen et al. (2006), and Botkin et al. (2007). Despite these limitations, these models can provide a first approximation of how a species might respond spatially to climate or land-use change at broad spatial scales.

14.4.3 Future Improvements for Applications

Several recommendations have been made for improving species distribution models for various conservation- and management-related applications (Pearson and Dawson 2004; Araújo et al. 2005b; Guisan and Thuiller 2005; Guisan et al. 2006; Heikkinen et al. 2006; Botkin et al. 2007). Among these are calls for: (1) models that are more strongly grounded in ecological theory, (2) more thorough model evaluations using more independent data sets, (3) more mechanistic modeling, (4) better and more species distribution data, (5) models that account for spatial autocorrelation and, (6) multi-scale modeling approaches. Below, we briefly discuss four additional advances in modeling capabilities that will allow practitioners to make better use of species distribution models for conservation applications.

The first of these is ensemble modeling. Ensemble modeling involves using multiple models and/or data sets to produce a suite of model predictions (Araújo and New 2007). The predictions from multiple models or from multiple input data sets are usually then combined in some way (e.g., averaged). Ensembles of models can be used when it is difficult to determine a priori which type of model or which model structure should produce the most accurate prediction. By using a number of different models, one can potentially reap the benefit of the more accurate models, although depending on how the predictions of the models are combined one runs the risk of diluting more accurate predictions with less accurate models. The random forest modeling approach used in the species range–shift modeling discussed above is in essence an ensemble-modeling approach. The random forest algorithm builds many individual classification (or regression) tree models using random subsets of the observations and the predictor variables in the data set. The predictions from these models are then combined (averaged in the case of regression trees, or combined using vote counting or averaging in the case of classification trees). Predictions based on ensembles of different input datasets allow the user to evaluate a range of potential predictions. Again, the range–shift projections described above for the three bird species are examples of this type of ensample approach in which output from ten different general circulation models were summarize to provide range shift projections that depict the level of agreement in projected range expansions and contractions across different climate projections.

The second major advance that we are likely to see in species distribution modeling is the combination of empirical and mechanistic approaches. Although a few studies have linked mechanistic and empirical models for making forecasts (Iverson et al. 2004), this is an area of research that has not yet received much attention. By combining empirical models that can provide a first approximation of species distributions with mechanistic models that better capture interspecific interactions, dispersal, or other mechanisms, we are likely to be able to more accurately capture species distributions and potential changes in those distributions. Iverson et al. (2004) combined empirical models that projected potential changes in the climatic conditions for trees in the eastern United States with a mechanistic model that simulated seed dispersal and establishment. Such combinations of approaches will greatly enhance our ability to predict species distributions.

The third advance that will result in improved predictions of species distributions involves the availability of data for model building and testing. Modeling efforts are often limited by the lack or the quality of data on species occurrences, natural history, and environmental conditions. Increased efforts to collect these data through coordinated national and international programs and citizen-science campaigns (e.g., Global Biodiversity Information Facility; http://www.gbif.org) will greatly expand the capacity to build and test species distribution models. In some regions of the globe much of the data required for building models already exists. The data, however, are often dispersed and inaccessible – often existing only in file cabinets in government offices or university labs. Thus, a massive effort to find, organize, preserve, and make these data publicly available is needed (Huettmann 2005, see also Zuckerberg et al., Chap. 4 this volume).

The fourth advance that holds particular promise for improving the application of species distribution modeling to conservation or management problems, is alternative scenario modeling. Scenario modeling involves evaluating multiple potential situations or conditions and has been used to investigate climate impacts (Nakicenovic et al. 2000; IPCC 2007b), alternative land-use plans (White et al. 1997), and alternative forest-management practices (Spies et al. 2007). Scenario modeling can be combined with ensemble approaches to synthesize the array of potential outcomes, but more appropriately, the different scenarios are often used to produce individual model predictions that are not combined, but rather compared and evaluated.

Species distribution models are a powerful tool for addressing conservation and natural resource management problems. Here, we have primarily discussed their use for conservation planning and ecological forecasting. However, these models are also useful tools for other management and conservation activities including species reintroductions, land-use zoning, and environmental risk assessments in general. The four advances discussed here – ensemble modeling, combining empirical and mechanistic techniques, improved data availability, and scenario modeling – when combined with the multitude of more specific recommended modeling advances, have the potential to greatly improve our ability to address difficult conservation and management questions.

Acknowledgments We are grateful for comments and suggestions from J. Roberts and two anonymous reviewers.

References

Alidina HM, Fischer DT, Steinback C, Ferdana Z, Lombana AV, Heuttmann F (2008) Assessing and managing data. In: Ardron JA, Possingham HP, Klein CJ (eds) Marxan good practices handbook. Pacific Marine Analysis and Research Association, Vancouver.
Andelman SJ, Ball I, Davis FW, Stoms DM (1999) SITES Version 1.0: an analytical toolbox for designing ecoregional conservation portfolios. The Nature Conservancy.
Araújo MB, Cabeza M, Thuiller W, Hannah L, Williams PH (2004) Would climate change drive species out of reserves? An assessment of existing reserve-selection methods. Glob Change Biol 10:1618–1626.
Araújo MB, New M (2007) Ensemble forecasting of species distributions. Trends Ecol Evol 22: 42–47.
Araújo MB, Pearson RG, Thuiller W, Erhard M (2005a) Validation of species–climate impact models under climate change. Glob Change Biol 11:1504–1513.
Araújo MB, Thuiller W, Pearson RG (2006) Climate warming and the decline of amphibians and reptiles in Europe. J Biogeogr 33:1712–1728.
Araújo MB, Whittaker RJ, Ladle R, Erhard M (2005b) Reducing uncertainty in projections of extinction risk from climate change. Glob Ecol Biogeogr 14:529–538.
Ardron J, Possingham H, Klein C (2008) Marxan good practices handbook. Pacific Marine Analysis and Research Association, Vancouver.
Battin J, Wiley MW, Ruckelshaus MH, Palmer RN, Korb E, Bartz KK, Imaki H (2007) Projected impacts of climate change on salmon habitat restoration. Proc Natl Acad Sci 104: 6720–6725.

Betts MG, Ganio LM, Huso MMP, Som NA, Huettmann F, Bowman J, Wintle BA (2009) Comment on "Methods to account for spatial autocorrelation in the analysis of species distributional data: a review." Ecography 32:374–378.

Botkin DB, Saxe H, Araújo MB, Betts R, Richard HWB, Cedhagen T, Chesson P, Dawson TP, Etterson JR, Faith DP, Ferrier S, Guisan A, Hansen AS, Hilbert DW, Loehle C, Margules C, New M, Sobel MJ, Stockwell DRB (2007) Forecasting the effects of global warming on biodiversity. Bioscience 57:227–236.

Boyce M, McDonald LL (1999) Relating populations to habitats using resource selection functions. Trends Ecol Evol 14:268–272.

Boyce MS, Vernier PR, Nielsen SE, Schmiegelow FKA (2002) Evaluating resource selection functions. Ecol Model 157:281–300.

Breiman L (2001) Random forests. Mach Learn 45:5–32.

Brooks TM, Bakarr MI, Boucher T, deFonesca GAB, Hilton-Taylor C, Hoekstra JM, Moritz T, Olivieri S, Parrish J, Pressey RL, Rodrigues ASL, Sechrest W, Stattersfield A, Strahm W, Stuart SN (2004) Coverage provided by the global protected areas system: is it enough. Bioscience 54:1081–1091.

Bugmann HKM (1996) A simplified forest model to study species composition along climate gradients. Ecology 77:2055–2074.

Cabeza M, Araújo MB, Wilson RJ, Thomas CD, Cowley MRJ, Moilanen A (2004) Combining probabilities of occurrence with spatial reserve design. J Appl Ecol 41:252–262.

Carroll C (2007) Interacting effects of climate change, landscape conversion, and harvest on carnivore populations at the range margin: marten and lynx in the Northern Appalachians. Conserv Biol 21:1092–1104.

Cutler DR, Edwards TC, Jr., Beard KH, Cutler A, Hess KT, Gibson J, Lawler JJ (2007) Random forests for classification in ecology. Ecol 88:2783–2792.

Davis MB, Shaw RG (2001) Range shifts and adaptive responses to Quaternary climate change. Sci 292:673–679.

Dormann CF, McPherson JM, Araújo MB, Bivand R, Bolliger J, Carl G, Davies RG, Hirzel A, Jetz W, Kissling WD, Kühn I, Ohlemüller R, Peres-Neto PR, Reineking B, Schröder B, Schurr FM, Wilson R (2007) Methods to account for spatial autocorrelation in the analysis of species distributional data: a review. Ecography 30:609–628.

Elith J, Graham CH, Anderson RP, Dudik M, Ferrier S, Guisan A, Hijmans RJ, Huettmann F, Leathwick JR, Lehmann A, Li J, Lohmann LG, Loiselle BA, Manion G, Moritz C, Nakamura M, Nakazawa Y, Overton J, Townsend Peterson A, Phillips SJ, Richardson K, Scachetti-Pereira R, Schapire RE, Soberon J, Williams S, Wisz MS, Zimmermann NE (2006) Novel methods improve prediction of species' distributions from occurrence data. Ecography 29:129–151.

Forman RTT (1995) Landscape mosaics: the ecology of landscapes and regions. Cambridge University Press, Cambridge.

Graham CH, Hijmans RJ (2006) A comparison of methods for mapping species ranges and species richness. Glob Ecol Biogeogr 15:578–587.

Guisan A, Edwards TC, Hastie T (2002) Generalized linear and generalized additive models in studies of species distributions: setting the scene. Ecol Model 157:89–100.

Guisan A, Lehmann A, Ferrier S, Austin M, Overton JMC, Aspinall R, Hastie T (2006) Making better biogeographical predictions of species' distributions. J Appl Ecol 43:386–392.

Guisan A, Thuiller W (2005) Predicting species distribution: offering more than simple habitat models. Ecol Lett 8:993–1009.

Guisan A, Zimmermann NE (2000) Predictive habitat distribution models in ecology. Ecol Model 135:147–186.

Habib LD, Wiersma YF, Nudds TD (2003) Effects of sampling bias on estimates of historical species richness and faunal relaxation of mammals in Canadian national parks. J Biogeogr 30:375–380.

Hannah L, Midgley GF, Andelman S, Araujo MB, Hughes G, Martinez-Meyer E, Pearson RG, Williams P (2007) Protected area needs in a changing climate. Front Ecol Environ 5:131–138.

Heikkinen RK, Luoto M, Araújo MB, Virkkala R, Thuiller W, Sykes MT (2006) Methods and uncertainties in bioclimatic envelope modelling under climate change. Prog Phys Geogr 30:751–777.

Higgins SI, Richardson DM, Cowling RM, Trinder-Smith TH (1999) Predicting the landscape-scale distribution of alien plants and their threat to plant diversity. Conserv Biol 13:303–313.

Howard PC, Davenport TRB, Kigenyi FW, Viskanic P, Baltzer MC, Dickinson CJ, Lwanga J, Matthrews RA, Mupada E (2000) Protected area planning in the tropics: Uganda's national system of forest nature reserves. Conserv Biol 14:858–875.

Huettmann F (2005) Databases and science-based management in the context of wildlife and habitat: towards a certified ISO standard for objective decision-making for the global community by using the internet. J Wildl Manage 69:466–472.

Huettmann F, Franklin SE, Stenhouse GB (2005) Predictive spatial modeling of landscape change in the Foothills Model Forest. For Chron 81:525–537.

Hurlbert AH, White EP (2005) Disparity between range map- and survey-based analyses of species richness: patterns, processes and implications. Ecol Lett 8:319–327.

IPCC (2007a) Climate change 2007: the physical science basis. Contribution of working group I to the fourth assessment report of the Intergovernmental Panel on Climate Change. Cambridge University Press, Cambridge.

IPCC (2007b) Climate change 2007: impacts, adaptation and vulnerability, contribution of working group II to the fourth assessment report of the Intergovernmental Panel on Climate Change. Cambridge University Press, Cambridge.

Iverson LR, Schwartz MW, Prasad AM (2004) Potential colonization of new available tree species habitat under climate change an analysis for five eastern US species. Landsc Ecol 19:787–799.

Jennings MD (2000) Gap analysis: concepts, methods, and recent results. Landsc Ecol 15:5–20.

Jetz W, Wilcove DS, Dobson AP (2007) Projected impacts of climate and land-use change on the global diversity of birds. PLoS Biol 5:e157.

Khan ML, Menon S, Bawa KS (1997) Effectiveness of the protected areas network in biodiversity conservation: a case-study of Meghalaya state. Biodiv Conserv 6:853–868.

Kiester AR, Scott JM, Csuti B, Noss RF, Butterfiel B, Sahr K, White D (1996) Conservation prioritization using GAP data. Conserv Biol 10:1332–1342.

Kirkpatrick JB (1983) An iterative model for establishing priorities for the selection of nature reserves: an example for Tasmania. Biol Conserv 25:127–134.

Lawler JJ, Aukema JE, Grant JB, Halpern BS, Kareiva P, Nelson CR, Ohleth K, Olden JD, Schlaepfer MA, Silliman BR, Zaradic P (2006a) Conservation science: a 20-year report card. Front Ecol Environ 4:473–480.

Lawler JJ, Shafer SL, White D, Kareiva P, Maurer EP, Blaustein AR, Bartlein PJ (2009) Projected climate-induced faunal change in the western hemisphere. Ecology 90:588–597.

Lawler SS, White D, Neilson RD, Blaustein AR (2006b) Predicting Climate-induced range shifts: model differences and model reliability. Global change Biology 12:1568–1584.

Lenoir J, Gegout JC, Marquet PA, de Ruffray P, Brisse H (2008) A significant upward shift in plant species optimum elevation during the 20th century. Science 320:1768–1771.

Lobo JM, Jimenez-Valverde A, Real R (2008) AUC: a misleading measure of the performance of predictive distribution models. Glob Ecol Biogeogr 17:145–151.

MacEachern A (2001) Natural selections: national parks in Atlantic Canada, 1935–1970. McGill-Queen's University Press, Montreal and Kingston.

Magness DR, Huettmann F, Morton JM (2008) Using Random Forests to provide predicted species distribution maps as a metric for ecological inventory and monitoring programs. In: Smolinski TG, Milanova MG, Hassanien AG (eds). Applications of computational intelligence in biology: current trends and open problems. Studies in computational intelligence. Springer-Verlag, Berlin, pp. 209–229.

Manel S, Dias J-M, Ormerod SJ (1999) Comparing discriminant analysis, neural networks and logistic regression for predicting species distributions: a case study with a Himalayan river bird. Ecol Model 120:337–347.

Margules CR, Nicholls AO, Pressey RL (1988) Selecting networks of reserves to maximise biological diversity. Biol Conserv 43:63–76.

Margules CR, Stein JL (1989) Patterns in the distributions of species and the selection of nature reserves: an example from Eucalyptus forests in south-eastern New South Wales. Biol Conserv 50:219–238.

McDonnell MD, Possingham HP, Ball IR, Cousins EA (2002) Mathematical models for spatially cohesive reserve design. Environ Model Assess 7:107–114.

Moilanen A, Kujala H (2008) Zonation spatial conservation planning framework and software v. 2.0, User manual. http://www.helsinki.fi/bioscience/consplan.

Nakicenovic N, Alcamo J, Davis G, Vries BD, Fenhann J, Gaffin S, Gregory K, Grübler A, Jung TY, Kram T, Rovere ELL, Michaelis L, Mori S, Morita T, Pepper W, Pitcher H, Price L, Riahi K, Roehrl A, Rogner H-H, Sankovski A, Schlesinger M, Shukla P, Smith S, Swart R, Rooijen SV, Victor N, Dadi Z (2000) Special report on emissions scenarios. a special report of working group III of the Intergovernmental Panel on Climate Change. Cambridge University Press, Cambridge.

Nelson E, Polasky S, Lewis DJ, Plantinga AJ, Lonsdorf E, White D, Bael D, Lawler JJ (2008) Efficiency of incentives to jointly increase carbon sequestration and species conservation on a landscape. Proc Natl Acad Sci 105:9471–9476.

New M, Hulme M, Jones PD (1999) Representing twentieth-century space-time climate variability. Part 1: development of a 1961-90 mean monthly terrestrial climatology. J Clim 12:829–856.

New M, Lister D, Hulme M, Makin I (2002) A high-resolution data set of surface climate over global land areas. Clim Res 21:1–25.

Nielsen SE, Stenhouse GB, Beyer HL, Huettmann F, Boyce MS (2008) Can natural disturbance-based forestry rescue a declining population of grizzly bears? Biol Conserv 141:2193–2207.

Olden JD, Jackson DA (2001) Fish-habitat relationships in lakes: gaining predictive and explanatory insight using artificial neural networks. Trans Am Fish Soc 130:878–897.

Olden JD, Jackson DA (2002) A comparison of statistical approaches for modeling fish species distributions. Freshwater Biol 47:1976–1995.

Olden JD, Lawler JJ, Poff NL (2008) Machine learning without tears: a primer for ecologists. Q Rev Biol 83:171–193.

Onyeahialam A, Huettmann F, Bertazzon S (2005) Modeling sage grouse: progressive computational methods for linking a complex set of local biodiversity and habitat data towards global conservation statements and decision support systems. Lecture Notes in Computer Science (LNCS) 3482, International Conference on Computational Science and its Applications (ICCSA) Proceedings Part III:152–161.

Parmesan C (2006) Ecological and evolutionary responses to recent climate change. Annu Rev Ecol Syst 37:637–669.

Parmesan C, Ryrholm N, Stefanescu C, Hill JK, Thomas CD, Descimon H, Huntley B, Kaila L, Kullberg J, Tammaru T, Tennent WJ, Thomas JA, Warren M (1999) Poleward shifts in geographical ranges of butterfly species associated with regional warming. Nature 399:579–583.

Parmesan C, Yohe G (2003) A globally coherent fingerprint of climate change impacts across natural systems. Nature 421:37–42.

Pearson RG, Dawson TP (2003) Predicting the impacts of climate change on the distribution of species: are climate envelope models useful? Glob Ecol Biogeogr 12:361–371.

Pearson RG, Dawson TP (2004) Bioclimate envelope models: what they detect and what they hide; response to Hampe (2004). Glob Ecol Biogeogr 13:471–473.

Pearson RG, Thuiller W, Araújo MB, Martinez-Meyer E, Brotons L, McClean C, Miles L, Segurado P, Dawson TP, Lees DC (2006) Model-based uncertainty in species range prediction. J Biogeogr 33:1704–1711.

Peterson AT (2003) Predicting the geography of species' invasions via ecological niche modeling. Q Rev Biol 78:419–433.

Peterson AT, Ortega-Huerta MA, Bartley J, Sanchez-Cordero V, Soberon J, Buddemeier RH, Stockwell DRB (2002) Future projections for Mexican faunas under global climate change scenarios. Nature 416:626–629.

Peterson AT, Robins CR (2003) Using ecological-niche modeling to predict barred owl invasions with implications for spotted owl conservation. Conserv Biol 17:1161–1165.

Peterson AT, Vieglais DA (2001) Predicting species invasions using ecological niche modeling: new approaches from bioinformatics attack a pressing problem. Bioscience 51:363–371.

Phillips SJ, Anderson RP, Schapire RE (2006) Maximum entropy modeling of species geographic distributions. Ecol Model 190:231–259.

Pimentel D, Lach L, Zuniga R, Morrison D (2000) Environmental and economic costs of nonindigenous species in the United States. Bioscience 50:53–65.

Prasad AM, Iverson LR, Liaw A (2006) Newer classification and regression tree techniques: bagging and random forests for ecological prediction. Ecosystems 9:181–199.

Pressey RL (1994) Ad hoc reservations: forward or backward steps in developing representative reserve systems? Conserv Biol 8:662–668.

Pressey RL, Cowling RM (2001) Reserve selection algorithms and the real world. Conserv Biol 15:275–277.

Pressey RL, Humphries CJ, Margules CR, Vane-Wright RI, Williams PH (1993) Beyond opportunism: key principles for systematic reserve selection. Trends Ecol Evol 8:124–128.

Pressey RL, Nicholls AO (1989) Application of a numerical algorithm to the selection of reserves in semi-arid New South Wales. Biol Conserv 50:263–278.

Pressey RL, Watts M, Ridges M, Barrrett T (2005) C_Plan Conservation Planning software. User manual. NSW Department of Environment and Conservation.

Rahbek C (2005) The role of spatial scale and the perception of large-scale species-richness patterns. Ecol Lett 8:224–239.

Rahbek C, Graves GR (2001) Multiscale assessment of patterns of avian species richness. Proc Natl Acad Sci 98:4534–4539.

Ridgely RS, Allnutt TF, Brooks T, McNicol DK, Mehlman DW, Young BE, Zook JR (2003) Digital distribution maps of birds of the Western Hemisphere, version 1.0. NatureServe: Arlington.

Robertson MP, Peter CI, Villet MH, Ripley BS (2003) Comparing models for predicting species' potential distributions: a case study using correlative and mechanistic predictive modelling techniques. Ecol Model 164:153–167.

Rodrigues ASL, Tratt R, Wheeler BD, Gaston KJ (1999) The performance of existing networks of conservation areas in representing biodiversity. Proc R Soc Lond B 266:1453–1460.

Root TL, Price JT, Hall KR, Schneider SH, Rosenzweig C, Pounds JA (2003) Fingerprints of global warming on wild animals and plants. Nature 421:57–60.

Runte A (1987) National parks: the American experience, Third edition. University of Nebraska Press, Lincoln.

Sala OE, Chapin FS, III, Armesto JJ, Berlow E, Bloomfield J, Dirzo R, Huber-Sanwald E, Huenneke LF, Jackson RB, Kinzig A, Leemans R, Lodge DM, Mooney HA, Oesterheld M, Poff NL, Sykes MT, Walker BH, Walker M, Wall DH (2000) Global biodiversity scenarios for the year 2100. Science 287:1770–1774.

Schumaker NH, Ernst T, White D, Haggerty P (2004) Projecting wildlife responses to alternative future landscapes in Oregon's Willamette Basin. Ecol Appl 14:381–400.

Scott JM, Davis F, Csuti B, Noss R, Butterfield B, Caicco S, Groves C, Edwards Jr TC, Ulliman J, Anderson H, D'Erchia F, Wright RG (1993) Gap analysis: a geographic approach to protection of biological diversity. Wildl Monogr 123:1–41.

Scott JM, Murray M, Wright RG, Csuti B, Morgan P, Pressey RL (2001) Representation of natural vegetation in protected areas: capturing the geographic range. Biodivers Conserv 10:1297–1301.

Sellars RW (1997) Preserving nature in the national parks: a history. Yale University Press, New Haven.

Spies TA, McComb BC, Kennedy RSH, McGrath MT, Olsen K, Pabst RJ (2007) Potential effects of forest policies in terrestrial biodiversity in a multi-ownership province. Ecol Appl 17:48–65.

Stehman SV (1997) Selecting and interpreting measures of thematic classification accuracy. Remote Sensing Environ 62:77–89.

Steiner FM, Schlick-Steiner BC, VanDerWal J, Reuther KD, Christian E, Stauffer C, Suarez AV, Williams SE, Crozier RH (2008) Combined modelling of distribution and niche in invasion biology: a case study of two invasive *Tetramorium* ant species. Divers Distrib 14:538–545.

Thomas CD, Cameron A, Green RE, Bakkenes M, Beaumont LJ, Collingham YC, Erasmus BFN, de Siqueira MF, Grainger A, Hannah L, Hughes L, Huntley B, Van Jaarsveld AS, Midgley GF, Miles L, Ortega-Huerta MA, Peterson AT, Phillips OL, Williams SE (2004) Extinction risk from climate change. Nature 427:145–148.

Thomas CD, Lennon JJ (1999) Birds extend their ranges northwards. Nature 399:213.

Thuiller W, Brotons L, Araújo MB, Lavorel S (2004) Effects of restricting environmental range of data to project current and future species distributions. Ecography 27:165–172.

Thuiller W, Lavorel S, Araújo MB, Sykes MT, Prentice IC (2005) Climate change threats to plant diversity in Europe. Proc Natl Acad Sci 102:8245–8250.

Underhill LG (1994) Optimal and suboptimal reserve selection algorithms. Biol Conserv 70:85–87.

Urban DL (2002) Prioritzing reserves for acquisition. In: Gergel SE, Turner MG (eds) Learning landscape ecology: a practical guide to concepts and techniques. Springer, New York, pp. 293–305.

van Jaarsveld AS, Freitag S, Chown SL, Muller C, Kock S, Hull H, Bellamy C, Krüger M, Endrödy-Younga S, Mansell MW, Sholtz CH (1998a) Biodiversity assessment and conservation strategies. Science 279:2106–2108.

Van Jaarsveld AS, Gaston K, Chown SL, Freitag S (1998b) Throwing biodiversity out with the binary data? S Afr J Sci 94:210–214.

Vander Zanden MJ, Olden JD (2008) A management framework for preventing the secondary spread of aquatic invasive species. Can J Fish Aquat. Sci 65:1512–1522.

White D, Minotti PG, Barczak MJ, Sifneos JC, Freemark KE, Santelmann MV, Steinitz CF, Kiester AR, Preston EM (1997) Assessing risks to biodiversity from future landscape change. Conserv Biol 11:349–360.

Wiens JA (1989) Spatial scaling in ecology. Funct Ecol 3:385–397.

Wiersma YF, Beechey TJ, Oosenbrug B, Meikle J (2005) Protected areas in northern Canada: designing with integrity. Canada Council on Ecological Areas, Ottawa, ON.

Wilcove DS, Rothstein D, Dubow J, Phillips A, Losos E (1998) Quantifying threats to imperiled species in the United States. Bioscience 48:607–615.

Wilson KA, Westphal MI, Possingham HP, Elith J (2005) Sensitivity of conservation planning to different approaches to using predicted species distribution data. Biol Conserv 122:99–112.

Chapter 15
Conclusion: An Attempt to Describe the State of Habitat and Species Modeling Today

C. Ashton Drew, Yolanda F. Wiersma, and Falk Huettmann

We set out to deliver a book that would prompt increased attention to the ecological theory and assess the relevant assumptions that underlie predictive landscape-scale species and habitat modeling. We invited international authors who are actively engaged in advancing the discipline of predictive modeling in landscape ecology to provide chapters that would not only highlight current developments and identify outstanding gaps, but which would also reflect on how methodological choices were informed by ecological theory. In this manner, we have provided readers not with a "how-to" guide that will rapidly become outdated as methods advance, but rather insights into the thought processes, reasoning, and current debates that are common across modeling projects and methods. Such extended reflections help to show multiple viewpoints and stimulate new ideas; they rarely find space in published research manuscripts. However, we believe these will offer valuable guidance to both novice and advanced modelers seeking to discern trade-offs between alternative modeling approaches.

Our introduction identified several themes that we believe represent trade-offs and issues that are addressed to some degree in virtually all modeling projects. Exploring, understanding and settling these themes is therefore critical to advancing the discussion of predictive landscape-scale species and habitat modeling, whether for a specific application or for the development of the research field more generally. Here we offer a summary of several key themes that emerged across chapters as authors reported and reflected upon the theoretical and technological underpinnings of their work.

C.A. Drew (✉)
North Carolina Fish and Wildlife Cooperative Research Unit, Department of Biology,
North Carolina State University, Raleigh, NC, 27695, USA
e-mail: cadrew@ncsu.edu

15.1 Emerging Themes: Theoretical

15.1.1 The Value of Realism Versus Utility

All models aim for high predictive accuracy and precision. However, comparison of chapter authors' insights reveals that there remains significant debate regarding how much realism (i.e., accurately representing the mechanisms and complexity of natural systems) contributes to model performance, how significantly uncertainty impairs the utility of models in applied settings, and how much error is acceptable. The various facets of the complexity versus simplicity and the mechanistic versus correlative debates quickly emerge in reading through this volume. Many of the chapter authors contend that models with high predictive power, but little underlying biological realism or theoretical support will "ultimately" prove to be less useful. This is because in the face of stochastic processes, there will be little information about ecological mechanisms to predict how species will respond to habitat and environmental changes. Other authors (Huettmann and Gottschalk, Chap. 9; and others) however, place a high value on the utility of model output. They feel that prediction should override explanation, and become a science in itself ("predictive science"). In either case, the central argument is a clear one: virtually all models get more use and public attention when the highest possible prediction accuracy gets achieved: model-prediction accuracies of over 90% are frequently reported and convincingly shown! Such model accuracy is obviously highly desirable. Virtually all authors agree that the scientific question to be asked still influences the research methods, including choice of available data, sampling design if new data are to be collected, statistical methods, and the desired degree of model simplicity or complexity.

Several authors highlight the trade-offs between the extremes of model parsimony and complexity (Hooten's section on naïve models in Chap. 2; Huettmann and Gottschalk, Chap. 9; Wiersma, Chap. 10; Lawler et al., Chap. 13; see specifically the discussion on links between assumptions and model simplicity/complexity in Laurent et al., Chap. 4). As some of the chapters illustrate, there are cases where a model has a very specific purpose and an overly simple model will not suffice. The example of modeling malaria risk given by Kerr et al. (Chap. 1) shows that complex models are generally more robust and suited for epidemiological modeling. Hooten (Chap. 2) highlights agent-based models as a viable, but complex, set of models for prediction, and Huettmann and Gottschalk (Chap. 9) emphasize their view that model complexity is essential to have a high and (ecologically) meaningful predictive performance, particularly when using nonlinear statistical algorithms.

With respect to the debate on mechanistic versus predictive models, some of our chapter authors feel that models do not need to be mechanistic to be predictive; others do. Hooten (Chap. 2) offers contrasting case studies. Several chapters articulate situations where one can make reasonable predictions without modeling mechanic processes or explicitly testing ecological hypotheses. For example, Kerr et al. (Chap. 1) highlight a case where a nonmechanistic model for chameleon in

Madagascar led to the discovery of a sister species in a previously unsurveyed area. Evans et al. (Chap. 7) provide an introduction to machine learning methods that do not require underlying ecological hypothesis to develop predictions, a sentiment echoed by Huettmann and Gottschalk (Chap. 9). Other chapters illustrate where inclusion of ecological mechanisms is essential. Fletcher et al. (Chap. 5) show how the incorporation of metapopulation theory and considerations of density-dependent habitat use can improve models for four bird species in Montana. Murphy and Evans (Chap. 8) give insights on how theories from the field of genetics can be linked to predictive models, and Lookingbill et al. (Chap. 6) demonstrate how accommodation of ecological mechanisms can improve model understanding. Models that are built on a high level of realism will make them easier to comprehend for the non-informed public, and carry strong educational value. They might achieve a wider buy-in that way. The notion of "what is truth?" and the need for benchmark studies to resolve this question and for a transparent model accuracy assessment are covered in the chapter by Huettmann and Gottschalk (Chap. 9).

The debate on these model-prediction issues is confounded by the lack of a clear consensus definition of what actually constitutes a model with high utility. Roberts et al. (Chap. 12) discuss model utility at some length, and point out that a model may not be considered "useful" if it fails for certain purposes, or at scales for which it was not designed. They point out, quite correctly we think, that a model may have high accuracy for one purpose (e.g., predicting species distribution at the coarse grain) but not for another. Many models have that feature. Such a phenomenon is not uncommon and should not be grounds for outright rejection of the model (e.g., "this model is, or is not, useful"). Laurent et al. (Chap. 4) feel that model utility is both limited and empowered by its assumptions. Kerr et al. (Chap. 1) state that model utility is measured by whether predictions are testable and tested, whereas Hooten (Chap. 2) suggests that model utility is determined by whether or not it provides progress, any progress, in understanding ecological phenomena. This is also echoed by Kerr et al., who ask whether models "fail in interesting ways" such that insights on ecological processes and avenues for further research can be inferred. Model utility for management decisions also hinges on the type of model generated (e.g., binary vs. probabilistic) and whether the species or process in question exhibits threshold dynamics. It is clear though: both modeling methods and results must be effectively communicated to end users for models to have maximum utility.

15.1.2 Modeling Uncertain Futures

The chapters in this volume suggest that we are good at knowing and quantifying the spatial uncertainty in our models, but that we are not doing as well when considering temporal dynamics. Although landscape ecology as a discipline is as much about temporal as spatial scales, we note that none of the chapters in this volume use temporal models (e.g., back-casting). However, temporal issues are not completely ignored – Kerr et al. (Chap. 1) discuss space-for-time assumptions in predictive species/habitat modeling and Laurent et al. (Chap. 4) discuss assumptions inherent with making temporal predictions. In addition, Fletcher et al. (Chap. 5), and briefly Huettmann and Gottschalk (Chap. 9) highlight how temporal replication may yield better model prediction than replications in space, especially for species with high temporal variability.

Related to these questions is the need for robust accuracy assessment of all models. Many of the chapter authors highlight the importance of accuracy assessment (Fletcher et al., Chap. 5; Huettmann and Gottschalk, Chap. 9; Wiersma, Chap. 10; Roberts et al., Chap. 12; Lawler et al., Chap. 13) and stress that the appropriate research design, free of autocorrelation and addressing scale, repeatability and transparency, should consider accuracy assessment right at the start of the research project. Roberts et al. also discuss issues with model accuracy that are linked to the species biology – specialists may be easier to model, but models for rare and/or cryptic species may have a high commission error. This emphasizes the challenges of conducting models and accuracy assessments for a wide range of species.

A final and important theme running through all the chapters is how models are ultimately applied to management questions and real-world issues. Lawler et al. (Chap. 13) highlight how predictive species–habitat models are incorporated into conservation planning, and emphasize the limitations of modeling on models. Huettmann and Gottschalk (Chap. 9), Drew and Perera (Chap. 11) and Roberts et al. (Chap. 12) suggest that species/habitat models can be applied within adaptive management frameworks. Implicit in these statements is an assumption that models be of high quality and utility in order to be an effective component of adaptive management.

15.2 Emerging Themes: Technical

15.2.1 *Effective Data Management is Essential*

The importance of effective and high-quality data management emerged as a clear theme across all chapters. A model's value is, in large part, determined by the suitability and quality of the input data. Zuckerberg et al. (Chap. 3) state this most strongly: the certainty of model inferences depends as much upon data management and maintenance, as upon the analytical approach. Similarly, in situations of "modeling on a model," as discussed by Huettmann and Gottschalk (Chap. 9) and Lawler et al. (Chap. 13), knowledge of the input data characteristics and quality of each component model (e.g., climate change model plus species–habitat model) is essential to define the overall strengths and limits of the final combined product. Put simply, metadata for underlying data and models, are the key to achieve good modeling practice. By demonstrating how the characteristics of species and habitat datasets often determine the validity and appropriateness of model assumptions, Laurent et al. (Chap. 4) reiterate the value of having clearly defined and documented data resources. This sentiment is echoed in the chapter on expert-based modeling by Drew and Perera (Chap. 11) and elsewhere, which advocates that the documentation and transparency of expert knowledge must occur with equal rigor as is typically applied when collecting empirical data.

The need for effective data management is likely to increase because, as Hooten (Chap. 2) argues, we are transitioning from a data-poor to a data-rich modeling context through advances in digitization, globalization, statistical techniques and increased effort to collect spatially referenced ecological data. Global data-mining efforts are well under way, and new data schemes are set up and growing exponentially. Unfortunately, the management of data resources requires knowledge, skill, and time at levels rarely allocated within modeling projects. Discussing this failure at length, Zuckerberg et al. (Chap. 3) challenge research, publishing, educational and management institutions to demonstrate leadership by requiring, funding, and rewarding data documentation and dissemination. However, they may face an uphill battle. Although many authors here advocate open access data sharing (Zuckerberg et al., Chap. 3; Huettmann and Gottschalk, Chap. 9; Roberts et al., Chap. 12), many scientists remain skeptical of open access resources and balk at the idea of participating in such programs. We are hopeful that in time, the majority of the research community will support Open Access data sharing (in part for their own gain).

15.2.2 *Transitioning from a Data-Poor to a Data-Rich Modeling Environment*

Historically, most ecological data were collected without adequate spatial and temporal reference to allow their effective use for landscape-scale predictive model development or assessment. However, with the development of GIS technology and novel spatial statistical and software techniques, there has been rapid growth in the availability of data suitable for predictive modeling. Three major trends are apparent in relation to this transition to a much more data-rich and statistically advanced modeling environment. First, we are reaching a point of "critical mass" in the number and variety of comparative studies that will allow us to begin to make generalizations that could provide clearer guidance for model development. This trend is evidenced probably most clearly within Wiersma's review of statistical models (Chap. 10), which identifies emerging guidelines relating data characteristics, statistical approach, and model performance. Second, there has been significant effort to research and develop sampling protocols that capture landscape pattern and process specifically for use in predictive modeling. Hooten (Chap. 2), Zuckerberg et al. (Chap. 3) and Huettmann and Gottschalk (Chap. 9) discuss some of the unique sampling considerations inherent to predictive species–habitat modeling and their algorithms. Several other chapters provide insights on alternative sampling designs for specific modeling applications. For example, Fletcher et al. (Chap. 5) outline methods for building models with samples through time, while Lookingbill et al. (Chap. 6) propose a sampling design to deal with scale-dependencies in predictive modeling. Third, with access to more data of higher quality, researchers are developing both new tools to explore these data (e.g., Random Forest models and Machine learning models, discussed in Evans et al., Chap. 7) and pushing toward new research frontiers. The contribution of Murphy and Evans (Chap. 8) exemplifies this expansion by their use of population genetics to infer landscape ecological processes, such as functional connectivity, that might better predict both patterns of species occurrence and habitat quality. It is interesting to note that the contrast between the foundational assumptions of these new tools reflect the theoretical debate of realism versus utility: some predict pattern in the absence of process hypotheses (data-mining approaches) while others specifically build such hypotheses (e.g., genetic flow and population dynamics approaches).

15.2.3 Recognizing the True Costs (and Benefits) of Good Models

Although rarely discussed in papers and texts about predictive species/habitat modeling, several chapters touch on issues of cost. Zuckerberg et al. (Chap. 3) discuss the costs that need to be allocated for data management; Roberts et al. (Chap. 12) mention costs of data acquisition and processing and discuss the trade-offs between increasing model accuracy and the costs of acquiring the data necessary to build more complex models; Hooten (Chap. 2) acknowledges that some GIS and statistical software packages may carry a cost; and Murphy and Evans (Chap. 8) include detailed cost estimates for predictive modeling in landscape genetics. Although not all chapters discuss costs explicitly, it is clear that a good project must have budget allocations that cover all steps, from model conception, through to data collection/management, model development, and accuracy assessment and (global/online) dissemination of results. A number of authors (Zuckerberg et al., Chap. 3; Huettmann and Gottschalk, Chap. 9; Roberts et al., Chap. 12) highlight the value of Open Access systems for data and models (see for example Open Modeller; http://openmodeller.sourceforge.net), but Zuckerberg et al. remind us that doing so also carries a cost and requires sustainable funding and a business plan. Open Access remains a contentious issue for some sectors, but is fully and globally established in most others. Although in principle, many scientists are supportive of the idea of sharing data and models widely, many researchers are hesitant when asked to publish their data and models in an open source environment, unless explicitly required to do so as a condition for publication and funding. Some of this reluctance may be due to a lack of infrastructure to facilitate it, or in a general unawareness of the benefits of Open Access.

Despite the apparent reluctance for investment in Open Access, paying for good data documentation and providing quality data delivery and management via open access systems offers great returns. Robust Open Access to data and models, with high-quality metadata affords many unknown opportunities for testing of new hypotheses via meta-analyses and data mining and global synthesis for instance. Bayesian methods can be widely improved with increased access to diverse data sets to facilitate setting of robust priors. A culture of open access also creates an identity of a network of experts and collaborators, which serve to improve modeling processes.

15.3 Summary and Future Directions

In summary, this volume highlights the current state of species and landscape ecological modeling as a large and rich area for research. It is clear that we have a wide array of technical tools and resources at our disposal. Although there are philosophical disagreements between chapter authors on certain issues (e.g., simplicity vs. complexity and mechanistic vs. predictive models), we believe these philosophical debates reflect the state of a discipline that has reached a certain stage of development. As long as there is progress, we do not intend for this volume to weigh definitively on one side of any of these debates. Rather we feel, as echoed by chapter authors, that the important take-home message of this book should be that modelers carefully consider and document *why* they are modeling, including thinking about their research hypothesis, management concerns, data sources, and model applications. Science is also about reflection and thinking. These chapters will not tell the reader which route to take, but will hopefully provide "food for thought" for a modeler considering these issues, and for the user who needs to interpret and use models. As we highlighted in the introduction, our initial motivation to consider ecological theory has proven to be important, but so too are issues of data type, data handling and model purpose. We have shown that there are a wide range of tools and techniques. No doubt there will be even more of these in just 5 years from now, but we hope that some of the philosophical questions raised here will continue to be considered by modelers when making choices about how to model, and finding the best models and solutions.

Clearly, there is still much to be discussed, debated and developed in the field of predictive species–habitat modeling. Several decades of modeling have yielded insights on different approaches, many of which are highlighted in this volume. Where do we envision going? Modeling got maturing, but clearly still represents a "frontier science" and is a discipline that will grow and evolve in a similar fashion as weather forecasting and remote sensing has. Models are here to stay, and they will be a key component of decision making; the use of models will likely increase worldwide.

We can safely say that there will continue to be some uncertainty; after all, uncertainty is a natural part of natural systems. However, this needs not impede model development or model application, but does needs to be critically defined, assessed, and accounted for. There is a need for continued effort to focus on links between theory and methods, because only with careful links between the two can we understand why some models predict better than others and from there, move forward to develop general guidelines. A focus on theory does not discount the value of simple, correlative models (evidenced by authors discussions that promote such approaches), but rather ensures that we move towards understanding when and why certain models are likely to be most appropriate.

Future research on modeling likely will focus on the twin issues of data and technology. Although there will always be cases where models to address specific questions are data-limited, it is worth noting that many of our chapters optimistically envision a data-rich future. How to effectively collect, organize, manage and share such data should be at the forefront of future work. Technologies are continually emerging that can facilitate ecological research, especially with respect to modeling. Research that investigates means to leverage these emerging technologies will be important, as modeling has always, and will continue to, depend heavily on the use of technology. We cannot anticipate the exact technology that modelers a decade from now will have at their disposal, but we hope that if modelers take time to consider some of the issues that have been raised in this volume, that future models will have a robust foundation upon which to develop predictions and tools to enhance knowledge of ecological systems and for a better sustainable management worldwide.

Author Bios
(*indicates past NASA–MSU awardee)

Adam Algar is a recent PhD graduate from the Canadian Facility for Ecoinformatics Research (CFER) in the Department of Biology at the University of Ottawa, Canada. He is currently an NSERC postdoctoral fellow at Harvard University.

Samuel A. Cushman* is a research landscape ecologist with the USFS Rocky Mountain Research Station in Flagstaff, AZ. His research interests are in statistics and wildlife biology. He is coeditor (with Falk Huettmann) of the recent book *Spatial Complexity, Informatics, and Wildlife Conservation*, published by Springer.

Michael Donovan is with the Michigan Department of Natural Resources, Wildlife Division.

Ashton Drew* is a postdoctoral researcher in the USGS North Carolina Cooperative Fish and Wildlife Research Unit in the Department of Biology at North Carolina State University. Her present work for the US Fish and Wildlife Service applies landscape ecology principles to step down regional conservation objectives to local scales. More broadly, her research interests focus on the role that models can play in support of adaptive monitoring and management programs.

Jeffrey S. Evans is a landscape ecologist for The Nature Conservancy, North America Science. His research is focused on quantifying biodiversity, species distribution, and landscape process using spatial statistics, niche modeling, landscape genetics, multiscale techniques, climate change, LiDAR and spectral remote sensing, and gradient modeling.

Robert Fletcher* is an assistant professor in the Department of Wildlife Ecology and Conservation at the University of Florida. His research broadly revolves around themes critical for understanding how space influences animal behavior, population dynamics, and community structure.

Thomas Gottschalk is a researcher in the Department of Animal Ecology at Justus-Liebig University Giessen. His research interests are in landscape ecology, GIS, remote sensing, ornithology, species distribution modeling, spatial scaling, and conservation ecology. Current projects investigate modeling climate-related changes of bird species distribution and assessing sustainability of agriculture.

Zachary A. Holden is with the USDA Forest Service, Northern Region, in Missoula, MT.

Richard L. Hutto is a professor of Biology and Wildlife Biology in the Division of Biology, University of Montana. His research interests revolve around habitat selection and the relative importance of landscape-level and local-scale variables in determining the suitability of sites to various landbird species.

Falk Huettmann* is an assistant professor in the Biology and Wildlife Department, Institute of Arctic Biology, University of Alaska-Fairbanks. His research interests are in wildlife habitat modeling, GIS and remote sensing, and data management issues worldwide. He is coeditor of the recent book *Spatial Complexity, Informatics, and Wildlife Conservation*, published by Springer.

Mevin Hooten* is an assistant professor in the Department of Mathematics and Statistics at Utah State University. His research interests include Bayesian methods; hierarchical models; ecological and environmental statistics; spatial, temporal, and spatiotemporal statistics.

Jeremy Kerr is an associate professor at the Canadian Facility for Ecoinformatics Research (CFER) in the Department of Biology at the University of Ottawa, Canada. His research interests include conservation biology, macroecology, global change and biodiversity. Current projects investigate causes of biodiversity decline, using remote sensing, field ecology, and computer-based modeling.

Manisha Kulkarni recently completed a postdoctoral fellowship at the Canadian Facility for Ecoinformatics Research (CFER) in the Department of Biology at the University of Ottawa, Canada. Her research focuses on the control of malaria in sub-Saharan Africa and the Amazon Basin. She is currently a program officer at the Canadian International Development Agency.

Edward J. Laurent* is a science coordinator at American Bird Conservancy. His interests include expanding the role of partnerships in collaboratively addressing conservation issues through the use of systems modeling, databases, GIS, and remote sensing technologies.

Josh J. Lawler is an associate professor in the College of the Environment at the University of Washington. His research interests generally lie in the fields of landscape ecology and conservation biology.

Todd R. Lookingbill* is an assistant professor in Geography and the Environment at the University of Richmond (VA). His research interests include protected areas management, natural resources monitoring design and assessment, and forest community ecology.

Brian A. Maurer is a professor in the Department of Fisheries and Wildlife at Michigan State University. His research interests are macroecology, biogeography, and quantitative ecology.

Melanie A. Murphy* is doing a postdoc in the Department of Biology at Colorado State University. Her research interests are in the integration of population genetics, quantitative landscape ecology, and conservation biology to develop robust spatial analysis techniques for analyzing functional landscape connectivity, particularly in fragmented systems.

Anna Noson is a GIS specialist and program coordinator at the Avian Science Center (ASC), Division of Biological Sciences, University of Montana, Missoula.

She develops and manages the ASC's riparian program, designing and implementing monitoring projects to inform restoration and management activities.

Ajith H. Perera is a landscape ecology research scientist with the Ontario Forest Research Institute. He focuses on understanding boreal forest landscape disturbance dynamics and developing landscape ecological models and user tools for forest policy and land use planning.

L. Jay Roberts* is an avian ecologist with PRBO Conservation Science in Petaluma, California. His research interests include ecological monitoring, biogeography, and landscape ecology.

Monique E. Rocca* is an assistant professor in the Department of Forest, Rangeland, and Watershed Stewardship at Colorado State University. Her research interests include forest fire management, plant community ecology, landscape ecology, spatial analysis, ecological restoration.

Christopher T. Rota is a PhD student in the School of Natural Resources at the University of Missouri.

Wayne Thogmartin is a research statistician with the US Geological Survey at the Upper Midwest Environmental Science Center. His research interests include avian ecology, conservation design, extinction risk, and spatial-temporal analysis of habitats.

Dean L. Urban is a professor in the Nicholas School of the Environment and currently the president of US-IALE. His research interests include landscape ecology and forest community ecology, conservation biology, spatial and multivariate analysis, and simulation modeling.

Yolanda F. Wiersma* is an assistant professor in Biology at Memorial University (St. John's, Canada). Her research interests are in boreal landscape ecology, with a focus on wildlife habitat interactions and land management (forestry and protected areas) issues. She also has a strong interest in the philosophy of science.

Jock S. Young was with the Landbird Monitoring Program at the Avian Science Centre at the University of Montana where he worked as data manager and program coordinator. He helped to develop LBMP habitat models as well as coordinate field work. He recently joined the Great Basin Bird Observatory (Reno, Nevada) as a wildlife statistician and research biologist.

Benjamin Zuckerberg* is a research associate at the Cornell Lab of Ornithology. His research interests include the effect of climate change and habitat loss on bird distributions. He is coauthor of the recent book *Monitoring Animal Populations and Their Habitats: A Practitioner's Guide*, published by CRC Press.

Index

A

Abundance, 17–21, 23, 32, 34, 35, 39, 72, 74, 80–85, 87, 114, 120, 124, 148, 149, 169, 197, 213, 220, 251, 252, 256, 272–274
Accuracy assessment, 4, 5, 79, 189, 199, 291, 292, 294
Acker, S.A., 122
Adams, J.R., 166
Adaptive management, 11, 86, 187, 197, 199, 201–202, 228, 229, 231, 251, 292
Admixture, 171, 178
AFLP. *See* Amplified fragment length polymorphism
Akaike information criterion (AIC), 101, 189, 215, 217, 219
Algar, A.C., 9, 14
Algorithmic methods, 139, 140
Alleles, 163, 166, 169–172, 175, 178, 179
Allendorf, F.W., 178–180
Alternative scenario modeling, 177, 282
American Woodcock, 82
Amplified fragment length polymorphism (AFLP), 163, 166, 167, 169, 171, 178
Amstrup, S.C., 202
Anderson, A.D., 167
Anderson, D.R., 51, 52, 194, 217
Anderson, J.R., 256
Anderson, R.P., 51, 221
Angerbjörn, A., 168
ANN. *See* Artificial neural networks
Antoni, O., 219
Application of ecological theory, 200
Arab, A., 37, 38
Araújo, M.B., 15, 281
Ardren, W.R., 167
Area under curve (AUC), 101–103, 145, 150, 189, 216, 219, 221–222, 254, 256–261, 265, 276

Artificial neural networks (ANN), 220, 273
Assignment tests, 165–167, 172, 174, 178
Assumptions, 1, 15, 16, 18, 22, 30–33, 71–87, 91, 92, 95, 119, 123, 137, 145, 167, 169, 172, 192, 194–195, 212, 213, 233, 238–240, 274, 289–292, 294
Atlases, 48, 50, 55, 271
AUC. *See* Area under curve
Aukema, J.E., 6
Austin, M.P., 46, 49, 71, 104, 114, 195
Autocorrelation, 2, 14, 30, 33, 34, 51, 60, 94, 112, 117–119, 125–127, 131, 137, 147, 174, 176, 179, 197, 199, 200, 218, 233, 274, 281, 292
Avise, J.C., 166, 178–180
Ayres, M.P., 12

B

Baddeley, M.C., 236
Bahn, V., 14
Ballou, J.D., 178, 179
Balloux, F., 166
Banerjee, S., 37
Bartlein, P.J., 6, 190, 279
Base-pair, 169, 178, 180
Bayesian, 31, 37, 104, 112, 118, 138, 174, 230, 236, 295
statistics, 32, 36, 39, 229, 241
Beard, K.H., 38
Beauliu, J., 219
Behavior, 1–2, 11–12, 15, 19, 21, 34–36, 39, 50, 60, 73, 77, 79–80, 92, 114, 127, 130, 153, 165, 196–197, 199, 237
Belanger, L., 219
Belkhir, K., 167
Bell, J.F., 199
Betts, R., 281

Bias, 14, 33, 45–47, 50, 54, 55, 59, 61, 62, 73–74, 81, 84, 100, 117, 141, 144, 166, 171, 174, 214, 217, 228, 233, 234, 236–242, 254, 274
Binary data, 33–34, 52, 97, 99
Biocomplexity, 78
Biological hypotheses, 10, 215, 216
Biological responses, 8, 12, 14, 15
Blackard, J.A., 219
Blackburnian Warbler *(Dendroica fusca)*, 279
Blanchong, J.A., 179
Blaustein, A.R., 6, 190, 279
Blazek, R., 218, 219
Bonhomme, F., 167
Booker, J.M., 237
Borcard, D., 118, 119
Bordage, D., 219
Botkin, D.B., 281
Bottlenecks, 166, 169–170, 178
Bray, J.R., 126
Breidt, E.J., 116
Breiman, L., 141, 144
Brennan, L.A., 217
Breshears, D.D., 111
Briscoe, D.A., 178, 179
Brotons, L., 15
Bruggeman, D.J., 179
Buckley, L.B., 20, 21
Burnham, K.P., 51, 194, 217
Bustamante, J., 57

C
Caldow, C., 220
Carl, G., 119
Carlin, B.P., 37
Carrying capacity, 74, 79, 81
CARTs. *See* Classification and regression trees
Castric, V., 167
Caswell, H., 202
Cavelli-Sforza, L.L., 167
Cedhagen, T., 281
Cerulean Warblers, 81
Chamberlain, T.C., 216
Changes in climate, 12, 120, 123, 124, 139, 153, 272, 280, 282
Chapin, F.S., 201
Chen, C., 144
Cherry, S., 216–217
Chesson, P., 281
Chistensen, J.D., 220
Clark, A.G., 179, 180
Clark, D.P., 178–180
Clark, J.S., 37

Clark's Nutcraker *(Nucigrafa Columbiana)*, 279
Classification and regression trees (CARTs), 52, 140–141, 174, 211, 214–215, 219–220, 222, 273, 279, 281
Classification uncertainty, 78
Clearly, D.F.R., 166
Climate change, 3, 11–12, 14, 50, 91, 113, 131, 138, 141, 153–154, 177, 187, 190, 228, 239, 269, 271–272, 274–276, 279–281, 292
Climate change-effects on species distribution, 11, 12, 50, 91, 131, 239
Climate-envelope models, 271
Climate-niche model, 149, 193, 196
Codes, 1–2, 37, 57, 61, 179
Coding, 178, 179
Cognitive limitations, 236–237
Cohen's Kappa, 145, 193, 276
Commission and omission, 74, 252–256, 260–262
Commission errors, 75, 81, 145, 253–256, 260, 262, 265
Community ecology, 19, 81, 139
Complexity, 1–6, 10, 17, 21, 33, 51, 73, 77–82, 85, 138–141, 176, 226, 227, 240
vs. simplicity, 5, 187–203, 290, 295
Confidence intervals, 32, 78, 126, 199, 230, 231
Connectivity, 77, 79–80, 160, 166, 170–177, 241, 294
Consensus, 138, 195, 230, 233, 236–239, 242, 291
Cooke, F., 190, 197, 220
Coral larvae, 77
Cornuet, J.M., 166
Corridors, 75, 80, 113, 173, 240
Cost, 9–10, 19–21, 47, 80, 131, 144, 163–166, 173–174, 187, 201, 216, 227, 228, 230, 249, 250, 258, 263–264, 266, 270, 272, 275, 278, 294–295
Costs of generating new data, 250
Coudun, C., 51
C-PLAN, 271
Craig, E., 192
Cressie, N.A.C., 37
Crookston, N.L., 147, 148
Currie, D.J., 5
Curtis, A., 236
Curtis, J.T., 126
Cushman, S.A., 137, 144, 149, 173
Cutler, A., 38
Cutler, D.R., 38

D

Dalén, L., 168
Darwin core, 56, 58, 64
Data, 1, 10, 29, 45, 71, 91, 111, 137, 159, 188, 208, 227, 249, 269, 290
Databases, 46–51, 53, 54, 56–58, 61–66, 97–99, 104, 105, 198, 209, 210, 231, 248–266
Data considerations, 4, 62, 63, 66, 74, 76, 86, 213, 238
Data dissemination, 45, 46, 55, 58, 62–65, 293, 294
Data management, 3–5, 39, 44–66, 227, 228, 251, 292–294
Data mining, 37, 39, 51–54, 137, 192, 293–295
Data quality, 45, 49, 54, 56, 65, 86, 164, 188, 213, 227, 250, 254, 263, 266, 269, 273, 282, 292, 294
Data storage, 45, 46, 64
Dawson, T.P., 15, 281
Decision support tools, 72, 265–266
deFrutos, A., 218, 219
Demmelle, E., 179, 180
DeYoung, R.W., 166
Dezzani, R., 179, 180
Dieringer, D., 167
Digital databases, 47, 49, 51, 57, 58, 63
Dijak, W.D., 190
Discriminant analysis, 219, 273
Dispersal, 1–2, 13, 17, 20–24, 35, 77, 80, 83, 84, 118, 120, 121, 123, 154, 168, 170–175, 208, 209, 251, 269–270, 274, 275, 282
Distance-based models, 214
Diversity, 9, 14, 18, 19, 46, 73, 81, 87, 111, 141, 160, 161, 166, 169–170, 177, 208, 211, 233, 256
Donovan, M., 249
Dorazio, R.M., 35
Dray, S., 118
Drew, C.A., 1, 71, 227, 289
Drift (neutral variation), 159
Drumm, D., 219
Dudík, M., 51, 221
Dynamical systems, 34, 35, 39
Dynamic or process-based models, 188
Dynamics, 1, 17, 18, 20, 21, 23, 35, 39, 48, 57, 77, 81–83, 85, 91–105, 116, 124, 168, 190–192, 207–209, 213, 229, 234, 236–237, 240, 251, 269, 274, 291, 294
Dzeroski, S., 219

E

Easements, 269
Eastman, J.R., 219
Ecological forecasting, 269–282
Ecological mechanisms, 3, 23, 24, 116, 202, 215, 290, 291
Ecological release, 81
Ecological risk assessment, 78
Ecosystem management, 125, 249
Ecotone, 112, 120–124, 131
Edwards, A.W.F., 167
Edwards, T.C., 38, 219
Effective population size, 169–170, 178
Elicitation, 229–231, 233–238, 241, 242
Elith, J., 51, 220, 221, 278
Elmhagen, B., 168
Emigration, 73, 83–84
Empirical models, 10, 11, 20, 22, 86, 227–228, 269–270, 272–275, 277, 279, 282
Environmental data, 15, 46, 52, 53, 71–72, 75, 77–79, 125, 127, 208, 249, 250
Environmental gradients, 15, 50, 54, 75, 91, 96, 114, 115, 125, 149, 152, 153, 239
Environmental variation, 50, 55, 75, 153
Epidemiological models, 22, 24, 36, 168, 290
Epperson, B.K., 179
Erickson, W.P., 194, 202
Error assessment, 78
Errors, 4, 32, 33, 38, 39, 45, 46, 49, 50, 52, 56, 59–61, 65, 73–75, 77, 78, 82, 84, 85, 94, 126, 139, 141–145, 150, 164, 165, 175–177, 193, 201, 219–222, 228, 235, 236, 242, 252–256, 259–262, 265, 274–276, 278, 290, 292
Etterson, J.R., 281
Evaluation, 1, 12, 71, 73, 78, 86, 91, 97, 101, 121, 123, 137, 138, 145, 150, 164, 214, 216, 219, 220, 222, 231, 242, 249, 250, 256, 258, 263, 276, 277, 281, 282
Evans, J.S., 5, 137, 143, 144, 148, 149, 159, 179, 180
Exclusion assumptions, 73, 74, 82, 85
Exclusion vs. inclusion, 73, 74
Expert, 6, 79, 171, 229, 232–240, 250–252, 258, 264
 judgment, 226, 228, 230, 232, 233, 236–237, 239, 240, 242
 knowledge, 13, 48, 78, 81, 86, 137, 173, 194, 219, 220, 226–242, 293
Expert-based models, 228–231, 234, 236, 240–242, 252, 266, 293
Expertise, 47–48, 66, 161–163, 227, 231–237, 239–242

Expert knowledge, 13, 48, 78, 81, 86, 137, 173, 194, 219, 220, 226–242, 293
Expert opinion, 6, 79, 171, 230, 250, 251, 258, 264
Explicit *vs.* implicit, 73, 111

F
Faith, D.P., 281
Fajardo, A., 118
Fauvelot, C., 166
Federal Geographic Data Committee (FGDC), 64, 65
Ferrier, S., 51, 194, 199, 221, 281
FGDC. *See* Federal Geographic Data Committee
Fielding, A.H., 199
Filion, B., 219
Finley, A.O., 147
Fire suppression, 113, 114
Fitness, 55, 81, 92, 93, 190
Flat database files, 63
Fletcher, R.J. Jr., 5, 91
Folke, C., 201
Forecasting, 3, 16, 30, 35, 39, 84, 93, 177, 187, 190–196, 198–199, 269–282, 295
Forest resource database, 249, 251, 255, 266
Fragmentation, 11, 17, 18, 75, 83, 170, 173, 228
Frair, J., 45
Frankham, R., 178, 179
Franklin, J., 219
Franklin, S.E., 202
Freeman, E.A., 219, 257, 265
Frequentist, 4–5, 31, 32, 137, 138, 143, 187, 192, 202, 216, 217
Frescino, T.S., 219
Friedman, J., 142
Fuller, W.A., 116
Functional connectivity, 79–80, 160, 166, 170–177, 294
Functional heterogeneity, 79
Fundamental, 2, 17–22, 31, 59, 83, 138, 159, 171, 207, 230
Furlanello, C., 218, 219
Furrer, R., 37
Fuzzy classification, 78–79
Fuzzy set theory, 78

G
Gabor, S., 219
Gaggiotti, O.E., 167, 178
GAMs. *See* Generalized additive models
Gap analysis, 6, 48, 255, 264, 277
Gardner, R.H., 111
GARP. *See* Genetic algorithms for rule-set production
Garton, E.O., 253
Garzon, M.B., 218, 219
Gégout, J.C., 51
Gelfand, A.E., 34, 37
Gene, 111, 143, 178, 179
Gene flow, 5, 77, 159, 160, 167, 171, 172, 174, 176, 179
Generalists, 221–222, 251, 252, 257, 259–261, 264
Generalized additive models (GAMs), 117–118, 215, 219–221, 273
Generalized linear mixed model, 34
Generalized linear modeling (GLM), 3, 33, 34, 52, 94, 99–100, 215, 218–222, 256, 273
Generation, 55, 140, 178, 179, 190, 197
Genetic algorithms, 13, 214, 222, 273
Genetic algorithms for rule-set production (GARP), 13, 20, 189, 214
Genetic drift, 178
Genetic patterns, 159–180
Genotypes, 163, 165, 169, 172, 175, 178, 179
Genton, M.G., 37
Geographic information systems (GIS), 1, 33, 37, 39, 48, 52, 53, 57, 62, 64, 74, 97–99, 161, 199, 209, 234, 238, 241, 264, 270, 293, 294
Geostatistics, 33, 117
Gibson, J., 38
GLM. *See* Generalized linear modeling
Global change, 8, 11, 15–16, 24, 76, 104, 113, 187, 190, 228
Global positioning system (GPS), 57, 60–62
Goldberg, C.S., 166, 179, 180
Goldenberg, N.E., 5
Gompper, M.E., 32
Gottschalk, T., 187
Gradient, 1, 11, 15, 24, 47, 50, 52, 54, 75–76, 78, 91, 96, 99, 100, 105, 113–115, 120, 121, 125, 139, 146–150, 152, 153, 167, 174, 219, 235, 239, 253, 277
Gradient models, 76, 120, 147–148
Graeber, R., 190
Graham, C.H., 51, 220, 221
Grain, 51, 71, 116, 149, 221, 234, 242, 274, 278, 291
Grant, J.B., 6
Graph theory, 77, 80, 173
Grenier, M., 219
Guisan, A., 47, 51, 59, 190, 220, 221, 281
Guthrey, F.S., 217

H

Habitat, 1, 11, 47, 71, 91, 111, 120, 168, 187, 207, 227, 249, 269, 289
  association, 2, 50, 55, 76, 80–84, 209, 229, 233–236, 239, 251, 252, 254, 255, 258, 264–266
  generalists, 249–250, 253, 258, 261, 263, 264
  quality, 71–87, 91–94, 96, 169, 170, 207, 213, 228, 250, 264–265, 292, 294
  selection, 17, 79, 81, 82, 91–94, 103, 104, 209, 236, 255–256
  specificity, 250, 252, 254, 258
Haight, R.G., 85
Halpern, B.S., 6
Halpin, P.N., 77, 113
Hansen, A.S., 281
Hardy, E.E., 256
Hargrove, W.W., 190
Harmon, M.E., 122
Hartl, D.L., 179, 180
Hastie, T., 142
Hawley, T.J., 166
Hayden, J., 173
Heglund, P.J., 253
Heikkinen, R.K., 195, 281
Herrick, J.E., 111
Hess, K.T., 38
Heuristics, 234, 236–237, 240, 242, 270
Hierarchical Bayesian models, 36, 37, 39, 112
Hierarchy theory, 115
Hijmans, R.J., 51, 220–222
Hilbert, D.W., 281
Hirzel, A., 47, 59
Historical data, 15, 32, 45, 50, 59, 63, 164
Holan, S., 38
Holden, Z.A., 137
Holder, M., 34
Holloway, G.L., 219
Home range, 232, 233, 238
Honeycutt, R.L., 166
Hooten, M.B., 5, 29, 34–37, 39
Horning, N., 13
Hrsak, V., 219
Huang, H.-C., 37
Hubbell, S.P., 18
Huettmann, F., 1, 45, 51, 187, 190, 192, 197, 202, 220, 221, 269, 289
Hunter, C.M., 202
Hutto, R.L., 91, 98, 253

Hypotheses, 3, 4, 10–12, 18, 30, 52–54, 71, 72, 74, 79, 86, 87, 115, 118, 125, 131, 132, 137–140, 143, 159, 162, 168, 172, 173, 175–177, 201, 215, 216, 231, 232, 239, 240, 242, 290–291, 294–295

I

Imbalanced data, 144
Immigration, 66, 73, 83, 84
Implicit assumptions, 78, 82, 86, 238–239
Inclusion assumptions, 74, 77
Integro-difference equations (IDE), 35
International Standardization Organization (ISO) standards, 56–58, 65, 66
Invasive species, 35, 81–83, 131, 168, 228, 269, 271, 272, 274
Isolation, 13, 75, 79–81, 94, 173, 176, 179, 233
Isolation by distance, 173, 176, 179
Iverson, L.R., 113, 282

J

Jelaska, S.D., 219
Jiménez-Valverde, A., 144
Jobin, B., 219
Johnsto, D.J., 217
Jones, A.G., 167

K

Kappa statistic, 101, 103, 145, 193
Kareiva, P., 6, 190, 279
Karl, J.W., 253
Kays, R.W., 32
Kearney, M., 21
Keating, K.A., 216–217
Keitt, T.H., 119
Kendall, K.C., 166
Kerr, J.T., 5, 9, 14
Kery, M., 37
Kharouba, H.M., 5, 14
Knutson, M.J., 82
Kofinas, G.P., 201
Kriging, 33, 37, 38, 117, 121, 123
Krizan, J., 219
Kuhn, I., 119
Kukal, O., 12
Kulkarni, M., 9

L

Land cover, 23, 24, 52, 53, 58, 74–79, 99, 115, 147, 175, 190, 197, 200, 228, 241, 249, 250, 255, 258, 263, 272, 277
Landscape fluidity, 76
Landscape genetics, 159–180, 294
Land-use change, 9, 15–16, 76, 95, 131, 228, 269, 271, 272, 275, 281
Lark Bunting *(Calamospiza melanocorys)*, 279, 280
Larsen, D.R., 34
Larson, M.A., 190
Latimer, A., 34
Latitudinal variation, 81
Laurent, E.J., 5, 71
Lawler, J.J., 6, 38, 190, 269, 279
Least-cost pathways, 80, 173, 174
Leathwick, J.R., 51, 221
Leberg, P.L., 166
Lee, C.Y., 179
Lees, D.C., 15
Legendre, P., 118, 119
Lehmann, A., 51, 220, 221
Levin, S.A., 115
Lewis, P.O., 34
Liaw, A., 144
Li, J., 51, 221, 222
Linear functions, 188
Linear regression, 32, 33, 95, 119, 170, 219–220, 273
Link, W.A., 217
Lippitt, C.D., 219
Lobo, J.M., 144
Local (fine-scale) wildlife habitat modeling, 249, 264, 265
Locus, 163–166, 169–175, 178, 179
Loehle, C., 281
Lohmann, L.G., 51, 221
Loiselle, B.A., 51, 220, 221
Lookingbill, T.R., 5, 111
Loucks, O.L., 115
Lugon-Moulin, N., 166
Luikart, G., 166, 178–180
Luoto, M., 195, 281
Lusk, J.J., 217
Luukkonen, D.R., 179

M

Machine learning, 5, 37–39, 104, 138, 143, 145, 154, 198, 201, 215, 221, 273, 279, 291, 294
Machine-learning-based tools, 5, 138, 273, 279

Machine learning methods, 5, 39, 138, 145, 154, 198, 215, 221, 273, 279, 291
Maisonneuve, C., 219
Ma, K.P., 119
Malcolm, J.R., 219
Management, 46–65, 92–97, 113–115, 160–164, 201–202, 212–217, 251–253, 270–272
 perspective, 218
Managers, 4, 30, 55, 57, 63, 113, 116, 159, 161, 164, 172, 173, 175–178, 189, 212, 218, 222, 228, 229, 231, 249–251, 269, 271, 272
Manel, S., 145, 167, 178, 199
Manion, G., 51, 221
Manly, F.J., 194, 202
Mantel, 118, 120, 125, 126, 174
Mantel test, 118, 119, 174
Margules, C.R., 277, 281
Markers, 159–180
Markov Chain models, 198
Markov models, 95, 104
Marmion, M., 195
MARS models. *See* Multivariate adaptive regression spline (MARS) models
Martínez-Meyer, E., 13, 15, 190
Mastro, V., 219
Matrix models, 35, 75, 173
Maurer, B.A., 249
Maurer, E.P., 6, 190, 279
Maximum entropy, 16, 23, 24, 188, 214, 220–222, 273
McClean, C., 15
McDonald, L.L., 194, 202
McDonald, L.T., 194, 202
McGill, B.J., 14
McIntire, E.J.B., 118
McKee, W.A., 122
McKelvey, K.S., 173
Mechanisms, 3, 10, 12–14, 16–17, 20, 22–24, 60, 115, 116, 121, 131, 138, 139, 166, 172, 193, 195, 202, 215, 279, 290, 291
Mechanistic, 9, 14, 52, 115, 117, 146
 modeling, 3, 4, 10–12, 15, 17, 19–24, 52, 92, 114, 120, 138, 145, 193, 269–270, 272–275, 277, 281, 282, 290, 295
 vs. predictive models, 3, 4, 9, 10, 12, 131–132, 138, 290, 295
Melick, D.R., 220
Menza, C., 220
Meta-analysis, 6, 193, 195, 200, 202, 211, 218
Metadata, 4, 46, 56–58, 61–66, 292, 294–295
Metapopulation, 17, 77, 79–80, 83, 91–95, 103, 104, 251, 291

Metapopulation theory, 83, 91–94, 103, 291
Meyer, M.A., 237
Microsatellite, 163, 164, 169–173, 175, 179
Migrants, 93, 171, 174, 179
Migration, 82, 113, 120, 124–125, 161, 171, 179, 197, 228, 270
Miles, L., 15
Miller, C.R., 166
Miller, J., 219
Millspaugh, J.J., 190
Mitochondrial DNA (mtDNA), 163, 165–168, 180
Mi, X.C., 119
Mixed models, 32, 34
Mladenoff, D.L., 85
Model, 1–6, 9–24, 29–39, 45–66, 73–74, 91–105, 117–119, 137–154, 159–177, 187–202, 207–222, 227–242, 253–263, 269–282, 289–296
  accuracy, 53, 73, 76, 101–103, 105, 149, 189, 194, 197, 218–222, 230, 249, 250, 254, 259–262, 264–265, 275–278, 290–292, 294
  assessment, 78, 80, 102, 103, 187, 189, 199, 200, 208, 211, 230, 242, 273, 276, 292, 293
  comparison, 100, 214, 216, 217, 221, 230, 231, 233, 255, 260, 262
  end-users, 72, 86
  fit, 30, 86, 91, 92, 100–101, 104, 117, 143, 144, 150, 173, 177, 187–202, 231, 252, 255–260, 262, 266
  objectives, 73, 77, 82, 85, 86
  performance, 54, 96, 98, 103, 143–145, 198, 200, 219, 221, 230–231, 250, 251, 254, 264, 276, 290, 293
  selection, 4, 52–53, 71, 96, 100–102, 143–144, 149, 171, 176, 209, 215–217, 219, 270, 273, 276–278
  sensitivity, 20, 78, 145, 150, 216
  utility, 72, 264, 291
Model-averaging, 188, 216, 258, 262, 273, 279
Model development and testing, 11, 73, 74, 76, 86, 97–101, 104, 105, 145, 234, 235, 255, 293–295
Modeling culture, 192, 201–202, 295
Moisen, G.G., 219, 257, 265
Monaco, M.E., 220
Monitoring, 37, 39, 51, 54, 58, 60, 62, 76, 82, 85, 86, 97, 98, 105, 116, 122, 164, 165, 169, 175, 227–230, 249–266
Morin, P.A., 180
Moritz, C., 51, 221
mtDNA. *See* Mitochondrial DNA

Multi-scale models, 111, 112, 117, 120, 139, 281
Multi-stage, 117, 120
Multi-stage sampling, 116, 120
Multivariate adaptive regression spline (MARS) models, 221, 222
Murphy, M.A., 5, 137, 143, 159, 166, 167, 179, 180
Myers, R., 195

N

Nakamura, M., 51, 221
Nakazawa, Y., 51, 221
National Biodiversity Information Infrastructure (NBII), 58, 65
Nature reserves, 270
NBII. *See* National Biodiversity Information Infrastructure
Nei, M., 167, 179
Nelson, C.R., 6
Neteler, M., 218, 219
Neutral, 16–19, 77, 159, 160, 163, 169, 179, 180
Neutral theory, 17–19
New, M., 281
Niche, 13, 16–24, 49, 79, 81, 83, 103, 124, 139, 147–149, 153, 168, 190, 192, 193, 196, 197, 200–202, 207, 214, 221, 232, 264, 271
  breadth, 81
Nikoli, T., 219
"No analog" future, 76
Non-equilibrium landscapes, 76
Non-parametric, 31, 37–38, 137, 139, 140, 146, 147, 152, 153, 174, 176
Non-stationary, 137
Northern Region Landbird Monitoring Program (NRLMP), 97–99, 104, 105
Noson, A., 91
Nussbaum, R.A., 13
Nusser, S.M., 116
Nychka, D., 37

O

Occupancy, 34, 84, 85, 92, 94–96, 98, 102, 104, 105, 174, 234–236, 238
Occurrence data, 48–57, 59, 96, 149, 164–165, 250, 252, 264, 266, 271, 272, 276, 277
Ogden, R.T., 119
Ohleth, K., 6
Olden, J.D., 6
Olea, P.P., 218, 219

Olivier, F., 220
Ollero, H.S., 218, 219
Online databases, 47, 48, 52, 53, 58
Open access, 4, 201, 202, 266, 293–295
Optimal sampling, 38
Ormerod, S.J., 145, 199
Ortega-Huerta, A., 13
Ouyang, Z.S., 119
Overton, J.M.C., 51, 220, 221

P
Pan, Y.W., 179
Papes, M., 144
Parametric, 31, 38, 118, 140, 141, 174, 192
Parsimony, 4, 32–33, 73, 95, 143, 188, 189, 193, 194, 227, 290
Partial differential equations (PDE), 35
Parviainen, M., 195
Patch characteristics, 75
Patch-matrix model, 75
PCRs. *See* Polymerase chain reactions
PDE. *See* Partial differential equations
Pearce, J., 194, 199
Pearson, R.G., 15, 281
Peck, R.M., 57
Pedrana, J., 57
Perera, A.H., 227
Peres-Neto, P.R., 118, 119
Permeability, 80, 175
Peters, D.P., 111
Peterson, A.T., 13, 51, 144, 190, 220–222
Peterson, E., 33
Peterson, M.J., 217
Phillips, S.J., 51, 221
Pittman, S.J., 220
Plazibat, M., 219
Plot-level field vegetation sampling, 255, 258, 264
Plot-level measurements, 255, 260, 264
Polymerase chain reactions (PCRs), 163–166, 175, 178, 180
Porter, W.P., 21
Possingham, H.P., 278
Posterior distribution, 36–37, 229
Prasad, A.M., 113, 282
Pratson, L.F., 77
Predictive accuracy, 3, 4, 17, 73, 76, 101–103, 105, 149, 188, 189, 192–194, 197, 198, 202, 218, 221–222, 290, 291
Predictive habitat, 4, 207–213, 216, 218–222, 277, 278
Predictive performance, 4, 51, 71, 143, 188, 189, 192, 200, 290

Prescribed fire, 125–130
Presence-only data, 59–61, 214, 220, 221
Presence-only database management, 61–62
Preserves, 55, 121, 170, 265, 269, 282
Prevalence, 24, 37, 39, 51, 54, 83, 101, 103, 165, 168, 199, 219, 230–231, 241, 250, 252, 254, 257–265
Primers, 164, 169, 178, 180
Prince, H.H., 179
Prior distribution, 31, 36–37, 229
Probability threshold, 215, 252–254, 256, 260, 265, 276, 278
Procopio, D., 57

R
Radio-telemetry, 60, 61, 171
Random forest (RF), 5, 38, 137–154, 174, 176, 188, 198–199, 215, 219, 273, 279, 281, 294
Range shifts, 16, 271, 279–281
Raxworthy, C.J., 13
Ray, J.C., 32
Realized niche, 20, 21, 49, 81, 83, 149
Rebelo, A.G., 34
Receiver operator characteristic (ROC), 101, 103, 145, 193, 199, 216, 254, 256–260, 265, 278
Receiver-operator characteristic curve (ROC/AUC), 145, 254, 256–260, 265, 278
Recombination, 178, 180
Recursive partitioning (RPART), 140, 255–258, 260, 263
Regehr, E.V., 202
Regression-based models, 211, 215–216
Rehfeldt, G.E., 148
Relational databases, 46, 55, 57, 58, 65
Remotely sensed imagery, 74
Ren, H.B., 119
Reserve-selection, 91, 270, 276–278
RF. *See* Random forest
Richard, H.W.B., 281
Richardson, K., 51, 221
Ritter, J., 192
Roach, J.T., 256
Roberts, L.J., 6, 249, 256
Robinson, D.H., 216
Robinson, S., 179, 180
ROC. *See* Receiver operator characteristic
ROC/AUC. *See* Receiver-operator characteristic curve
Rocca, M.E., 111
Rodríguez, A., 57
Rogan, J., 219

Rota, C.T., 91
Rousset, F., 167
Royle, J.A., 34, 35, 37
RPART. *See* Recursive partitioning
Runge, M.C., 202
Rushin, J., 39
Russell, L.D., 178–180

S
Sadler, J., 38
Sample size, 47–48, 50, 51, 56, 59, 116, 117, 130, 131, 144, 164, 171, 213, 215, 220–222, 250, 255
Sampling, 39, 54–61, 74, 92, 95, 112, 115–117, 122, 125, 149, 165, 167, 171, 172, 178, 199, 220, 227, 231, 235–237, 251, 252, 264
  designs, 2, 4, 5, 38, 39, 46–52, 54, 57–59, 97, 105, 115, 116, 120, 121, 138, 220, 233, 241, 290, 293–294
Sanchez de Dios, R., 218, 219
Sangermano, F., 219
Sauer, J.R., 82
Sawyer, A., 219
Saxe, H., 281
Scachetti-Pereira, R., 51, 221
Scale, 2, 3, 5, 22, 23, 33, 34, 36, 38, 47, 54, 58, 60, 73, 77, 91, 99, 111–132, 137–139, 148, 149, 161, 163, 164, 167–169, 172, 180, 195, 227–229, 234, 239–242, 249–250, 259, 274, 276–277, 281, 289, 292–294
Schapire, R.E., 51, 221
Schlaepfer, M.A., 6
Schlötterer, C., 167
Schneider, G.E., 13
Schoennagel, T., 125
Schwartz, M.K., 173
Schwartz, M.W., 113, 282
Scott, J.M., 253
Scriber, J.M., 12
Scribner, K.T., 179
Segurado, P., 15
Semantic framework, 72–74
Semi-parametric, 37–38
Sensitivity analysis, 78, 230, 242
Sensitivity and specificity, 20, 59, 78, 103, 139, 140, 145, 150, 193, 201, 216, 219–221, 230, 242, 250, 252, 254, 255, 257, 258, 265, 278
Sequence, 162–164, 166, 168, 169, 180
Shafer, S.L., 6, 190, 279
Sheriff, S., 39

Shifley, S.R., 190
Shi, T., 37
Shorey, R.I., 179
Sickley, T.A., 85
Silander, J.A., 34
Silliman, B.R., 6
Simplicity (parsimony), 4
Simplifying assumptions, 17, 77, 79, 82, 83
Single nucleotide polymorphisms (SNPs), 163, 165, 169, 171, 180
SITES/MARXAN, 271
Skidmore, A.K., 220
Slatkin, M., 166
Smith, T.M., 114
Smithwick, A.H.E., 125
Sobel, M.J., 281
Soberón, J., 51, 144, 221
Spatial, 1, 9, 11, 14–16, 23, 29–39, 46, 49, 51, 54, 59, 71, 73–79, 82, 85, 91, 105, 111–127, 137, 143, 147–149, 154, 159, 161, 163, 165, 167, 168, 174, 179, 187–202, 208, 220, 227, 232, 238, 242, 249, 264, 269, 274, 277, 291, 293
  autocorrelation, 14, 30, 33, 34, 51, 94, 112, 117, 118, 125, 131, 137, 147, 179, 199, 233, 274, 281
  extent, 30, 46, 49–51, 54–55, 71, 85, 105, 112, 201, 230, 271, 274
  scale, 73, 77, 91, 92, 97, 111–113, 115, 117–121, 125, 127, 131, 149, 174, 208, 227, 228, 239, 274, 277, 281, 291
  and temporal domain, 39, 73, 114
Spatio-temporal, 9, 29–39, 93, 94, 103
Spear, S.F., 179, 180
Specialists, 72, 221, 250, 252, 253, 257–261, 263, 264, 292
Specialization, 2, 74, 75, 84, 163, 165, 211, 229, 251
Species, 1, 9, 34, 45, 71, 91, 111, 137, 160, 187, 207, 227, 249, 269, 289
  distribution, 1, 2, 4, 12, 13, 15, 16, 19, 22, 45–66, 74, 84, 91–105, 111–113, 117–120, 123, 125–126, 131, 137–154, 187, 235, 258, 269–282, 291
  interactions, 80, 81, 269–270
  location data, 79, 251, 264
  ranges, 14, 15, 22, 79, 81, 116, 220, 271, 274, 276, 280, 281
Species distributions-cryptic species, 84, 165, 292
Species-habitat relationships, 72, 78, 79, 82, 85, 95, 231, 252
Spies, T.A., 122
Stand-level forest inventory database, 264

Stand-level vegetation measurements, 255, 264
Stationarity, 79, 82, 119
Statistical, 1–6, 29–39, 45–47, 51, 54, 64, 71, 74, 81, 86, 92–97, 111, 117, 119, 130, 131, 138–140, 167, 178, 192, 198–199, 202, 207–222, 229, 241, 251–253, 256, 258, 262, 265, 266, 270–273, 290, 293, 294
  hypotheses, 4, 53, 216
  'noise', 174, 197
Stein, J.L., 277
Stenhouse, G.B., 202
Stirling, I., 202
Stockwell, D.R.B., 281
Stone, K., 38
Storfer, A.S., 143, 179, 180
Subjectivity, 31, 63–65, 84, 86, 99, 140, 143, 189, 221, 228–230, 234, 242
Suitability, 4, 24, 45, 51, 74, 75, 80, 83–84, 86, 93, 94, 104–105, 116–117, 147–148, 154, 164, 173–174, 209, 220–221, 227, 234–235, 238, 252–254, 258, 260, 263–265, 269, 272, 292
Suitable habitat, 17, 55, 82–85, 93, 113, 174, 218, 220, 249, 254
Sykes, M.T., 281

T
Telemetry, 60–62, 85, 171, 213
Temporal, 2, 9, 14–15, 22, 29–39, 60, 73, 74, 76–78, 83, 91–105, 114, 119, 168, 179, 190–192, 196–200, 230, 232, 233, 242, 291–293
  effects, 200
  processes, 5, 76
  and spatial scale, 23, 34, 38, 73, 77, 83, 91–93, 97, 111–113, 115, 117–121, 125, 127, 131, 148, 149, 153, 163, 174, 208, 227, 228, 239, 242, 274, 277, 281, 291
  and thematic resolutions, 74, 75
  variation, 81, 92, 94, 95, 98, 101–105
Thematic precision, 75
Thematic resolution, 74, 75
Theobald, D., 33
Theoretical *vs.* empirical models, 10
Thogmartin, W.E., 71, 82
Thomas, D.L., 194, 202
Thompson, F.R., 190

Threshold in probability of occurrence, 252–254, 265
Thuiller, W., 15, 190, 195, 281
Tibshirani, R., 142
Toledana, J., 219
Traditional knowledge (TK), 198, 229
Travaini, A., 57
Treml, E.A., 77
Turner, M.G., 125

U
Uncertainty, 1–6, 29–31, 34, 38, 39, 78, 100, 150–152, 187–203, 220, 222, 228–230, 235, 238, 239, 241–242, 278, 290, 291, 295
Uncertainty analysis, 78
Underlined, 161
Under-parametrized, 73
Unintended consequences, 73
Urban, D.L., 77, 111, 119

V
Van Loon, E.E., 166
Variable and model selection, 4, 52–53, 71, 96, 100–102, 138, 141, 143–144, 149, 171, 176, 209, 215–219, 256, 270, 273, 276–278
Variation, 207–222
Vera, R., 218, 219
Ver Hoef, J.M., 33
Vierling, L., 179, 180
Virkkala, R., 281
Visualization, 39, 58, 64, 66, 145–147, 149, 232, 238

W
Wainer, H., 216
Waits, L.P., 166, 179, 180
Wang, J., 166
Wang, S., 38
Waples, R.S., 167, 178
Warwell, M.V., 148
Wavelet analysis, 119, 120, 127
Wayne, R., 180
Weight, 53, 60, 61, 80, 101, 117, 144, 145, 147, 149–152, 173, 216, 217, 239, 241
Weir, B.S., 167
Wei, W., 119

Westphal, M.I., 278
White, D., 6, 190, 279
Whittaker, R.H., 114
Wickert, C., 190
Wiersma, Y.F., 1, 5, 207, 269, 289
Wikle, C.K., 34–37, 39
Williams, B.H., 5
Williams, H.C., 145, 199
Williams, S., 51, 221
Wilson, K.A., 278
Winterstein, S.R., 179
Wisz, M.S., 51, 221, 222
Witmer, R.E., 256
Wood, R.A., 236
Worm, B., 195
Wotherspoon, S.J., 220
Wright, N.M., 253
Wright, S., 166
Wu, J., 115
Wu, S., 34
Wydeven, A.P., 85

X
Xu, J., 220

Y
Yang, X., 220
Yen, P.P.W., 190, 197, 220
Young, E.M., 14
Young, J.S., 91, 98

Z
Zaniewski, A.E., 220
Zapata, S., 57
Zaradic, P., 6
Zhou, Z., 220
Zhu, J., 37
Zimmermann, N.E., 51, 219, 221
Ziv, Y., 190
Zonation, 271
Zuckerberg, B., 5, 45

Lightning Source UK Ltd.
Milton Keynes UK
UKOW06f0223310315

248760UK00005B/16/P